A CENTURY OF NATURE

A CENTURY OF nature

TWENTY-ONE DISCOVERIES

THAT CHANGED **SCIENCE** AND THE **WORLD**

EDITED BY LAURA GARWIN & TIM LINCOLN

WITH A FOREWORD BY STEVEN WEINBERG

THE UNIVERSITY OF CHICAGO PRESS / CHICAGO AND LONDON

Laura Garwin is at the Bauer Center for Genomics Research at Harvard University. She was formerly the Physical Sciences editor and North American editor of *Nature*.

Tim Lincoln is the News and Views editor of *Nature*.

The University of Chicago Press, Chicago 60637
The University of Chicago Press, Ltd., London

Printed in the United States of America

12 11 10 09 08 07 06 05 04 03   1 2 3 4 5

ISBN: 0-226-28413-1 (cloth)
ISBN: 0-226-28415-8 (paper)

Library of Congress Cataloging-in-Publication Data

A century of Nature : twenty-one discoveries that changed science and the world / edited by Laura Garwin and Tim Lincoln.
p. cm.
Includes bibliographical references and index.
ISBN 0-226-28413-1 (cloth : alk. paper) — ISBN 0-226-28415-8 (pbk. : alk. paper)
1. Nature (London, England). 2. Science—History. 3. Science—Social aspects—History. 4. Science and civilization. I. Garwin, Laura. II. Lincoln, Tim.
Q125 .C384 2003
500—dc22

2003014899

The paper used in this publication meets the minimum requirements of the American National Standard for Information Sciences—Permanence of Paper for Printed Library Materials, ANSI Z39.48-1992.

Contents

Foreword

Steven Weinberg

Few scientists and even fewer nonscientists will ever have read the original reports of classic scientific discoveries. For instance, in my life I have met only two physicists who have read Newton's *Principia* (and I am not one of them). And why should they? Science, unlike theology or the arts, is a cumulative enterprise. We now understand things much better than those who preceded us. Any competent graduate student in theoretical physics today understands the general theory of relativity better than Einstein did, so why should he or she read the 1915 papers in which Einstein struggled to understand gravitation, space, and time?

Still, it would be a pity to cut ourselves off completely from our past. The history of science is as interesting as any other branch of history, and for scientists one of the gratifications of our work is the sense of continuing a great historical tradition, of carrying forward the work of our predecessors and preparing the ground for our successors. Indeed, in learning science, we all absorb a certain amount of scientific history—Darwin and Wallace did this, and then Mendel did that, and so on. The trouble with this potted history is that it never captures the extreme difficulty of taking a new step, and it is often just plain wrong. For an understanding of the history of science, there is no substitute for actually reading some of the great works of past scientists. It is therefore cause for celebration that Laura Garwin and Tim Lincoln have assembled this collection of classic scientific papers from *Nature,* with explanatory essays to make the papers accessible to a general audience.

The collection is remarkable, not only for the importance of the individual articles, but also for the fact that they all appeared in the same journal. It used to be assumed that natural philosophers (only later called "scientists") would be able to read and understand reports of new discoveries in any area of science, and could often make contributions to many of them. Think of Charles-François de Cisternay Du Fay, who in the early eighteenth century wrote papers on geometry, fire pumps, artificial gems, phosphorescence, slaked lime, plants, and dew. Now we are terribly specialized, and can read papers only in our own sub-sub-sub-specialties. Aside from the proceedings of learned academies, there are only two English-language scientific journals that keep up the old tradition of encompassing all of science: *Science* in the U.S., and *Nature* in Britain. In this collection of articles from *Nature* one can read seminal articles in every field of twentieth-century science. Every field, that is, except my own: elementary particle theory. For some reason, particle physicists rarely submit their work to *Science* or *Nature*. (But *Nature* had its chance: Enrico Fermi submitted to *Nature* his great 1932 paper on the theory of beta decay, which founded the modern theory of weak interactions, but it was rejected.) Looking over this collection, I begin to feel that we should. We would be in such good company.

Department of Physics, University of Texas at Austin

Preface

The book you are holding contains the original reports of some of the most explosive science of the twentieth century. Some explosions came literally, as a by-product of the discovery of nuclear fission. Others came in the thunderous debate that followed the announcement of the successful cloning of an adult sheep. But most of the chain reactions set off by the twenty-one discoveries presented here took the form of waves of revelation spreading through a community of scientists, providing impetus to further scientific or technological progress.

Textbooks present much of our understanding of the world as if it has always been known: the nuclei of atoms contain protons and neutrons; the structure of DNA is a double helix. But there was a time—not so long ago—when these basic facts were at the sharp edge of scientific discovery. What did it feel like to make such a discovery, or to read about it for the first time? As editors at the weekly science journal *Nature,* we have been among the first to know of world-changing discoveries (not that their impact was always immediately apparent). With this book, we hope to share that experience.

Here, a disclaimer is in order. This is not a collection of papers announcing the twenty-one most important scientific discoveries of the twentieth century—although, arguably, all of them would place in the top fifty. (Readers wishing to decide for themselves can find a wider selection in the chronology beginning on page xv.) For one thing, we restricted our choice to papers published in *Nature.* The reason is simply pragmatic. As editors at *Nature,* we are surrounded by our own journal's history—easily accessible in bound volumes that contain the weekly contents of decade upon decade of scientific endeavor, all neatly packaged for publication. That history goes back to 4 November 1869, when the first issue appeared. But it was not until well into the following century that *Nature* really began to hit its stride as a vehicle for original research reports. Hence the choice of articles from the twentieth century: the first published in 1925, and the last in 1997. In all, twenty-two papers are included, because two announced a simultaneous finding.

So are these the most important papers published in *Nature* during the twentieth century? We would not make that claim either. A good case could be made for inclusion of the paper that cracked the nature of the genetic code, by which the letters of the DNA sequence are translated, by threes, into the building blocks of proteins. (A first draft of the complete human genome sequence escapes on a technicality, having been published in the twenty-first century.) Perhaps the invention of holography, or of monoclonal antibodies, both of which won Nobel prizes for their authors, should have been included. For editors of anthologies such as this, second-guessing comes with the territory. We hope readers will have fun picking holes in our choices.

Fortunately, however, we did not set ourselves the task of choosing *Nature*'s "greatest" papers. What we do claim is that these are all papers that transformed their subject—even, in some cases, the world. Some did this by reporting the discovery of unexpected phenomena: the first pulsar, for example, or the "hole" in the ozone layer over Antarctica. Others, such as the explanation of magnetic stripes on the sea floor, provided a theoretical framework in which previously mysterious observations could be interpreted. Some gave the world a useful new technology, such as the laser. And still others reported the implementation of ingenious techniques that drove innovative science: DNA sequencing, and the "patch clamp" for recording electrical signals from individual cells.

Our selection method was thorough, but not exhaustive. One starting point was a list of "classic" papers started many years ago by *Nature*'s editors, and extended as new candidates have cropped up in reference lists or award citations. Classic papers do keep coming to light, even from long ago. For example, it was only in 1998 that we rediscovered a flurry of seminal *Nature* papers on superfluidity that had been published sixty years previously, thanks to a commentary in *Nature* that recalled some of them.

We also picked the brains of our colleagues and advisers, in part to help improve the balance among different disciplines. In many cases the Nobel committee had done our work for us: twelve of the discoveries presented here propelled their authors to Stockholm. But the Nobel Prize isn't the final arbiter of significance—not least because there is no Nobel for paleontology, geology, or observational astronomy, fields in which five of the remaining nine papers lie. Nevertheless, we are conscious of remaining gaps in the subject spread, for example in evolutionary biology, ecology, and mainstream chemistry.

A collection of papers on its own would be a poor thing—hence the specially commissioned explanatory essays that accompany each paper. We asked the authors of these essays, distinguished scientists and commentators in their own right, to place the discoveries in context by describing the state of knowledge at the time the paper was published, the immediate effect of the report of the discovery, and the abundance of science or technology that ensued. Each of the essayists has a close intellectual connection to the work in question, and many have a personal one as well. Remarkably, the eyewitness accounts extend even to some of the discoveries from the 1920s and '30s, reminding us of how far our understanding has come in the course of a lifetime.

The Nobel Prize–winning biologist Sir Peter Medawar once famously asked if the scientific paper is a fraud, as the conventional format for such papers, logically proceeding from facts to explanation, "misrepresents the processes of thought that accompanied or gave rise to the work." In the same way, textbook accounts of the progress of science are too often forced to leave out the messy details of the human side of discovery. In asking our authors to cast their commentaries as narratives, our hope was that they would give a flavor of what it feels like to experience the thrill (or, often, endure the frustration) of pioneering science. And in asking them to write for the general reader, the intention was that in each case the significance of the work would be clear for such an audience, complementing the technicality of the original papers.

With the advances of the twentieth century, dilemmas over the practice of science itself and its applications have of course arisen. Exploring such issues is not the purpose of this book. Instead, we hope that this collection will illuminate how the best science is done and reported, and how its influence spreads, both immediately and in the longer term. You will see examples of the role of serendipity, and the importance of the prepared mind; the rewards that come from following up unexpected observations instead of ignoring them; and the benefits to be had from searching for needles in haystacks. You will see that no discovery, no matter how startlingly original, stands alone: each of the papers included here derives from previous work, and in turn provides support for what follows. In that sense, they constitute bricks in an edifice. The edifice is subject to modification or even destruction in the light of later knowledge, but we suspect that these twenty-one bricks from the twentieth century will continue to provide durable components of science through the twenty-first century and beyond.

If further evidence is needed that science is a collective enterprise, one need only consider the supreme importance of the scientific paper in the lives of scientists: in a very real sense, a discovery does not exist until it has been published, and validated by one's peers. Electronic publication, which looks set to become the norm, is a change of medium from the printed format in which these papers first appeared. But it is unlikely to change that essential message.

Creating a book is also a collective enterprise. This book would not have happened without the vision and commitment of our colleague Richard Nathan at Macmillan and Christie Henry of the University of Chicago Press. Particular appreciation is also due to Radha Clelland for the graphics, Susan Boobis for the index, and A. A. Bene for help when it was most needed. Finally, our thanks in addition to other staff at the University of Chicago Press, especially Michael Koplow and Renate Gokl.

Chronology of twentieth-century science

The twenty-one discoveries discussed in this volume are highlighted in bold type.

1900 Quantum theory proposed / *Planck*
1901 Discovery of human blood groups / *Landsteiner*
1905 Wave-particle duality of light / *Einstein*
1905 Special theory of relativity / *Einstein*
1906 Existence of vitamins proposed / *Hopkins*
1906 Evidence that Earth has a core / *Oldham*
1908 Synthesis of ammonia from its elements / *Haber*
1909 Idea of genetic disease introduced / *Garrod*
1909 Boundary between Earth's crust and mantle identified / *Mohorovičić*
1909 Discovery of Burgess Shale: ancient invertebrate fossils / *Walcott*

1910 First mapping of a gene to a chromosome / *Morgan and others*
1911 Discovery of the atomic nucleus / *Rutherford*
1911 Superconductivity discovered / *Onnes*
1912 Discovery of cosmic rays / *Hess*
1912 Idea of continental drift presented / *Wegener*
1914 First steps in elucidating chemical transmission of nerve impulses: neurotransmitters / *Dale; Barger; Loewi*
1914 Astronomical theory of climate change / *Milankovitch*
1915 General theory of relativity / *Einstein*
1918 onwards Synthesis of genetics with the theory of evolution by natural selection: neodarwinism / *Fisher; Haldane; Wright*

1921 Isolation of insulin / *Banting & Best*
1923 Nature of galaxies discovered / *Hubble*
1925 Description of *Australopithecus africanus* / *Dart*
1925–26 Matrix and wave formulations of quantum mechanics / *Heisenberg; Schrödinger*
1927 Matter is proved to be wavelike / *Davisson & Germer*
1928 Discovery of penicillin / *Fleming*
1929 Expansion of the Universe established / *Hubble*
1929 First suggestion that Earth's magnetic field reverses / *Matuyama*

1930 First absolute geological timescale / *Holmes*
1930s Theory of chemical bonds developed / *Pauling*
1930s onward Establishment of the scientific study of animal behavior / *von Frisch; Lorenz; Tinbergen*
1931 Birth of radioastronomy / *Jansky*
1931 First electron microscope / *Ruska*
1932 Discovery of the neutron / *Chadwick*
1932 Discovery of the positron, first antimatter particle / *Anderson*
1935 Magnitude scale for earthquakes / *Richter*
1935 Theory of the nuclear force / *Yukawa*
1937 Discovery of the citric acid cycle / *Krebs*
1938 Nuclear reactions in stars / *Bethe; von Weizsäcker*
1938 First observation of superfluidity / *Kapitza*
1939 Discovery of nuclear fission / *Meitner & Frisch*

1943 Mutations in bacteria identified / *Luria & Delbrück*
1944 Evidence in bacteria that DNA is the genetic material / *Avery, MacLeod, and McCarty*
1944 Start of Mexican wheat improvement program, leading to the "green revolution" / *Borlaug*
1945 Formulation of the one-gene, one-enzyme hypothesis / *Beadle & Tatum*
1946 Radiocarbon dating / *Libby*
1946 Initial elucidation of the reactions involved in photosynthesis / *Calvin*
1947 Invention of the transistor / *Shockley, Bardeen, and Brattain*
1948 Big Bang theory for origin of the Universe / *Gamow, Alpher, and Herman*

1948 Quantum electrodynamics / *Feynman; Schwinger; Tomonaga*
1949 Immunological tolerance hypothesis proposed / *Burnet*

1951 Presentation of the idea of gene transposition: "jumping genes" / *McClintock*
1952 First polio vaccine / *Salk*
1952 Theory of nerve-cell excitation announced / *Hodgkin & Huxley*
1953 Production of amino acids in "early Earth" conditions / *Miller & Urey*
1953 First determination of the amino-acid sequence of a protein / *Sanger* et al.
1953 Structure of DNA: the double helix / *Watson & Crick*
1956 Discovery of the neutrino / *Cowan & Reines*
1957 Superconductivity explained / *Bardeen, Cooper, and Schrieffer*
1958 Quantum tunneling of electrons in semiconductors / *Esaki*
1958 First three-dimensional protein structure published / *Kendrew* et al.

1960 First laser / *Maiman*
1960 onward Discoveries of fossils of early *Homo* in East Africa / *Leakeys and others*
1961 Nature of the genetic (triplet) code proposed / *Crick* et al.
1963 Deterministic chaos: the butterfly effect / *Lorenz*
1963 Discovery of quasars / *Schmidt*
1963 Explanation for magnetic stripes on the sea floor: seafloor spreading / *Vine & Matthews*
1964 Existence of quarks proposed / *Gell-Mann; Zweig*
1964 Genetic explanations proposed for animal social behavior / *Hamilton*
1965 Discovery of cosmic microwave background radiation / *Penzias & Wilson*
1967 First warning of an anthropogenic "greenhouse effect" / *Manabe & Wetherald*
1967 Theory of plate tectonics / *McKenzie & Parker; Morgan*
1967 Electroweak theory, first unification of fundamental forces / *Weinberg; Glashow; Salam*
1967 Proposal that certain cell organelles are descended from free-living bacteria / *Margulis*
1968 Pulsars discovered / *Hewish* et al.
1968 Theory of random molecular evolution (the neutral theory) proposed / *Kimura*

1970 **Reverse transcriptase discovered** / *Baltimore; Temin & Mizutani*
1973 Gamma-ray bursts from outer space / *Klebesadel, Strong, and Olsen*
1973 Advent of genetic engineering techniques / *Cohen, Boyer, and Berg*
1973 **Invention of magnetic resonance imaging** / *Lauterbur*
1974 Identification of CFCs as threat to ozone layer / *Molina & Rowland*
1974 **Principles of cell-mediated immunity unveiled** / *Zinkernagel & Doherty*
1974 Discovery of "Lucy," *Australopithecus afarensis* / *Johanson & Taieb*
1974 First Grand Unified Theory of particle physics / *Georgi & Glashow*
1975 Monoclonal antibodies created / *Köhler & Milstein*
1976 **Patch-clamp technique for studying single ion channels** / *Neher & Sakmann*
1977 **First complete DNA sequence of an organism** / *Sanger* et al.
1977 Discovery of deep-sea hydrothermal vents / *Corliss* et al.
1978 Observation of astronomical dark matter / *Rubin*

1980 **Unveiling of genetic controls on animal body-plan development** / *Nüsslein-Volhard & Wieschaus*
1980 First human oncogene ("cancer gene") identified / *Weinberg*
1980 Impact hypothesis for extinctions at the Cretaceous/Tertiary boundary / *Alvarez* et al.
1981 Superstring theory / *Green & Schwarz*
1982 Prion hypothesis proposed / *Prusiner*
1983 AIDS virus identified / *Barré-Sinoussi* et al.
1985 Genetic fingerprinting invented / *Jeffreys*
1985 **Ozone hole discovered** / *Farman* et al.
1985 **Discovery of buckminsterfullerene** / *Kroto* et al.
1986 First high-temperature superconductor / *Bednorz & Müller*
1987 Formulation of the "Out of Africa" hypothesis of human evolution using molecular data / *Cann, Stoneking, and Wilson*

1995 Bose-Einstein condensation of trapped atoms / *Cornell & Wieman*
1995 **First extrasolar planet identified** / *Mayor & Queloz*
1997 **Dolly the sheep created by cloning** / *Wilmut* et al.

A CENTURY OF NATURE

1925

Australopithecus africanus: The Man-Ape of South Africa.

By Prof. RAYMOND A. DART, University of the Witwatersrand, Johannesburg, South Africa.

VARDS the close of 1924, Miss Josephine Salmons, student demonstrator of anatomy in iversity of the Witwatersrand, brought to me silised skull of a cercopithecid monkey which, h her instrumentality, was very generously to the Department for de... G. Izod, of the Rand Mines ... lea is valuable fossil had been ... ne cliff formation—at a vert... dep... horizontal depth of 200 feet at Taungs, ... miles north of Kimberley on the m... to sia, in Bechuanaland, by operati... the rn Lime Company. Important stratigraphical e has been forthcoming recently from this dis- ncerning the succession of stone ages in South (Neville Jones, Jour. Roy. Anthrop. Inst., 1920), e feeling was entertained that this lime deposit, at of Broken Hill in Rhodesia, might contain emains of primitive man.

mediately consulted Dr. R. B. Young, professor ogy in the University of the Witwatersrand, about the discovery, and he, by a fortunate coincidence, was called down to Taungs almost synchronously to investigate geologically the lime deposits of an adjacent farm. During his visit to Taungs, Prof. Young was enabled, through the courtesy of Mr. A. F. Campbell, general manager of the Northern Lime Company, to inspect the site of the discovery and to select further samples of fossil material for me from the same formation. These included a natural cercopithecid endocranial a second and larger cast, and some rock nts disclosing portions of bone. Finally, Dr. D. Laing, senior lecturer in anatomy, obtained hrough his friend Mr. Ridley Hendry, of another e skull from the same cliff. This cercopithecid he possession of Mr. De Wet, of the Langlaagte Mine, has also been liberally entrusted by him to partment for scientific investigation.

Norma facialis of *Australo-us africanus* aligned on the kfort horizontal.

cercopithecid remains placed at our disposal ly represent more than one species of catarrhine The discovery of Cercopithecidæ in this area is vel, for I have been informed that Mr. S. ton has in the press a paper discussing at least ecies of baboon from this same spot (Royal of South Africa). It is of importance that, of the famous Fayüm area, primate deposits een found on the African mainland at Oldaway Reck, *Sitzungsbericht der Gesellsch. Naturforsch.* land, for these discoveries lend promise to the exp tion that a tolerably complete story of higher pri evolution in Africa will yet be wrested from our r

In manipulating the pieces of rock brought bac Prof. Young, I found that the larger natural e ... exactly by its fractured fr ... nother piece of rock in which bro... ower ... terior margin of the left side ... was vi... After cleaning the rock m ... of the ... er and lower part of the f skeleton came i... w. Careful development of solid limesto... which it was embedded fr revealed the almost entire face depicted in the ac panying photographs.

It was apparent when the larger endocranial cast first observed that it was specially important, fo size and sulcal pattern revealed sufficient simila with those of the chimpanzee and gorilla to demonst that one was handling in this instance an anthro and not a cercopithecid ape. Fossil anthropoids h not hitherto been recorded south of the Fayüm Egypt, and living anthropoids have not been discov in recent times south of Lake Kivu region in Bel Congo, nearly 2000 miles to the north, as the c flies.

All fossil anthropoids found hitherto have b known only from mandibular or maxillary fragme so far as crania are concerned, and so the gen appearance of the types they represented has b unknown; consequently, a condition of affairs w virtually the whole face and lower jaw, replete w teeth, together with the major portion of the b pattern, have been preserved, constitutes a specime unusual value in fossil anthropoid discovery. H as in *Homo rhodesiensis*, Southern Africa has prov documents of higher primate evolution that are amo the most complete extant.

Apart from this evidential completeness, the sp men is of importance because it exhibits an ext race of apes *intermediate between living anthropoids man*.

In the first place, the whole cranium disp *humanoid* rather than anthropoid lineaments. I markedly dolichocephalic and leptoprosopic, and m fests in a striking degree the *harmonious relatio* calvaria to face emphasised by Pruner-Bey. Topinard says, "A cranium elongated from be backwards, and at the same time elevated, is alre in harmony by itself; but if the face, on the o hand, is elongated from above downwards, and narro the harmony is complete." I have assessed roug the difference in the relationship of the glab gnathion facial length to the glabella-inion calva length in recent African anthropoids of an age c parable with that of this specimen (depicted in Du worth's "Anthropology and Morphology," sec edition, vol. i.), and find that, if the glabella-in length be regarded in all three as 100, then the glab gnathion length in the young chimpanzee is app

ONE

Raymond Dart and our African origins

C. K. Brain

In the early twentieth century, the prevailing view was that humans had originated in Eurasia. In 1925, the first evidence of an early fossil link between the apes and man was published, and Africa was proposed as the cradle of humanity. This bold claim was largely dismissed at the time. But with further finds, especially in the eastern part of the continent, Africa has since remained at the center of the search for human origins.

In 1924, the fossilized skull of a child, half-ape, half-human, found its way without warning into the hands of a young anatomist in Johannesburg, South Africa. He was in an excellent position to interpret it and, in the subsequent paper in *Nature,*[1] to challenge the accepted concepts of the time. This man was Raymond Dart; his insight shows the value of the prepared mind (see box 1.1).

In 1923, Dart and his wife Dora traveled from Britain to South Africa, where Dart was to take up a new post. He was thirty years old and not enamored of the prospect. He later recalled, "I hated the idea of uprooting myself from what was then the world's center of medicine [University College, London] . . . to take over the anatomy department at Johannesburg's new and ill-equipped University of the Witwatersrand. I felt I had lived a pioneer's life for quite long enough in my younger days." But what was to happen there the following year was surely beyond his wildest dreams.

Dart wished to establish an anatomical museum in his new department, and his attention was drawn to fossilized baboon skulls that were being unearthed in a lime mine at Taung in the northern Cape (fig. 1.1). In *Adventures with the Missing Link,*[2] Dart relates how two boxes of fossils from Taung

Box 1.1 Raymond Dart: a prepared mind

Dart was born in 1893, in rural Queensland, Australia. He won a scholarship to study medicine at the University of Sydney, and it was there that his interest in neurology began—an interest further stimulated by Professor Grafton Elliot Smith, an Australian visiting his homeland from Britain, who lectured on brain evolution. Dart qualified as a doctor during the First World War, serving in both England and France, before joining Elliot Smith's department at University College, London. Here he acquired specialized knowledge of human neuroanatomy.

In 1920, Dart was awarded a fellowship, allowing him to further his studies at Washington University, Saint Louis. He returned briefly to Britain in 1921, before taking up the post in South Africa where he was to make his momentous discovery. In all, Dart had spent ten years in the acquisition of neuroanatomical insight, making his mind very much prepared.

In his later years I worked closely with him in developing the new discipline of African cave taphonomy—deducing past behavior from the piles of bones in caves. Dart was more than just a pioneering thinker: he was intuitive and humble. In our work, he was much more interested in revealing truths than in any status he might derive from them. He was one of the great figures of twentieth-century science.

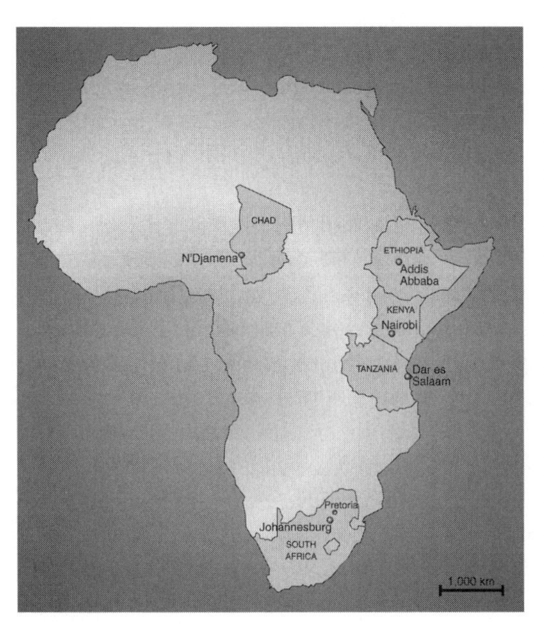

Fig. 1.1. Places and fossil sites in Africa. Taung, where the fossil of *Australopithecus africanus* described by Dart was found, lies some 300 km southwest of Johannesburg. The Sterkfontein caves are about 30 km northwest of Johannesburg. Since 1959, and the discovery of a fossil hominid at Olduvai Gorge in northern Tanzania, the center of attention in paleoanthropology has shifted to East African sites in Tanzania, Kenya, and Ethiopia. With the most recent discoveries, fossil sites in Central Africa, including Chad[7], have also entered the picture.

were delivered to his house one Saturday afternoon in 1924, just as he was dressing for a wedding reception to be held there. Unable to contain his curiosity, he wrenched open the boxes in the driveway. The first did not seem to contain anything of interest. But when he looked into the second, he later recalled:

> a thrill of excitement shot through me. On the very top of the rock heap was what was undoubtedly an endocranial cast or mold of the interior of the skull. Had it been only the fossilised brain cast of any species of ape it would have ranked as a great discovery, for such a thing had never before been reported. But I knew at a glance that what lay in my hands was no ordinary anthropoidal brain. Here in lime-consolidated sand was the replica of a brain three times as large as that of a baboon and considerably bigger than that of an adult chimpanzee. The startling image of the convolutions and furrows of the brain and the blood vessels of the skull were plainly visible.
>
> It was not big enough for primitive man, but even for an ape it was a big bulging brain and, most important, the forebrain was so big and had grown so far backward that it completely covered the hindbrain.
>
> But was there anywhere among this pile of rocks, a face to fit the brain? I ransacked feverishly through the boxes. My search was rewarded, for I found a large stone with a depression into which the cast fitted perfectly. . . .
>
> I stood in the shade holding the brain as greedily as any miser hugs his gold, my mind racing ahead. Here I was certain was one of the most significant finds ever made in the history of anthropology.
>
> Darwin's largely discredited theory that man's early progenitors probably lived in Africa came back to me. Was I to be the instrument by which his 'missing link' was found?
>
> These pleasant daydreams were interrupted by the bridegroom himself tugging at my sleeve. 'My God, Ray,' he said, striving to keep the nervous urgency out of his voice. 'You've got to finish dressing immediately—or I'll have to find another best man. The bridal car should be here any moment'.
>
> Reluctantly, I replaced the rocks in the boxes, but I carried the endocranial cast and the stone from which it had come along with me and locked them away in my wardrobe.

For the next three months Dart used every spare moment to patiently chip away the matrix from the skull, using his wife's sharpened knitting needles. Then, two days before Christmas, the rock parted and the face of a child emerged, with a full set of milk teeth and its permanent molars in the process of erupting. Dart wrote: "I doubt if there was any parent prouder of his offspring than I was of my Taungs baby on that Christmas of 1924."

Dart wasted no time in preparing his report for submission to *Nature.*[1] In essence, it pointed out that while the skull, teeth, and jaw of this child had been "humanoid," rather than anthropoid or apelike, this was undoubtedly a small-brained hominid, or member of the human family—the first of its kind to be described. He pointed out that the forward position of the foramen magnum, where the spinal cord attached to the skull, clearly indicated that this hominid had walked upright, with its hands free for the manipulation of tools and weapons in an open environment far to the south of the equatorial forests inhabited by chimpanzees and gorillas. Finally, Dart asserted that *Australopithecus africanus,* the southern ape of Africa, as he called it, provided clear evidence that Africa had been the cradle of mankind.

Although Charles Darwin had predicted[3] that human ancestors must have lived in Africa, subsequent finds of large-brained fossil humans in Europe had swung scientific opinion in favor of Eurasia as the birthplace of humanity. These included numerous Neanderthal remains, those of the modern-looking Cro-Magnon man from the Dordogne region of France, discovered in 1868, and the Piltdown skull from southern England in 1912, whose large brain and apelike jaw fulfilled the expectations of the time—until it was shown to be a hoax. In fact Piltdown had seemed far more acceptable than had *Pithecanthropus* (now *Homo*) *erectus,* a fossil hominid with a relatively small brain but upright stature whose remains were found in 1893 by Eugene Dubois in river gravels of Java after a five-year search. The reception this discovery received was so disappointing that Dubois locked the remains away for twenty-five years in a Dutch museum before they became available for others to study.

So it is not surprising that Dart's child from Taung, presented as the "missing link" from Africa, met a chilly reception in Europe. The authorities dismissed it as, at best, a relative of the chimpanzee or gorilla with little relevance to human ancestry, stressing that until an adult specimen was available, the matter was hardly worth discussing.

This attitude prevailed even though Dart took the specimen to Britain in 1931 and exhibited it at scientific gatherings. At this time, the Taung child had a strange experience: by mistake, Dora Dart left it in the back of a London taxi. After a prolonged tour of London, the box was opened by the taxi driver who, alarmed at seeing a skull inside, took it straight to a police station. Here a distraught Dora was reunited with the child.

Although Dart's claims endured severe criticism overseas, in South Africa they enjoyed the unwavering support of Robert Broom, a paleontologist known for his work on the evolution of mammals from reptiles. In his later years, while he was based at the Transvaal Museum in Pretoria, Broom started a deliberate search for an adult fossil of *Australopithecus.* His attention was drawn by several of Dart's students to the Sterkfontein caves near Krugersdorp, where lime mining had unearthed fossil baboon skulls as at Taung.

In August 1936, on his second visit to Sterkfontein, Broom was handed the endocranial cast of an adult ape-man by the quarry manager and in the next few days he found much of the rest of the skull. One month later his report on *Australopithecus transvaalensis,* as he named the new find, appeared in *Nature*[4] and *The Illustrated London News.* The initial discovery was followed by many others during the next few years, leaving no doubt as to the hominid status of this African ape-man.

Not content with this, in 1938 Broom described a second kind of hominid from the nearby cave of Kromdraai as *Paranthropus robustus,* with a wide flat face and extremely large molar teeth. Subsequent work has shown that this "robust" ape-man lineage arose from an *africanus*-like stock about 2.5 million years ago and then coexisted with early humans until about a million years ago, when it became extinct.

With fossils of adult ape-men now available for study, Dart's concept of *Australopithecus* as an African ancestor of later humans was generally accepted. Heartened and relieved, Dart reentered the emotional field of hominid paleontology and started a long-term investigation of the Makapansgat Limeworks cave, 250 kilometers northeast of Johannesburg. Here miners had blasted out a vast accumulation of fossil bones, and among them Dart identified and described several new *Australopithecus* specimens. Most of the other fossils came from antelope and Dart speculated as to how all of these bones had found their way to the cave. In a long series of publications he argued that the ape-men had been mighty hunters that underwent a "predatory transition from ape to man," bringing back to their cave those bones from their kills that could serve as useful tools and weapons. Using dramatic and provocative prose, Dart presented his view of "the blood-bespattered archives of humanity" and provoked further research on the ways that bones accumulate in African caves.

East Africa came into the paleontological spotlight in 1959, when Mary Leakey[5] found a very complete robust ape-man skull at Olduvai Gorge, Tanzania; this has been followed by numerous other finds in Tanzania, Kenya, Ethiopia, and elsewhere. Many of these fossils come from lake bed sediments, which can be dated from the volcanic ash beds laid down with them. It appears now that more than four million years ago, small upright-walking hominids such as *Ardipithecus ramidus* and *Australopithecus anamensis* were living in forest-edge habitats of northeast Africa; in time they were succeeded by small *Australopithecus afarensis* individuals, known now by many fossils including the skeleton of "Lucy" immortalized by Don Johanson.[6] These appear to have been the ancestors of Dart's *Australopithecus africanus,* which could have given rise to both our own and the robust ape-man lineages (fig. 1.2).

Today paleoanthropology is a rapidly evolving field. New discoveries and interpretations confirm Africa's place as an evolutionary center. Attention has lately shifted to Chad, in the central part of the continent, with the announcement[7] of the discovery of a six- to seven-million-year-old hominid

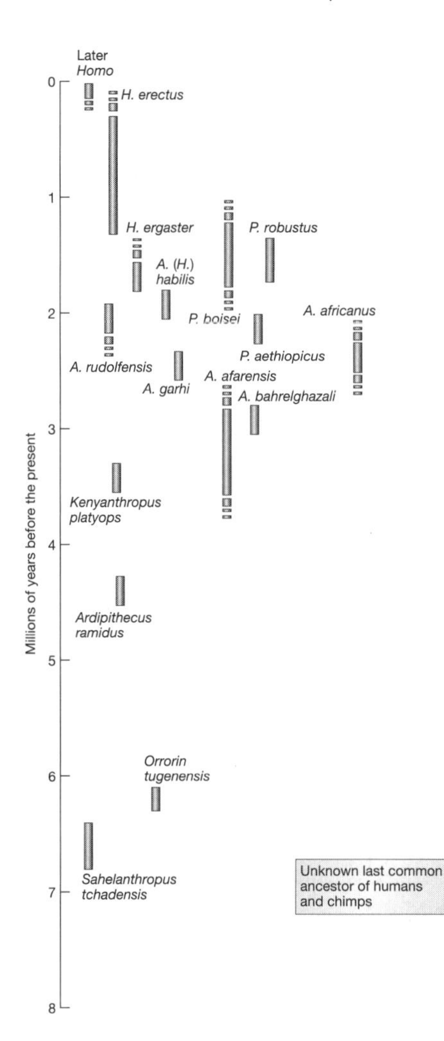

Fig. 1.2. Currently recognized hominid groups. There is considerable uncertainty and disagreement over the names and relationships of the various species, but shown here are the main groups (genera) now usually recognized. Broken bars indicate the uncertainty in the time spans over which each species existed. The generic *H.* indicates *Homo* (of which we, *Homo sapiens,* are the only surviving species); *P.* indicates *Paranthropus; A.* indicates *Australopithecus.* The "robust" ape-man lineage, which was an evolutionary dead end, is represented by *Paranthropus* species. The dating and interpretation of the three oldest species discovered to date—*Ardipithecus ramidus, Orrorin tugenensis,* and *Sahelanthropus tchadensis*—especially remain subjects of debate. In discovering *Australopithecus africanus,* Dart identified the first piece in the puzzle of human evolution in Africa.

skull there. But Asia should not be ignored, as the latest evidence of early *Homo erectus* technology in Japan and China has emphasized. Moreover, fossil discoveries are now not the only way to investigate human origins: molecular techniques, which involve the tracing of our ancestry through analysis of genetic material in living humans, and even in Neanderthals, provide a further tool.[8] Dart would be delighted with the expansion of his vision.

References

1. Dart, R. A. *Australopithecus africanus:* the man-ape of South Africa. *Nature* **115,** 195–199 (1925).
2. Dart, R. A. & Craig, D. *Adventures with the Missing Link,* pp. 5–6 (Hamish Hamilton, London, 1959; also published by Better Baby Press, Philadelphia, 1982).
3. Darwin, C. *The Descent of Man and Selection in Relation to Sex* (John Murray, London, 1871).
4. Broom, R. A new fossil anthropoid skull from South Africa. *Nature* **138,** 486–488 (1936).
5. Leakey, L. S. B. A new fossil skull from Olduvai. *Nature* **184,** 491–493 (1959).
6. Johanson, D. C., White, T. D. & Coppens, Y. A new species of the genus *Australopithecus* (Primates : Hominidae) from the Pliocene of eastern Africa. *Kirtlandia* (Cleveland, Ohio) **28,** 1–14 (1978).
7. Brunet, M. *et al.* A new hominid from the Upper Miocene of Chad, Central Africa. *Nature* **418,** 145–151 (2002).
8. Stringer, C. & McKie, R. *African Exodus: The Origins of Modern Humanity* (Cape, London, 1996).

Further reading

Dart, R. A. The osteodontokeratic culture of *Australopithecus prometheus* (Transvaal Museum Memoir No. 10, Pretoria, 1957).

Johanson, D. C. & Edey, M. A. *Lucy: The Beginnings of Humankind* (Simon & Schuster, New York, 1981).

Leakey, R. E. *The Making of Mankind* (Michael Joseph, London, 1981).

Lewin, R. *Bones of Contention: Controversies in the Search for Human Origins* (Simon & Schuster, New York, 1987).

Lewin, R. *Human Evolution: An Illustrated Introduction* (Freeman, New York, 1984).

Tattersall, I. *Becoming Human: Evolution and Human Uniqueness* (Harcourt Brace, San Diego, 1998).

Wheelhouse, F. & Smithford, K. S. *Dart: Scientist and Man of Grit* (Transpareon Press, Sydney, 2001).

1925

Australopithecus africanus: the man-ape of South Africa

Raymond A. Dart

Towards the close of 1924, Miss Josephine Salmons, student demonstrator of anatomy in the University of the Witwatersrand, brought to me the fossilised skull of a cercopithecid monkey which, through her instrumentality, was very generously loaned to the Department for description by its owner, Mr. E. G. Izod, of the Rand Mines Limited. I learned that this valuable fossil had been blasted out of the limestone cliff formation—at a vertical depth of 50 feet and a horizontal depth of 200 feet—at Taungs, which lies 80 miles north of Kimberley on the main line to Rhodesia, in Bechuanaland, by operatives of the Northern Lime Company. Important stratigraphical evidence has been forthcoming recently from this district concerning the succession of stone ages in South Africa (Neville Jones, Jour. Roy. Anthrop. Inst., 1920), and the feeling was entertained that this lime deposit, like that of Broken Hill in Rhodesia, might contain fossil remains of primitive man.

I immediately consulted Dr. R. B. Young, professor of geology in the University of the Witwatersrand, about the discovery, and he, by a fortunate coincidence, was called down to Taungs almost synchronously to investigate geologically the lime deposits of an adjacent farm. During his visit to Taungs, Prof. Young was enabled, through the courtesy of Mr. A. F. Campbell, general manager of the Northern Lime Company, to inspect the site of the discovery and to select further samples of fossil material for me from the same formation. These included a natural cercopithecid endocranial cast, a second and larger cast, and some rock fragments disclosing portions of bone. Finally, Dr. Gordon D. Laing, senior lecturer in anatomy, obtained news, through his friend Mr. Ridley

Hendry, of another primate skull from the same cliff. This cercopithecid skull, the possession of Mr. De Wet, of the Langlaagte Deep Mine, has also been liberally entrusted by him to the Department for scientific investigation.

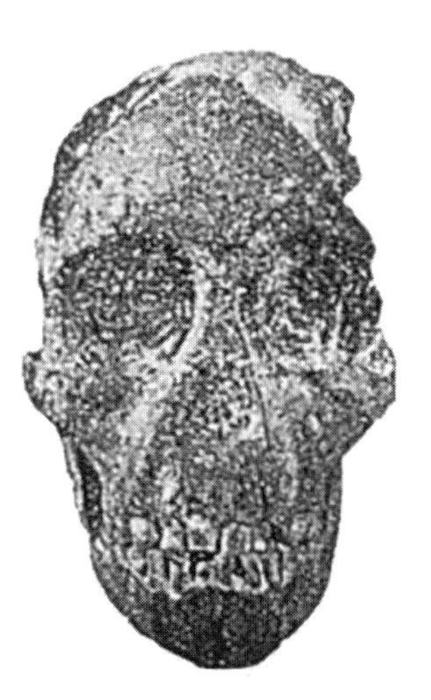

Fig. 1. Norma facialis of *Australopithecus africanus* aligned on the Frankfort horizontal.

The cercopithecid remains placed at our disposal certainly represent more than one species of catarrhine ape. The discovery of Cercopithecidæ in this area is not novel, for I have been informed that Mr. S. Haughton has in the press a paper discussing at least one species of baboon from this same spot (Royal Society of South Africa). It is of importance that, outside of the famous Fayüm area, primate deposits have been found on the African mainland at Oldaway (Hans Reck, *Sitzungsbericht der Gesellsch. Naturforsch. Freunde,* 1914), on the shores of Victoria Nyanza (C. W. Andrews, *Ann. Mag. Nat. Hist.,* 1916), and in Bechuanaland, for these discoveries lend promise to the expectation that a tolerably complete story of higher primate evolution in Africa will yet be wrested from our rocks.

In manipulating the pieces of rock brought back by Prof. Young, I found that the larger natural endocranial cast articulated exactly by its fractured frontal extremity with another piece of rock in which the broken lower and posterior margin of the left side of a mandible was visible. After cleaning the rock mass, the outline of the hinder and lower part of the facial skeleton came into view. Careful development of the solid limestone in which it was embedded finally revealed the almost entire face depicted in the accompanying photographs.

It was apparent when the larger endocranial cast was first observed that it was specially important, for its size and sulcal pattern revealed sufficient similarity with those of the chimpanzee and gorilla to demonstrate that one was handling in this instance an anthropoid and not a cercopithecid ape. Fossil anthropoids have not hitherto been recorded south of the Fayüm in Egypt, and living anthropoids have not been discovered in recent times south of Lake Kivu region in Belgian Congo, nearly 2000 miles to the north, as the crow flies.

All fossil anthropoids found hitherto have been known only from mandibular or maxillary fragments, so far as crania are concerned, and so the general appearance of the types they represented has been unknown; consequently, a condition of affairs where virtually the whole face and lower jaw, replete with teeth, together with the major portion of the brain pattern, have been preserved, constitutes a specimen of unusual value in fossil anthropoid discovery. Here, as in *Homo rhodesiensis,* Southern Africa has provided documents of higher primate evolution that are amongst the most complete extant.

Apart from this evidential completeness, the specimen is of importance because it exhibits an extinct race of apes *intermediate between living anthropoids and man.*

In the first place, the whole cranium displays *humanoid* rather than anthropoid lineaments. It is markedly dolichocephalic and leptoprosopic, and manifests in a striking degree the *harmonious relation* of calvaria to face emphasised by Pruner-Bey. As Topinard says, "A cranium elongated from before backwards, and at the same time elevated, is already in harmony by itself; but if the face, on the other hand, is elongated from above downwards, and narrows, the harmony is complete." I have assessed roughly the difference in the relationship of the glabella-gnathion facial length to the glabella-inion calvarial length in recent African anthropoids of an age comparable with that of this specimen (depicted in Duckworth's "Anthropology and Morphology," second edition, vol. i.), and find that, if the glabella-inion length be regarded in all three as 100, then the glabella-gnathion length in the young chimpanzee is approximately 88, in the young gorilla 80, and in this fossil 70, which proportion suitably demonstrates the enhanced relationship of cerebral length to facial length in the fossil (Fig. 2).

The glabella is tolerably pronounced, but any traces of the salient supra-orbital ridges, which are present even in immature living anthropoids, are here entirely absent. Thus the relatively increased glabella-

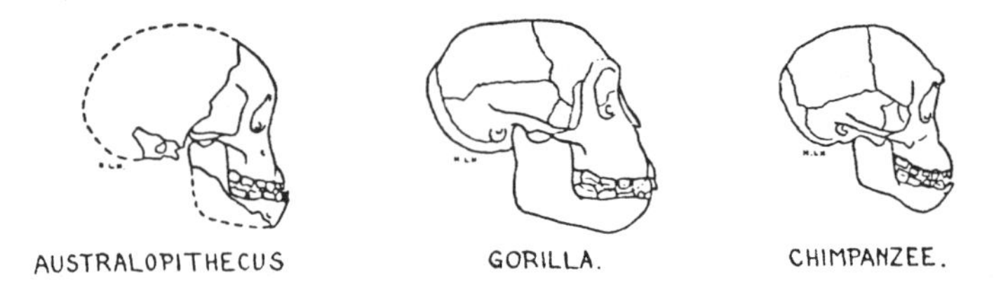

Fig. 2. Cranial form in living anthropoids of similar age (after Duckworth) and in the new fossil. For this comparison, the fossil is regarded as having the same calvarial length as the gorilla.

inion measurement is due to brain and not to bone. Allowing 4 mm. for the bone thickness in the inion region, that measurement in the fossil is 127 mm.; *i.e.* 4 mm. less than the same measurement in an adult chimpanzee in the Anatomy Museum at the University of the Witwatersrand. The orbits are not in any sense detached from the forehead, which rises steadily from their margins in a fashion amazingly human. The interorbital width is very small (13 mm.) and the ethmoids are not blown out laterally as in modern African anthropoids. This lack of ethmoidal expansion causes the lacrimal fossæ to face posteriorly and to lie relatively far back in the orbits, as in man. The orbits, instead of being subquadrate as in anthropoids, are almost circular, furnishing an orbital index of 100, which is well within the range of human variation (Topinard, "Anthropology"). The malars, zygomatic arches, maxillæ, and mandible all betray a delicate and humanoid character. The facial prognathism is relatively slight, the gnathic index of Flower giving a value of 109, which is scarcely greater than that of certain Bushmen (Strandloopers) examined by Shrubsall. The nasal bones are not prolonged below the level of the lower orbital margins, as in anthropoids, but end above these, as in man, and are incompletely fused together in their lower half. Their maximum length (17 mm.) is not so great as that of the nasals in *Eoanthropus dawsoni*. They are depressed in the median line, as in the chimpanzee, in their lower half, but it seems probable that this depression has occurred post-mortem, for the upper half of each bone is arched forwards (Fig. 1). The nasal aperture is small and is just wider than it is high (17 mm. × 16 mm.). There is no nasal spine, the floor of the nasal cavity being continuous with the anterior aspect of the alveolar portions of the maxillæ, after the fashion of the chimpanzee and of certain New Caledonians and negroes (Topinard, *loc. cit.*).

In the second place, the dentition is *humanoid* rather than anthropoid. The specimen is juvenile, for the first permanent molar tooth only has erupted in both jaws on both sides of the face; *i.e.* it corresponds anatomically with a human child of six years of age. Observations upon the milk dentition of living primates are few, and only one molar tooth of the deciduous dentition in one fossil anthropoid is known (Gregory, "The Origin and Evolution of the Human Dentition," 1920). Hence the data for the necessary comparisons are meagre, but certain striking features of the milk dentition of this creature may be mentioned. The tips of the canine teeth transgress very slightly (0.5–0.75 mm.) the general margin of the teeth in each jaw, *i.e.* very little more than does the human milk canine. There is no diastema whatever between the premolars and ca-

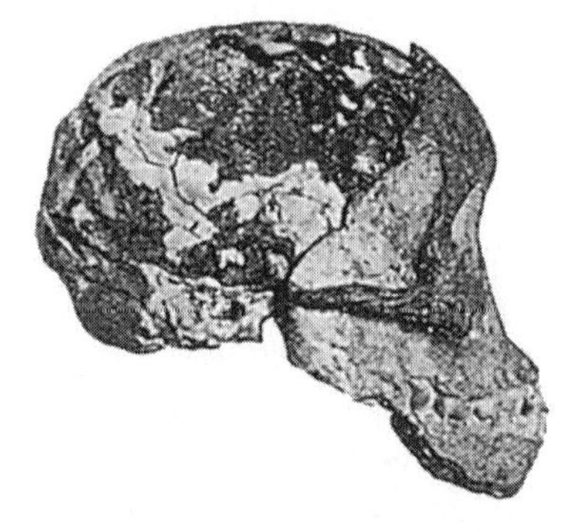

Fig. 3. Norma lateralis of *Australopithecus africanus* aligned on the Frankfort horizontal.

nines on either side of the lower jaw, such as is present in the deciduous dentition of living anthropoids; but the canines in this jaw come, as in the human jaw, into alignment with the incisors (Gregory, *loc. cit.*). There is a diastema (2 mm. on the right side, and 3 mm. on the left side) between the canines and lateral incisors of the upper jaw; but seeing, first, that the incisors are narrow, and, secondly, that diastemata (1 mm.–1.5 mm.) occur between the central incisors of the upper jaw and between the medial and lateral incisors of both sides in the lower jaw, and, thirdly, that some separation of the milk teeth takes place even in mankind (Tomes, "Dental Anatomy," seventh edition) during the establishment of the permanent dentition, it is evident that the diastemata which occur in the upper jaw are small. The lower canines, nevertheless, show wearing facets both for the upper canines and for the upper lateral incisors.

The incisors as a group are irregular in size, tend to overlap one another, and are almost vertical, as in man; they are not symmetrical and well spaced, and do not project forwards markedly, as in anthropoids. The upper lateral incisors do project forwards to some extent and perhaps also do the upper central incisors very slightly, but the lateral lower incisors betray no evidence of forward projection, and the central lower incisors are not even vertical as in most races of mankind, but are directed slightly backwards, as *sometimes* occurs in man. Owing to these remarkably human characters displayed by the deciduous dentition, when contour tracings of the upper jaw are made, it is found that the jaw and the teeth, as a whole, take up a parabolic arrangement comparable only with that presented by mankind amongst the higher primates. These facts, together with the more minute anatomy of the teeth, will be illustrated and discussed in the memoir which is in the process of elaboration concerning the fossil remains.

In the third place, the mandible itself is *humanoid* rather than anthropoid. Its ramus is, on the whole, short and slender as compared with that of anthropoids, but the bone itself is more massive than that of a human being of the same age. Its symphyseal region is virtually complete and reveals anteriorly a more vertical outline than is found in anthropoids or

Fig. 4. Norma basalis of *Australopithecus africanus* aligned on the Frankfort horizontal.

even in the jaw of Piltdown man. The anterior symphyseal surface is scarcely less vertical than that of Heidelberg man. The posterior symphyseal surface in living anthropoids differs from that of modern man in possessing a pronounced posterior prolongation of the lower border, which joins together the two halves of the mandible, and so forms the well-known *simian shelf* and above it a deep genial impression for the attachment of the tongue musculature. In this character, *Eoanthropus dawsoni* scarcely differs from the anthropoids, especially the chimpanzee; but this new fossil betrays no evidence of such a shelf, the lower border of the mandible having been massive and rounded after the fashion of the mandible of *Homo heidelbergensis.*

That hominid characters were not restricted to the face in this extinct primate group is borne out by the relatively forward situation of the foramen magnum. The position of the basion can be assessed within a few millimetres of error, because a portion of the right exoccipital is present alongside the cast of the basal aspect of the cerebellum. Its position is such that the basi-prosthion measurement is 89 mm., while the basi-inion measurement is at least 54 mm. This relationship may be expressed in the form of a "head-balancing" index of 60.7. The same index in a baboon provides a value of 41.3, in an adult chimpanzee 50.7, in Rhodesian man 83.7, in a dolichocephalic European 90.9, and in a brachycephalic European 105.8. It is significant that this index, which indicates in a measure the poise of the skull upon the vertebral column, points to the assumption by this fossil group of an attitude appreciably more erect than that of modern anthropoids. The improved poise of the head, and the better posture of the whole body framework which accompanied this alteration in the angle at which its dominant member was supported, is of great significance. It means that a greater reliance was being placed by this group upon the feet as organs of progression, and that the hands were being freed from their more primitive function of accessory organs of locomotion. Bipedal animals, their hands were assuming a higher evolutionary rôle not only as delicate tactual, examining organs which were adding copiously to the animal's knowledge of its physical environment, but also as instruments of the growing intelligence in carrying out more elaborate, purposeful, and skilled movements, and as organs of offence and defence. The latter is rendered the more probable, in view, first, of their failure to develop massive canines and hideous features, and, secondly, of the fact that even living baboons and anthropoid apes can and do use sticks and stones as implements and as weapons of offence ("Descent of Man," p. 81 *et seq.*).

Lastly, there remains a consideration of the endocranial cast which was responsible for the discovery of the face. The cast comprises the right cerebral and cerebellar hemispheres (both of which fortunately meet the median line throughout their entire dorsal length) and the anterior portion of the left cerebral hemisphere. The remainder of the cranial cavity seems to have been empty, for the left face of the cast is clothed with a picturesque lime crystal deposit; the vacuity in the left half of the cranial cavity was probably responsible for the fragmentation of the specimen during the blasting. The cranial capacity of the specimen may best be appreciated by the statement that the length of the cavity could not have been less than 114 mm., which is 3 mm. greater than that of an adult chimpanzee in the Museum of the Anatomy Department in the University of the Witwatersrand, and only 14 mm. less than the greatest length of the cast of the endocranium of a gorilla chosen for casting on account of its great size. Few data are available concerning the expansion of brain matter which takes place in the living anthropoid brain between the time of eruption of the first permanent molars and the time of their becoming adult. So far as man is concerned, Owen ("Anatomy of Vertebrates," vol. iii.) tells us that "The brain has advanced to near its term of size at about ten years, but it does not usually obtain its full development till between twenty and thirty years of age." R. Boyd (1860) discovered an increase in weight of nearly 250 grams in the brains of male human beings after they had reached the age of seven years. It is therefore reasonable to believe that the adult forms typified by our present specimen possessed brains which were larger than that of this juvenile specimen, and equalled, if they did not actually supersede, that of the gorilla in absolute size.

Whatever the total dimensions of the adult brain may have been, there are not lacking evidences that the brain in this group of fossil forms was distinctive in type and was an instrument of greater intelligence than that of living anthropoids. The face of the endocranial cast is scarred unfortu-

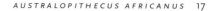

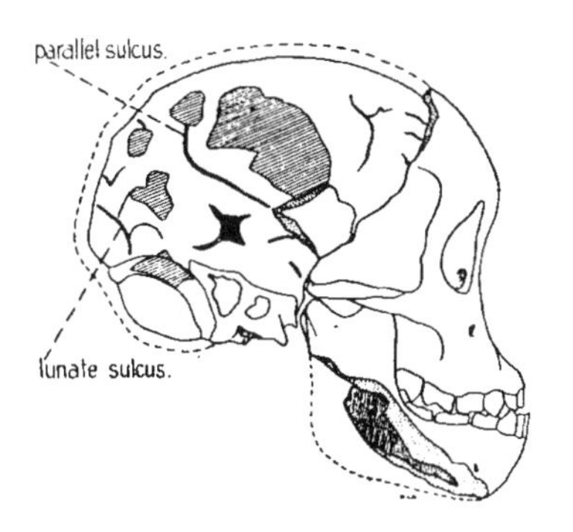

Fig. 5. Dioptographic tracing of *Australopithecus africanus* (right side), X1/3.

nately in several places (cross-hatched in the dioptographic tracing—see Fig. 5). It is evident that the relative proportion of cerebral to cerebellar matter in this brain was greater than in the gorilla's. The brain does not show that general pre- and post-Rolandic flattening characteristic of the living anthropoids, but presents a rounded and well-filled-out contour, which points to a symmetrical and balanced development of the faculties of associative memory and intelligent activity. The pithecoid type of parallel sulcus is preserved, but the sulcus lunatus has been thrust backwards towards the occipital pole by a pronounced general bulging of the parieto-temporo-occipital association areas.

To emphasise this matter, I have reproduced (Fig. 6) superimposed coronal contour tracings taken at the widest part of the parietal region in the gorilla endocranial cast and in this fossil. Nothing could illustrate better the mental gap that exists between living anthropoid apes and the group of creatures which the fossil represents than the flattened atrophic appearance of the parietal region of the brain (which lies between the visual field on one hand, and the tactile and auditory fields on the other) in the former and its surgent vertical and dorso-lateral expansion in the latter. The expansion in this area of the brain is the more significant in that it explains the posterior *humanoid* situation of the sulcus lunatus.

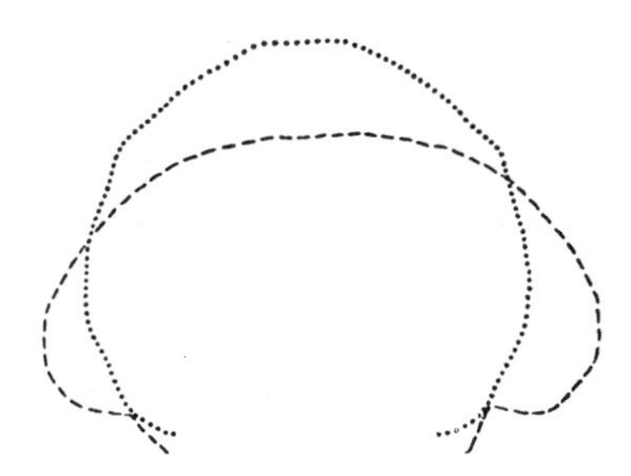

Fig. 6. Contour tracings of coronal sections through the widest part of the parietal region of the endocranial casts in Australopithecus ...... and in a gorilla -------.

It indicates (together with the narrow interorbital interval and human characters of the orbit) the fact that this group of beings, having acquired the faculty of stereoscopic vision, had profited beyond living anthropoids by setting aside a relatively much larger area of the cerebral cortex to serve as a storehouse of information concerning their objective environment as its details were simultaneously revealed to the senses of vision and touch, and also of hearing. They possessed to a degree unappreciated by living anthropoids the use of their hands and ears and the consequent faculty of associating with the colour, form, and general appearance of objects, their weight, texture, resilience, and flexibility, as well as the significance of sounds emitted by them. In other words, their eyes saw, their ears heard, and their hands handled objects with greater meaning and to fuller purpose than the corresponding organs in recent apes. They had laid down the foundations of that discriminative knowledge of the appearance, feeling, and sound of things that was a necessary milestone in the acquisition of articulate speech.

There is, therefore, an ultra-simian quality of the brain depicted in this immature endocranial cast which harmonises with the ultra-simian features revealed by the entire cranial topography and corroborates the various inferences drawn therefrom. The two thousand miles of territory which separate this creature from its nearest living anthropoid cousins is indirect testimony to its increased intelligence and mastery of its environment. It is manifest that we are in the presence here of a pre-human stock, neither chimpanzee nor gorilla, which possesses a series of differential characters not encountered hitherto in any anthropoid stock. This complex of characters exhibited is such that it cannot be interpreted as belonging to a form ancestral to any living anthropoid. For this reason, we may be equally confident that there can be no question here of a primitive anthropoid stock such as has been recovered from the Egyptian Fayüm. Fossil anthropoids, varieties of Dryopithecus, have been retrieved in many parts of Europe, Northern Africa, and Northern India, but the present specimen, despite its youth, cannot be confused with anthropoids having the dryopithecid dentition. Other fossil anthropoids from the Siwalik hills in India (Miocene and Pliocene) are known which, according to certain observers, may be ancestral to modern anthropoids and even to man.

Whether our present fossil is to be correlated with the discoveries made in India is not yet apparent; that question can only be solved by a careful comparison of the permanent molar teeth from both localities. It is obvious, meanwhile, that it represents a fossil group distinctly advanced beyond living anthropoids in those two dominantly human characters of facial and dental recession on one hand, and improved quality of the brain on the other. Unlike Pithecanthropus, it does not represent an ape-like man, a caricature of precocious hominid failure, but a creature well advanced beyond modern anthropoids in just those characters, facial and cerebral, which are to be anticipated in an extinct link between man and his simian ancestor. At the same time, it is equally evident that a creature with anthropoid brain capacity, and lacking the distinctive, localised temporal expansions which appear to be concomitant with and necessary to articulate man, is no true man. It is therefore logically regarded as a man-like ape. I propose tentatively, then, that a new family of *Homo-simiadæ* be created for the reception of the group of individuals which it represents, and that the first known species of the group be designated *Australopithecus africanus,* in commemoration, first, of the extreme southern and unexpected horizon of its discovery, and secondly, of the continent in which so many new and important discoveries connected with the early history of man have recently been made, thus vindicating the Darwinian claim that Africa would prove to be the cradle of mankind.

It will appear to many a remarkable fact that an ultra-simian and prehuman stock should be discovered, in the first place, at this extreme southern point in Africa, and, secondly, in Bechuanaland, for one does not associate with the present climatic conditions obtaining on the eastern fringe of the Kalahari desert an environment favourable to higher primate life. It is generally believed by geologists (*vide* A. W. Rogers, "Post-Cretaceous Climates of South Africa," *South African Journal of Science,* vol. xix., 1922) that the climate has fluctuated within exceedingly narrow limits in this country since Cretaceous times. We must therefore conclude that it was only the enhanced cerebral powers possessed by this group which made their existence possible in this untoward environment.

In anticipating the discovery of the true links between the apes and man in tropical countries, there has been a tendency to overlook the fact that, in the luxuriant forests of the tropical belts, Nature was supplying with profligate and lavish hand an easy and sluggish solution, by adaptive specialisation, of the problem of existence in creatures so well equipped mentally as living anthropoids are. For the production of man a different apprenticeship was needed to sharpen the wits and quicken the higher manifestations of intellect—a more open veldt country where competition was keener between swiftness and stealth, and where adroitness of thinking and movement played a preponderating rôle in the preservation of the species. Darwin has said, "no country in the world abounds in a greater degree with dangerous beasts than Southern Africa," and, in my opinion, Southern Africa, by providing a vast open country with occasional wooded belts and a relative scarcity of water, together with a fierce and bitter mammalian competition, furnished a laboratory such as was essential to this penultimate phase of human evolution.

In Southern Africa, where climatic conditions appear to have fluctuated little since Cretaceous times, and where ample dolomitic formations have provided innumerable refuges during life, and burial-places after death, for our troglodytic forefathers, we may confidently anticipate many complementary discoveries concerning this period in our evolution.

In conclusion, I desire to place on record my indebtedness to Miss Salmons, Prof. Young, and Mr. Campbell, without whose aid the discovery would not have been made; to Mr. Len Richardson for providing the photographs; to Dr. Laing and my laboratory staff for their willing assistance; and particularly to Mr. H. Le Helloco, student demonstrator in the Anatomy Department, who has prepared the illustrations for this preliminary statement.

University of the Witwatersrand, Johannesburg, South Africa

[Published 7 February 1925]

Letters to the Editor.

*[ditor does not hold himself responsible for
ons expressed by his correspondents. Neither
e undertake to return, nor to correspond with
riters of, rejected manuscripts intended for this
y other part of* Nature. *No notice is taken
nymous communications.]*

cattering of Electrons by a Single Crystal of Nickel.

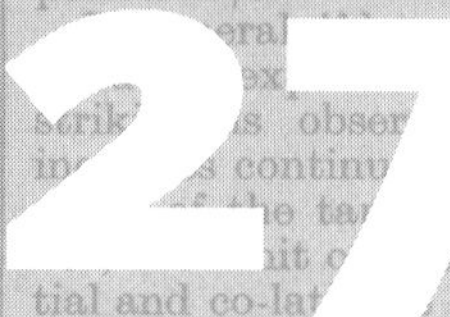

series of experiments now ... ro... , w...
g a narrow beam of electr... or... a...
cut from a single crysta... nic...
ng the intensity of sca... g (numb...
s per unit solid angle with speeds near ... of
nbarding electrons) in various d... in
the target. The experimental arrangement
hat the intensity of scattering can be measured

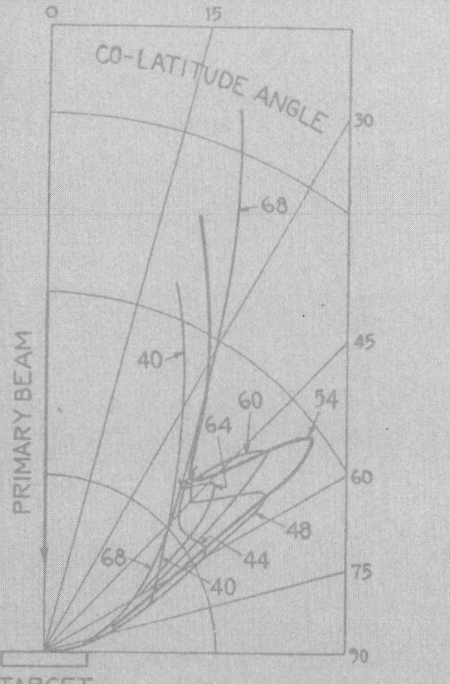

—Intensity of electron scattering vs. co-latitude angle for various bombarding voltages—azimuth-{111}-330°.

latitude from the equator (plane of the target) in 20° of the pole (incident beam) and in any n.

face of the target is cut parallel to a set of lanes of the crystal lattice, and etching by ation has been employed to develop its surface 11}-facets. The bombardment covers an area t 2 mm.[2] and is normal to these facets.

ewed along the incident beam the arrangement s in the crystal exhibits a threefold symmetry. 100}-normals equally spaced in azimuth emerge ne crystal in latitude 35°, and, midway in between these, three {111}-normals emerge de 20°. It will be convenient to refer to the of any one of the {100}-normals as a {100}-, and to that of any one of the {111}-normals 11}-azimuth. A third set of azimuths must specified; this bisects the dihedral angle adjacent {100}- and {111}-azimuths and

target. There are six such azimuths, and any one these will be referred to as a {110}-azimuth. It follo from considerations of symmetry that if the intens of scattering exhibits a dependence upon azimuth we pass from a {100}-azimuth to the next adjace {111}-azimuth (60°), the same dependence must exhibited in the reverse order as we continue through 60° to the next following {100}-azimu Dependence on azimuth must be an even function period $2\pi/3$.

...eral ... ding potential and azimuth ... ex... made in latitude, nothing v... strik... s obser... The intensity of scatter... in... s continu... and regularly from zero in ... of the ta... o a highest value in co-latitu... ...it o... ervations. If bombarding pote... tial and co-lat... are fixed and exploration is ma... in azimuth, ... ation in the intensity of scatteri... of the type to be expected is always observed, but general this variation is slight, amounting in so... cases to not more than a few per cent. of the avera... intensity. This is the nature of the scattering bombarding potentials in the range from 15 volts near 40 volts.

At 40 volts a slight hump appears near 60° in co-latitude curve for azimuth-{111}. This hu... develops rapidly with increasing voltage into a stro... spur, at the same time moving slowly upward tow... the incident beam. It attains a maximum intens... in co-latitude 50° for a bombarding potential 54 volts, then decreases in intensity, and disappea... in co-latitude 45° at about 66 volts. The growth a... decay of this spur are traced in Fig. 1.

A section in azimuth through this spur at maximum (Fig. 2—Azimuth-330°) shows that it sharp in azimuth as well as in latitude, and that forms one of a set of three such spurs, as was to expected. The width of these spurs both in latitu... and in azimuth is almost completely accounted for the low resolving power of the measuring device. *T... spurs are due to beams of scattered electrons which ... nearly if not quite as well defined as the primary bea...* The minor peaks occurring in the {100}-azimuth ... sections of a similar set of spurs that attains maximum development in co-latitude 44° for a bo... barding potential of 65 volts.

Thirteen sets of beams similar to the one j... described have been discovered in an exploration the principal azimuths covering a voltage range fr... 15 volts to 200 volts. The data for these are set do... on the left in Table I. (columns 1-4). Small correctio... have been applied to the observed co-latitude ang... to allow for the variation with angle of the 'ba... ground scattering,' and for a small angular displa... ment of the normal to the facets from the incide... beam.

If the incident electron beam were replaced by beam of monochromatic X-rays of adjustable wa... length, very similar phenomena would, of course, observed. At particular values of wave-length, sets three or of six diffraction beams would emerge fr... the incident side of the target. On the right Table I. (columns 5, 6 and 7) are set down data the ten sets of X-ray beams of longest wave-leng... which would occur within the angular range of o... observations. Each of these first ten occurs in o... of our three principal azimuths.

Several points of correlation will be noted betwe... the two sets of data. Two points of difference w... also be noted; the co-latitude angles of the electr... beams are not those of the X-ray beams, and t... three electron beams listed at the end of the Ta... appear to have no X-ray analogues.

TWO

Electrons make waves

Akira Tonomura

At the atomic scale, the distinction between particles and waves breaks down. The realization, early in the twentieth century, that light consists of discrete packages, or "photons," prompted the converse suggestion that particles could behave like waves. Confirmation of this idea came in 1927, from an experiment involving the scattering of electrons from a crystal. Today, "wave-particle duality" is central to quantum mechanics, and the wave properties of subatomic particles are widely used in microscopy and crystallography.

Accident plays a larger part in the history of scientific discovery than is generally acknowledged. When Clinton Davisson and Lester Germer, at the Western Electric Company laboratories in New York, bombarded a nickel target with an electron beam in 1924, their aim was to investigate the atomic structure of nickel. But an accident in the laboratory, and a fortuitous trip by Davisson to a conference in England, led them to realize that they were onto something much more important. In fact, their experiments, reported in *Nature* in 1927,[1] provided the first evidence that electrons could behave like waves. This set the seal on the concept of wave-particle duality—one of the pillars of the quantum revolution that shook physics in the early part of the twentieth century.

Our story starts in 1900, when Max Planck explained a puzzling property of the radiation emitted by heated bodies by proposing that light is emitted and absorbed in packets, or "quanta," of energy. Each quantum of light would have an amount of energy proportional to the frequency of the light—a relationship expressed in the equation $E = h\nu$, where E is energy, ν (the Greek letter "nu") is frequency, and h is a constant of proportionality, now known as the Planck constant. In 1905, Albert Einstein famously used the idea of light quanta—later to be named "photons"—to explain the process by which light ejects electrons from atoms (the "photoelectric effect") and some other phenomena that could not be explained in the framework of classical physics.

While light waves were being rediscovered as particles, the reverse fate awaited electrons. J. J. Thomson had identified electrons as negatively charged particles that comprised the radiation from cathode ray tubes, and had deduced that they were elementary constituents of all atoms. When, in 1911, Ernest Rutherford discovered the atomic nucleus, he proposed a "planetary" model for the atom, in which negatively charged electrons orbit the positively charged nucleus. But according to classical physics, the electrons would be expected to radiate energy as they orbited, causing them to spiral down into the nucleus; in such a picture, no atom could survive for longer than about a hundred trillionths of a second.

Two years later, Niels Bohr addressed this problem by introducing Planck's quantum into atoms. Bohr postulated that electrons were allowed to occupy only certain orbits, with specific ("quantized") values of energy. Electrons could move from one orbit to another by absorbing or emitting a quantum of energy, but they could not radiate energy continuously. These ideas, like Einstein's proposal for photons, were in blatant violation of the laws of classical physics. Bohr's model seemed to be on the right track, because it accounted for the discrete frequencies of light emitted from hydrogen atoms. But his quantum condition remained mysterious, having no supporting physical picture.

Such a picture was provided by Louis de Broglie, who suggested that if light could behave as particles, then why shouldn't electrons behave as waves? In his 1924 doctoral thesis for the Sorbonne, he proposed that electrons of velocity v should be associated with a wave of wavelength $\lambda = h/mv$ (where m is the electron mass). Bohr's quantum condition could then be understood as a requirement that the electron wave form a standing wave along the electron's path around the nucleus. In other words, the electron wavelength should fit a whole number of times into the circumference of the orbit.

De Broglie's hypothesis implied that electrons outside atoms should also exhibit wavelike properties, such as interference and diffraction. But there was no evidence for such behavior: this is what Davisson and Germer's experiments would eventually provide.

Davisson and his colleague Charles Kunsman, who were working for the engineering department of Western Electric (later to become the Bell Telephone Laboratories), had started an experiment in the early 1920s to study the secondary electrons emitted from nickel surfaces bombarded by electron beams. The original purpose of the experiment was to provide ammunition for a patent dispute with the General Electric Company, relating to the behavior of vacuum tubes.

Much to Davisson and Kunsman's surprise, some electrons were reflected back without changing their energies. Davisson tried hard to relate this behavior to the structure of atoms in much the same way as Rutherford had analyzed the back-scattering of α-particles (see p. 40) by an atomic nucleus. He failed to do so, however, and returned to work on the emission of electrons from the heated filaments of vacuum tubes.

In 1924, Davisson and Germer restarted the electron bombardment experiments. The following year, they had what can only be described as an accidental breakthrough. A bottle of liquefied air exploded in the laboratory: the glass tube containing the nickel target cracked, causing the nickel to become oxidized. To Davisson and Germer's surprise, the angular distribution of reflected electrons measured after this accident was completely different: instead of varying smoothly with angle, the number of electrons showed strong peaks at particular angles (fig. 2.1).

Puzzled by this result, they cut the glass tube open and found that the heating process used to remove the oxidized layer had caused the nickel target to recrystallize from many small crystals into a few large ones. The resulting change in the electrons' behavior showed that they were responding to the crystal structure of the target, not the atomic structure of nickel. Moreover—although Davisson and Germer did not recognize it until later—the presence of peaks at special angles was reminiscent of the diffraction of X-rays from a crystal (fig. 2.2). Clearly, they were onto something quite new.

After this, they decided to use single-crystal nickel for the target, instead of a sample containing crystals of various orientations. After a year of sample preparation, they resumed the experiments in April 1926. But the first results obtained were similar to those before the accident: they were not able to reproduce the sharp peaks. The reason for this isn't known, but it may be that impurity atoms adsorbed on the nickel surface affected the behavior of the rebounding electrons.

That summer, a disappointed Davisson sailed to England with his wife for the holidays. While there he attended the annual meeting of the British Association for the Advancement of Science, in Oxford, and was surprised to hear Max Born say that Davisson and Kunsman's results—the ones without the strong peaks—might provide evidence for the electron waves that

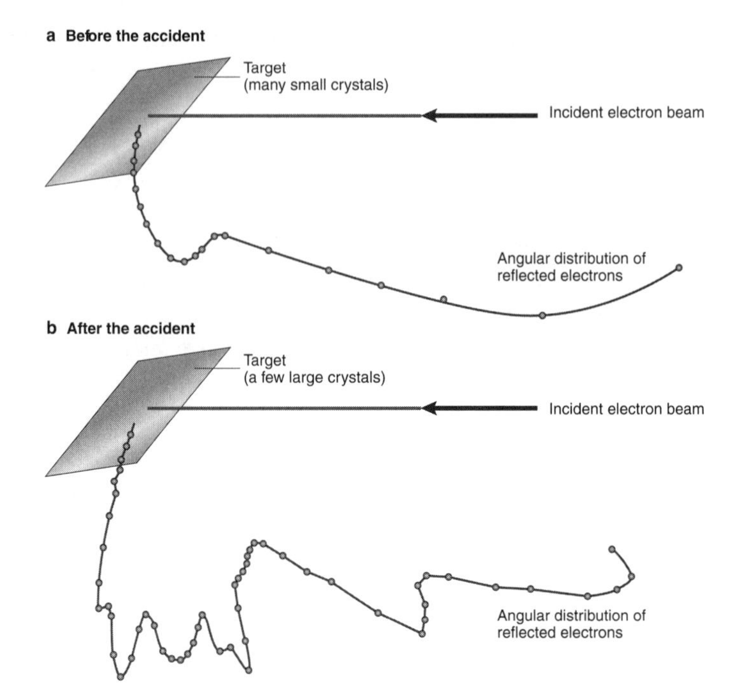

Fig. 2.1. Davisson and Germer's accidental breakthrough: the clue that led to the discovery of electron waves. a, In Davisson and Germer's initial experiments, the intensity of electrons reflected from their nickel target varied smoothly with angle. b, After the accident in the lab, the reflected electron intensity showed strong peaks at particular angles. Davisson and Germer realized that, in the later experiments, the nickel target had recrystallized into a few large crystals, and the electrons were influenced by the crystal structure. As they concluded after further experiments, the strong peaks are due to constructive and destructive interference of electron waves reflected from the regular array of atoms in the nickel crystal (see fig. 2.2).

de Broglie had predicted. Davisson also learned that an assertion similar to Born's had already been published[2] by Born's student Walter Elsasser. But Davisson did not believe Born and Elsasser's interpretations, because the earlier results had been obtained with polycrystalline samples, which would not have produced interference peaks.

Once Davisson was back in the United States, he and Germer pursued their experiments with renewed vigor: their objective now was to test the wave nature of electrons. They calculated the diffraction angles that would result from the de Broglie wavelength of their electrons (calculated from the electron energy) and compared these to the positions of the weak peaks that Germer had managed to produce during Davisson's absence. They did not find a one-to-one match, and soon found that a precise comparison was not possible, owing to misalignment of the apparatus.

They improved the apparatus, and finally, in early 1927, they obtained sharp peaks (see p. 33, figure 2). These peaks could be identified as arising from beams diffracted at specific angles from the regularly arranged atoms in the nickel surface (fig. 2.2).

Davisson and Germer submitted their results to *Nature* in early March, and their paper[1] was published on 16 April. It contains a detailed comparison between their reflected beams and those that would be produced by illuminating the crystal with X-rays (see p. 34, table 1). There were thirteen reflected beams in all, of which ten corresponded to those seen in X-ray diffraction patterns, although the authors had to introduce a "contraction factor" to obtain precise agreement of the diffraction angles. This discrepancy was

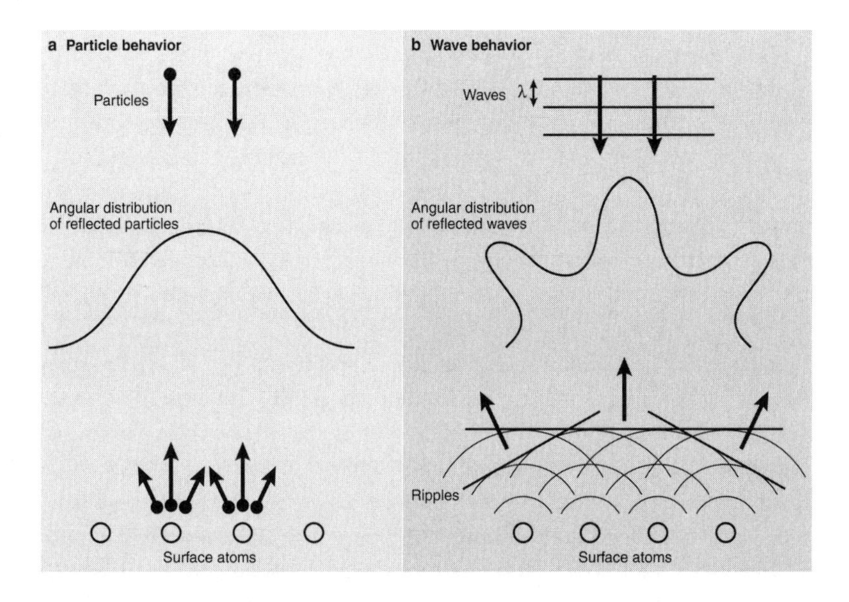

Fig. 2.2. Particles and waves reflecting from a crystal. a, When particles are reflected from a regular array of atoms on a solid surface, the angular distribution of the reflected particles is given by the sum of the particles scattered by each atom. b, But when waves, such as X-rays, impinge on the surface, each atom produces ripples of reflected radiation. The intensity of the reflected wave is greater in the directions in which the crests of the ripples overlap to interfere constructively and is less in the directions in which the crest of one ripple and the trough of another overlap to interfere destructively. This constructive and destructive interference produces the diffraction peaks, which occur at angles related to the wavelength of the impinging wave (λ) and the spacing of the atoms in the crystal.

later attributed[3,4] to the effect of the acceleration of electrons by the electric field at the nickel surface: when the electrons entered the crystal, their velocity increased, and so their de Broglie wavelength decreased.

The remaining three peaks for which Davisson and Germer could not find any correspondence with X-ray data were later identified as being due to diffraction from atoms adsorbed on the surfaces of the target crystal,[5] because the peaks vanished after the crystal was heated enough to vaporize the adsorbed atoms.

Davisson and Germer used their diffraction data to calculate wavelengths for the electrons of different energies, and these wavelengths agreed with de Broglie's formula. This led to the inescapable conclusion that electrons behave like waves. At each atom the incident electrons produce wave ripples, which can propagate only in those directions in which the crests and troughs of many ripples overlap to interfere constructively, thus forming strong diffracted beams (fig. 2.2).

Just two months after Davisson and Germer's paper appeared, George Thomson and Alexander Reid of the University of Aberdeen reported more evidence for the wave nature of electrons.[6] Their experiment also involved diffraction, but whereas Davisson and Germer's electrons had been reflected from a single crystal of nickel, Thomson and Reid's electron beam passed straight through a thin polycrystalline film of celluloid. Thomson—the son of J. J. Thomson, the discoverer of the electron—had attended the same Oxford meeting that Davisson had, after which he had immediately started his experiment, and obtained results quite quickly.

Because Thomson and Reid's film was made of many small crystals, they saw diffraction rings, rather than discrete spots. But one year later Seishi Kikuchi, at the Institute of Physical and Chemical Research in Tokyo, obtained clear electron diffraction spots from a thin single crystal of mica, providing welcome additional confirmation. Davisson and Thomson shared the 1937 Nobel Prize in Physics for their discovery of electron diffraction.

By 1927, when Davisson and Germer's paper appeared, the essential framework of quantum mechanics had already been constructed by Erwin Schrödinger and others. But the direct evidence for the wave nature of electrons accelerated further developments in theoretical physics.

Davisson and Germer's discovery also had practical implications, for the study of matter using particles instead of light. The inventor of the electron microscope, Ernst Ruska, was already deeply involved in the development of magnetic lenses for focusing electron beams when he learned about the wave nature of electrons in 1932. Ruska calculated the de Broglie wavelength of the electrons in his thirty-kilovolt beams as less than one hundredth of a nanometer (about one tenth the size of an atom), and immediately recognized the possibility that electron microscopes could resolve atomic-scale structures.

Electron microscopes have since become indispensable tools for observing objects that are too small to be resolved by light microscopes, as recognized by the award of the 1986 Nobel Prize in Physics to Ruska. The wave nature of electrons is not directly used in conventional electron microscopes, but more recently "electron holography" microscopes have been developed, which use the interference between coherent electron waves to measure electric and magnetic fields inside and outside solids.

Beyond microscopy, electron diffraction techniques are widely used to characterize atomic arrangements on solid surfaces and inside thin films and to investigate the structures of molecules in gases. And electrons are not the only particles whose wave behavior has been put to use: neutrons (see chapter 3) are also used to probe the structure of matter and the collective motions of the atoms that comprise it.

The dual nature of waves and particles is part of the "spookiness" of quantum mechanics, which cannot be explained in any classical way. It used to be that one could only imagine an experiment that would simultaneously demonstrate the particle and wave nature of electrons—an experiment that Nobel laureate Richard Feynman said would exhibit "the heart of quantum mechanics," but would require "an impossibly small" apparatus. But in 1989, thanks to the availability of very bright, coherent electron beams, such an experiment was performed.[7] We watched single electrons arrive at a video monitor, their locations seemingly distributed at random. But after enough of them had arrived, we could see that their distribution formed an interference pattern, which could have been produced only by waves. Sixty years after Davisson and Germer's paper, we were able to reap the full harvest from their discovery.

References

1. Davisson, C. & Germer, L. H. The scattering of electrons by a single crystal of nickel. *Nature* **119,** 558–560 (1927).
2. Elsasser, W. Bemerkungen zur Quantenmechanik freier Elektronen. *Naturwiss.* **13,** 711 (1925).
3. Eckart, C. The reflection of electrons from crystals. *Proc. Natl Acad. Sci. USA* **13,** 460–462 (1927).
4. Bethe, H. Theorie der Beugung von Elektronen an Kristallen. *Ann. Phys.* **87,** 55–129 (1928).
5. Davisson, C. & Germer, L. H. Diffraction of electrons by a crystal of nickel. *Phys. Rev.* **30,** 705–740 (1927).
6. Thomson, G. P. & Reid, A. Diffraction of cathode rays by a thin film. *Nature* **119,** 890 (1927).

7. Tonomura, A., Endo, J., Matsuda, T., Kawasaki, T. & Ezawa, H. Demonstration of single-electron buildup of an interference pattern. *Am. J. Phys.* **57,** 117–120 (1989).

Further reading

Gehrenbeck, R. K. "Davisson and Germer," in *Fifty Years of Electron Diffraction* (ed. Goodman, P.), 12–27 (Reidel, Dordrecht, 1981).

Nobel e-Museum. *The Nobel Prize in Physics 1937* (http://www.nobel.se/physics/laureates/1937/). (Also physics prizes in 1918, 1921, 1922, 1929, and 1986.)

Tonomura, A. *The Quantum World Unveiled by Electron Waves* (World Scientific, Singapore, 1998).

1927

The scattering of electrons by a single crystal of nickel

C. Davisson and L. H. Germer

In a series of experiments now in progress, we are directing a narrow beam of electrons normally against a target cut from a single crystal of nickel, and are measuring the intensity of scattering (number of electrons per unit solid angle with speeds near that of the bombarding electrons) in various directions in front of the target. The experimental arrangement is such that the intensity of scattering can be measured in any latitude from the equator (plane of the target) to within 20° of the pole (incident beam) and in any azimuth.

The face of the target is cut parallel to a set of {111}-planes of the crystal lattice, and etching by vaporisation has been employed to develop its surface into {111}-facets. The bombardment covers an area of about 2 $mm.^2$ and is normal to these facets.

As viewed along the incident beam the arrangement of atoms in the crystal exhibits a threefold symmetry. Three {100}-normals equally spaced in azimuth emerge from the crystal in latitude 35°, and, midway in azimuth between these, three {111}-normals emerge in latitude 20°. It will be convenient to refer to the azimuth of any one of the {100}-normals as a {100}-azimuth, and to that of any one of the {111}-normals as a {111}-azimuth. A third set of azimuths must also be specified; this bisects the dihedral angle between adjacent {100}- and {111}-azimuths and includes a {110}-normal lying in the plane of the target. There are six such azimuths, and any one of these will be referred to as a {110}-azimuth. It follows from considerations of symmetry that if the intensity of scattering exhibits a dependence upon azimuth as we pass from a {100}-azimuth to the next adjacent {111}-azimuth (60°), the same depen-

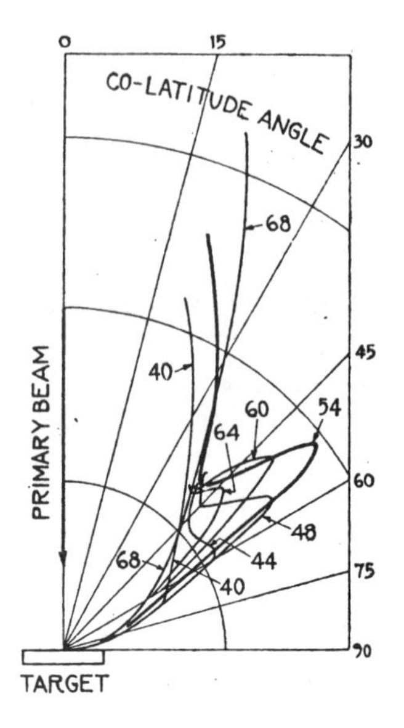

Fig. 1. Intensity of electron scattering vs. co-latitude angle for various bombarding voltages—azimuth-{111}-330°.

dence must be exhibited in the reverse order as we continue on through 60° to the next following {100}-azimuth. Dependence on azimuth must be an even function of period $2\pi/3$.

In general, if bombarding potential and azimuth are fixed and exploration is made in latitude, nothing very striking is observed. The intensity of scattering increases continuously and regularly from zero in the plane of the target to a highest value in co-latitude 20°, the limit of observations. If bombarding potential and co-latitude are fixed and exploration is made in azimuth, a variation in the intensity of scattering of the type to be expected is always observed, but in general this variation is slight, amounting in some cases to not more than a few per cent. of the average intensity. This is the nature of the scattering for bombarding potentials in the range from 15 volts to near 40 volts.

At 40 volts a slight hump appears near 60° in the co-latitude curve for azimuth-{111}. This hump develops rapidly with increasing voltage into a strong spur, at the same time moving slowly upward toward the incident beam. It attains a maximum intensity in co-latitude 50° for a bombarding potential of 54 volts, then decreases in intensity, and disappears in co-latitude 45° at about 66 volts. The growth and decay of this spur are traced in Fig. 1.

A section in azimuth through this spur at its maximum (Fig. 2—Azimuth-330°) shows that it is sharp in azimuth as well as in latitude, and that it forms one of a set of three such spurs, as was to be expected. The width of these spurs both in latitude and in azimuth is almost completely accounted for by the low resolving power of the measuring device. *The spurs are due to beams of scattered electrons which are nearly if not quite as well defined as the primary beam.* The minor peaks occurring in the {100}-azimuth are sections of a similar set of spurs that attains

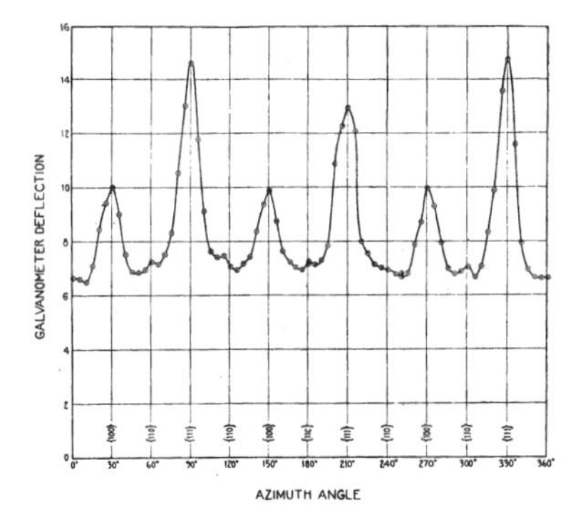

Fig. 2. Intensity of electron scattering vs. azimuth angle—54 volts, co-latitude 50°.

its maximum development in co-latitude 44° for a bombarding potential of 65 volts.

Thirteen sets of beams similar to the one just described have been discovered in an exploration in the principal azimuths covering a voltage range from 15 volts to 200 volts. The data for these are set down on the left in Table I. (columns 1–4). Small corrections have been applied to the observed co-latitude angles to allow for the variation with angle of the 'background scattering,' and for a small angular displacement of the normal to the facets from the incident beam.

If the incident electron beam were replaced by a beam of monochromatic X-rays of adjustable wave-length, very similar phenomena would, of course, be observed. At particular values of wave-length, sets of three or of six diffraction beams would emerge from the incident side of the target. On the right in Table I. (columns 5, 6 and 7) are set down data for the ten sets of X-ray beams of longest wave-length which would occur within the angular range of our observations. Each of these first ten occurs in one of our three principal azimuths.

Several points of correlation will be noted between the two sets of data. Two points of difference will also be noted; the co-latitude angles of the electron beams are not those of the X-ray beams, and the three electron beams listed at the end of the Table appear to have no X-ray analogues.

The first of these differences is systematic and may be summarised quantitatively in a simple manner. If the crystal were contracted in the direction of the incident beam by a factor 0.7, the X-ray beams would be shifted to the smaller co-latitude angles θ' (column 8), and would then agree in position fairly well with the observed electron beams—the average difference being 1.7°. Associated in this way there is a set of electron beams for each of the first ten sets of X-ray beams occurring in the range of observations, the electron beams for 110 volts alone being unaccounted for.

These results are highly suggestive, of course, of the ideas underlying the theory of wave mechanics, and we naturally inquire if the wave-length of the X-ray beam which we thus associate with a beam of electrons is

Table I

Azimuth	Electron Beams			X-ray Beams						
	Bomb. Pot. (volts)	Co-lat. θ	Intensity	Reflections	$\lambda \times 10^8$ cm.	Co-lat. θ	Co-lat. θ'	$v \times 10^{-8}$ cm./sec.	$n\lambda \times 10^8$ cm.	$n\left\{\frac{\lambda m v}{h}\right\}$
{111}	54	50°	0.5	{220}	2.03	70.5	52.7	4.36	1.65	0.99
	100	31	0.5	{331}	1.49	44.0	31.6	5.94	1.11	0.91
	174	21	0.9	{442}	1.13	31.6	22.4	7.84	0.77	0.83
	174	55	0.15	{440}	1.01	70.5	52.7	7.84	1.76	2(0.95)
{100}	65	44	0.5	{311}	1.84	59.0	43.2	4.79	1.49	0.98
	126	29	1.0	{422}	1.35	38.9	27.8	6.67	1.04	0.95
	190	20	1.0	{533}	1.04	28.8	20.4	8.19	0.74	0.83
	159	61	0.4	{511}	1.05	77.9	59.0	7.49	1.88	2(0.97)
{110}	138	59	0.07	{420}	1.22	78.5	59.5	6.98	1.06	1.02
	170	46	0.07	{531}	1.04	57.1	41.7	7.75	0.89	0.95
{111}	110	58	0.15	. .	. .	. .	. .	6.23	1.82	1.56
{100}	110	58	0.15	. .	. .	. .	. .	6.23	1.82	1.56
{110}	110	58	0.25	. .	. .	. .	. .	6.23	1.05	0.90

in fact the h/mv of L. de Broglie. The comparison may be made, as it happens, without assuming a particular correspondence between X-ray and electron beams, and without use of the contraction factor. Quite independently of this factor, the wave-lengths of all possible X-ray beams satisfy the optical grating formula $n\lambda = d \sin \theta$, where d is the distance between lines or rows of atoms in the surface of the crystal—these lines being normal to the azimuth plane of the beam considered. For azimuths-{111} and -{100}, $d = 2.15 \times 10^{-8}$ cm. and for azimuth-{110}, $d = 1.24 \times 10^{-8}$ cm. We apply this formula to the electron beams without regard to the conditions which determine their distribution in co-latitude angle. The correlation obtained by this procedure between wave-length and electron speed v is set down in the last three columns of Table I.

In considering the computed values of $n(\lambda mv/h)$, listed in the last column, we should perhaps disregard those for the 110-volt beams at the bottom of the Table, as we have had reason already to regard these beams as in some way anomalous. The values for the other beams do, indeed, show a strong bias toward small integers, quite in agreement with the type of phenomenon suggested by the theory of wave mechanics. These integers, one and two, occur just as predicted upon the basis of the correlation between electron beams and X-ray beams obtained by use of the contraction factor. The systematic character of the departures from integers may be significant. We believe, however, that this results from imperfect alignment of the incident beam, or from other structural deficiencies in the apparatus. The greatest departures are for beams lying near the limit of our co-latitude range. The data for these are the least trustworthy.

Bell Telephone Laboratories, Inc., New York, N.Y.

Mar. 3 [Published 16 April 1927]

Letters to the Editor

Editor does not hold himself responsible for
ions expressed by his correspondents. Neither
he undertake to return, nor to correspond with
riters of, rejected manuscripts intended for this
ny other part of Nature. *No notice is taken*
nonymous communications.]

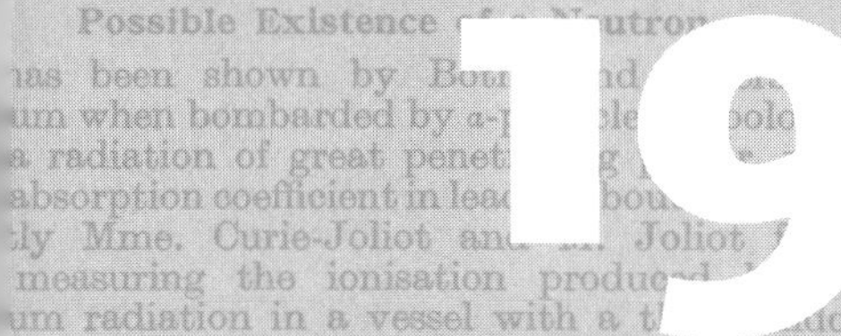

Possible Existence of a Neutron

1932

has been shown by Bo... nd ... at
um when bombarded by α-p... cle... olo... n
a radiation of great penet... g ...
absorption coefficient in lea... bou...
ly Mme. Curie-Joliot and M. Joliot ...,
measuring the ionisation produc... his
um radiation in a vessel with a th... dow,
e ionisation increased when matter containing
gen was placed in front of the window. The
appeared to be due to the ejection of protons
elocities up to a maximum of nearly 3 × 10⁹ cm.
c. They suggested that the transference of
to the proton was by a process similar to the
on effect, and estimated that the beryllium radia-
d a quantum energy of 50 × 10⁶ electron volts.
ve made some experiments using the valve
r to examine the properties of this radiation
l in beryllium. The valve counter consists of
l ionisation chamber connected to an amplifier,
e sudden production of ions by the entry of a
e, such as a proton or α-particle, is recorded
deflexion of an oscillograph. These experi-
have shown that the radiation ejects particles
ydrogen, helium, lithium, beryllium, carbon,
d argon. The particles ejected from hydrogen
, as regards range and ionising power, like
s with speeds up to about 3·2 × 10⁹ cm. per sec.
articles from the other elements have a large
g power, and appear to be in each case recoil
of the elements.
e ascribe the ejection of the proton to a Compton
from a quantum of 52 × 10⁶ electron volts,
he nitrogen recoil atom arising by a similar
s should have an energy not greater than about
0 volts, should produce not more than about
ions, and have a range in air at N.T.P. of
1·3 mm. Actually, some of the recoil atoms
rogen produce at least 30,000 ions. In col-
tion with Dr. Feather, I have observed the
atoms in an expansion chamber, and their
estimated visually, was sometimes as much
m. at N.T.P.
se results, and others I have obtained in the
of the work, are very difficult to explain on
ssumption that the radiation from beryllium
uantum radiation, if energy and momentum
be conserved in the collisions. The difficulties
ear, however, if it be assumed that the radia-
onsists of particles of mass 1 and charge 0, or
ns. The capture of the α-particle by the
ucleus may be supposed to result in the
tion of a C^{12} nucleus and the emission of the
n. From the energy relations of this process
elocity of the neutron emitted in the forward
ion may well be about 3 × 10⁹ cm. per sec.
llisions of this neutron with the atoms through
it passes give rise to the recoil atoms, and the
ed energies of the recoil atoms are in fair
ent with this view. Moreover, I have ob-
l that the protons ejected from hydrogen by the
ion emitted in the opposite direction to that of
citing α-particle appear to have a much smaller

This again receives a simple explanation on
neutron hypothesis.
If it be supposed that the radiation consist
quanta, then the capture of the α-particle by
Be^9 nucleus will form a C^{13} nucleus. The
defect of C^{13} is known with sufficient accurac
show that the energy of the quantum emitted in
process cannot be greater than about 14 × 10⁶ v
It is difficult to make such a quantum respon
for the effects observed.
... to be ... ed that many of the effects
... in ... hrough matter should reser
th... a quant... f high energy, and it is not
... ch the ... decision between the two h
... U... resent, all the evidence
fav... ... while the quantum hypot
ca... be upheld if the conservation of energy
...ntum be relinquished at some point.

J. Chadwic

Cavendish Laboratory,
Cambridge, Feb. 17.

The Oldoway Human Skeleton

A letter appeared in Nature of Oct. 24, 1
signed by Messrs. Leakey, Hopwood, and Reck
which, among other conclusions, it is stated
"there is no possible doubt that the human skel
came from Bed No. 2 and not from Bed No. 4".
must be taken to mean that the skeleton is t
considered as a natural deposit in Bed No. 2, whi
overlaid by the later beds Nos. 3 and 4, and tha
consideration of human interment is ruled out.
If this be true, it is a most unusual occurrence.
skeleton, which is of modern type, with filed te
was found completely articulated down even to
phalanges, and in a position of extraordinary
traction. Complete mammalian skeletons of
age are, as field palæontologists know, of great ra
When they occur, their perfection can usually
explained as the result of sudden death and immed
covering by volcanic dust. Many of the mor
less perfect skeletons which may be seen in muse
have been rearticulated from bones found somev
scattered as the result of death from floods, or in
neighbourhood of drying water-holes. We kno
no case of a perfect articulated skeleton being fo
in company with such broken and scattered rem
as appear to be abundant at Oldoway. Either
skeletons are all complete, as in the *Stenomylus* qu
at Sioux City, Nebraska, or are all scattered
broken in various degrees, as in ordinary bone b
The probability, therefore, that the Oldoway skel
represents an artificial burial is thus one that
occur to palæontologists.
The skeleton was exhumed in 1913, and publis
photographs show that the excavation made fo
disinterment was extensive. It is, therefore,
difficult to believe that in 1931 there can be reli
evidence left at the site as to the conditions u
which it was deposited. If naturally deposite
Bed No. 2, the skeleton is of the highest pos
importance, because it would be of pre-Mouste
age, and would be in the company of *Pithecanthr*
and the Piltdown, Heidelberg, and Peking men
of whose remains are fragmentary to the last deg
Of the few other human remains for which
antiquity is claimed, the Galley Hill skeleton and
Ipswich skeleton are, or apparently were, comp
The first of these was never seen *in situ* by
trained observer, and the latter has, we believe, b
withdrawn by its discoverer. The other fragme
found long ago, are entirely without satisfac

THREE

The atom completed

Maurice Goldhaber

In the 1920s, common wisdom held that atoms were made from only two constituents: positively charged protons and negatively charged electrons. When, in 1932, James Chadwick reported the existence of a third, electrically neutral particle—the neutron—it completed the inventory of atomic building blocks and marked the birth of modern nuclear physics. Today, neutrons are used to probe the structure of matter, and neutron physics underpins much of our understanding of astrophysics and cosmology.

Great discoveries rarely come out of the blue. Often they are the culmination of previous observations and thoughts, suddenly clearing the fog of many "nonsensical observations"—although this description is often inappropriate. When observations are unexpected but correct, they have not betrayed our senses; rather, thinking in grooves may make the observations seem paradoxical. James Chadwick's discovery of the neutron,[1] published in *Nature* on 27 February 1932, is a powerful example of the benefits to be had from thinking outside the groove.

The neutron has a fascinating prehistory. Ernest Rutherford, who had discovered the atomic nucleus and was a senior colleague of Chadwick's at Cambridge, speculated on the existence of a neutron as early as 1920, in his famous Bakerian lecture to the Royal Society.[2] He had made one of his rare mistakes, interpreting some particles "liberated from nitrogen and oxygen by bombardment with α-particles" (see box 3.1) as helium-3 (^{3}He) nuclei. Although Rutherford did not like to speculate in his publications, he allowed his imagination free rein in talks. He started from the belief, then quite general, that electrons were present inside atomic nuclei, as well as

Box 3.1 Probing the atomic nucleus

Many early discoveries concerning the structure and composition of atoms were made by bombarding different materials with energetic particles and observing the result. The most important such projectiles were **α-particles,** which are emitted by radioactive elements such as uranium. In 1911, Rutherford famously deduced the existence of atomic nuclei by interpreting the back-scattering of α-particles from gold foils.

We now know that α-particles consist of two protons and two neutrons. But in 1920, when Rutherford gave his Bakerian lecture,[2] all he knew was that they had a mass four times that of the proton (an **atomic mass** of 4) and a positive charge twice that of the proton (a charge of +2). He therefore assumed that they comprised four protons and two electrons (see figure). Similarly, he suggested that a ^{3}He nucleus (a particle of mass 3 and charge +2; see box 3.2) might comprise three protons and one electron, and a heavy hydrogen, or ^{2}H, nucleus (mass 2 and charge +1) might have two protons and one electron. Today the ^{3}He nucleus is known to comprise two protons and a neutron, and the ^{2}H (deuterium) nucleus, also known as the **deuteron** (and for a while called the "diplon" by Rutherford), consists of a single proton and a neutron. Tritium, or ^{3}H, the heaviest form of hydrogen, has one proton and two neutrons in its nucleus.

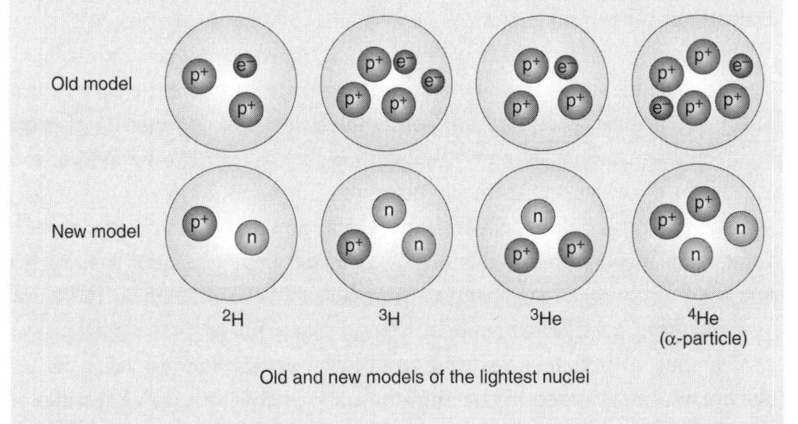

Old and new models of the lightest nuclei

orbiting around them, and he conjectured that if the ^{3}He nucleus consists of three protons and one electron, might not a combination of two protons and one electron (which he thought of as heavy hydrogen) also exist? Then he went further, discussing the possible existence of what he called an atom of mass 1, with zero nuclear charge, adding:[2]

> Such an atomic structure seems by no means impossible. On present views, the neutral hydrogen atom is regarded as a nucleus of unit charge with an electron attached at a distance, and the spectrum of hydrogen is ascribed to the movements of this distant electron. Under some conditions, however, it may be possible for an electron to combine much more closely with the H nucleus, forming a kind of neutral doublet. Such an atom would have very novel properties . . . it should be able to move freely through matter . . . it may be impossible to contain it in a sealed vessel. On the other hand, it should enter readily the structure of atoms. . . .
>
> The existence of such atoms seems almost necessary to explain the building up of the nuclei of heavy elements; for unless we suppose the production of charged particles of very high velocities it is difficult to see how any positively charged particle can reach the nucleus of a heavy atom against its intense repulsive field.

It took another decade, during which many new insights were developed (including the theory of quantum mechanics) before it was realized that the electrons presumed to exist inside nuclei appeared to have become ghosts, having lost all of their properties except electric charge. The negative charge (but negligible mass) carried by electrons seemed necessary to explain the charge and mass of atomic nuclei, which, with the single exception of ordinary hydrogen, carry less charge than would be expected if their mass were carried solely by protons. At that time, physicists were too conservative (unlike at later times!) to postulate the existence of a new particle.

At about the same time as Rutherford, William D. Harkins, a chemistry professor at the University of Chicago, also suggested the existence of a neutron.[3] He took as his starting point the existence of isotopes—forms of the same element that have different atomic mass. Although we now know that Harkins was right—different isotopes of the same element are indeed distinguished by the number of neutrons in the nucleus (see box 3.2)—his suggestion did not have much influence, because he did not discuss the possibility of free neutrons and their likely behavior.

Rutherford, with his uncanny ability to visualize the interactions of particles, helped to prepare Chadwick's mind, often refreshed by their discussions while they waited together in the dark until their eyes were adapted and sensitive enough to record the scintillations produced on a zinc sulfide screen by nuclear disintegrations. Although Rutherford soon corrected his mistaken ^{3}He interpretation (he had in fact seen α-particles liberated from nitrogen and oxygen), the idea of the neutron lived on in his mind. Late in his life, in his last experiment, he had the satisfaction of discovering the real ^{3}He—as well as ^{3}H, the heaviest form of hydrogen, also known as tritium.

Box 3.2 Naming nuclei

Chemical elements are identified by their **atomic number,** which is equal to the number of protons in the nucleus. Thus, hydrogen, with one proton in its nucleus, has atomic number 1, and helium, with two protons, has atomic number 2. To make a neutral atom, the number of electrons orbiting the nucleus must be the same as the number of protons; thus, the atomic number also specifies the number of orbiting electrons, and hence the element's chemical behavior. But a given element can exist in different forms, or **isotopes,** which differ in their **mass number**—the number of protons plus neutrons in the nucleus.

Because the symbol for a chemical element also specifies its atomic number (H = 1, He = 2, etc.), isotopes are usually denoted by the chemical symbol and the mass number: thus, ^{3}He (helium-3) and ^{4}He (helium-4) each have two protons in the nucleus (which is what makes them helium), but the ^{3}He nucleus has one neutron and the ^{4}He nucleus (or α-particle) has two.

A few specific experiments were the final trigger for Chadwick's discovery. First, there was the observation by Walter Bothe and Hans Becker[4] of γ-rays (very energetic photons) emitted from a beryllium target bombarded by α-particles. Then H. C. Webster,[5] a student of both Rutherford and Chadwick, found that these γ-rays were preferentially emitted in the forward direction, rather than back toward the source of the α-particles. This should have opened eyes to the fact that something strange was happening, as γ-rays were expected to be emitted more uniformly in all directions. By contrast, particles hit by other particles behave like billiard balls, and mostly keep traveling in the direction of the particle that hit them.

At about the same time, the husband-wife team of Irène Curie and Frédéric Joliot, in Paris, studied the interaction of the γ-rays with hydrogen. They reported[6] that the γ-rays were able to eject protons from the hydrogen atom. An analogous interaction, called the Compton effect, was already known to occur between γ-rays and electrons. The ability to eject protons led Curie and Joliot to deduce an energy for these γ-rays of about fifty million electron volts (MeV).

When Chadwick read the issue of *Comptes Rendus* containing Curie and Joliot's paper,[6] his pipe supposedly fell out of his mouth! From Albert Einstein's equation $E = mc^2$, the energy released in a nuclear reaction is given by the difference in mass between the starting materials and the products of the reaction. Chadwick knew nuclear masses well enough to realize that the proposed reaction between beryllium and α-particles (Be + α) to form carbon could not possibly release enough energy to produce 50-MeV γ-rays. What had been interpreted as γ-rays must instead be particles—which carry more momentum than photons for a given energy and can transmit it to another particle in a collision. The difference is striking: if a γ-ray needed 50 MeV of energy to eject the protons seen by Curie and Joliot, a particle with the mass of the proton would need only 5 MeV to accomplish the same task. Chadwick must have felt that here, at last, was the neutron.

Immediately, he set out to prove his suspicion with decisive experiments, studying the atomic recoils produced when hydrogen, helium, and nitrogen were exposed to the radiation from the Be + α reaction. His measurements of the energies of these recoil atoms in an ionization chamber led him to conclude that the Be + α radiation was best explained as "particles of mass 1 and charge 0, or neutrons." The only alternative was to forsake one of the most fundamental laws of physics—either the conservation of energy or of momentum. Nevertheless, he gave his *Nature* paper[1] the cautious title "Possible existence of a neutron."

Chadwick followed this with a more detailed paper,[7] now more confidently called "The existence of a neutron." Using the nuclear reaction $^{11}B + \alpha \rightarrow {}^{14}N + n$, in which bombardment of boron by α-particles leads to the production of nitrogen nuclei and neutrons, he determined the mass of the neutron as 1.0067 atomic mass units. This is less than the mass of the hydrogen atom (an electron orbiting a proton), satisfying Rutherford's expectation that the neutron was a compound of a proton and an electron. Because of the equivalence of mass and energy, a tightly bound compound (the proposed neutron) has less mass than a more loosely bound compound of the same constituents (the hydrogen atom).

Theoreticians then happily replaced the electrons that had mysteriously lost most of their properties inside nuclei by neutrons, and developed theories of nuclear structure with protons and neutrons (considered to be proton-electron compounds) as constituents. The rapid acceptance of the neutron was signaled by the award of the 1935 Nobel Prize in Physics to Chadwick for his discovery.

In 1934, Chadwick invited me to join him as one of the last students with whom he worked before he left Cambridge a year later to assume the Lyon Jones chair of physics at Liverpool University. We investigated the photodisintegration of the deuteron (deuteron + γ-ray $\rightarrow$ proton + neutron), a simple reaction, which allowed us to measure the neutron's mass more accurately than had previously been possible. We found that the neutron was in fact heavier than the hydrogen atom.[8] Thus, like the proton, it had to be considered a fundamental particle in its own right, rather than as a proton-electron compound. Ironically, the neutron is now again considered to be a compound of more elementary particles, called quarks, as is the proton.

As befits one of the fundamental building blocks of matter, neutrons have proved to be important in many different fields of science. In astrophysics, neutrons and their reactions are critical to our understanding of events immediately following the Big Bang, and all of the chemical elements heavier than iron are thought to have been formed by neutron capture in supernovae, and in certain types of star. Here on Earth, neutrons produced in nuclear reactors (see chapter 5) or by particle accelerators have become invaluable tools for probing the atomic structure and behavior of solids, liquids, and complex molecules. In the technique known as neutron scattering, neutrons collide with the atoms of a sample under examination, and are deflected in ways that can reveal both the positions of the atoms and their collective motions. In biologically important molecules such as proteins, the positions of hydrogen atoms, nearly invisible to X-rays because of their low charge, can be studied easily by neutron scattering because low-energy neutrons interact very strongly with protons. Neutron scattering techniques, which were the subject of the 1994 Nobel Prize in Physics, are possible because neutrons, like other elementary particles, behave like waves as well as particles (see chapter 2).

Many steps led to the discovery of the neutron. Little wonder that science historians often emphasize the importance of different routes to the top. One vital lesson for all of us is that experimenters should draw a clear line between observations and interpretations, as Chadwick did when he realized there was another way of interpreting the "γ-rays" that had been reported by others. Especially today, when expectations may become promises used to attract funding for research, we should remember Chadwick and not confuse our expectations with what we have really seen.

References

1. Chadwick, J. Possible existence of a neutron. *Nature* **129,** 312 (1932).
2. Rutherford, E. Nuclear constitution of atoms. *Proc. R. Soc. Lond.* A**97,** 374–400 (1920).
3. Harkins, W. D. The nuclei of atoms and the new periodic system. *Phys. Rev.* **15,** 73–94 (1920).
4. Bothe, W. & Becker, H. Eine Kern-γ-Strahlung bei leichten Elementen. *Naturwiss.* **18,** 705 (1930).
5. Webster, H. C. The artificial production of nuclear γ radiation. *Proc. R. Soc. Lond.* A**136,** 428–453 (1932).

6. Curie, I. & Joliot, F. Emission de protons de grande vitesse par les substances hydrogénées sous l'influence des rayons γ très pénétrants. *C. R. Acad. Sci. Paris* **194,** 273–275 (1932).
7. Chadwick, J. The existence of a neutron. *Proc. R. Soc. Lond.* A**136,** 692–708 (1932).
8. Chadwick, J. & Goldhaber, M. A 'nuclear photo-effect': disintegration of the diplon by γ-rays. *Nature* **134,** 237–238 (1934).

Further reading

Close, F., Marten, M. & Sutton, C. *The Particle Explosion* (Oxford University Press, 1987).
Nobel e-Museum. *The Nobel Prize in Physics 1935* (http://www.nobel.se/physics/laureates/1935/). (Also physics prize in 1994.)
't Hooft, G. *In Search of the Ultimate Building Blocks* (Cambridge University Press, 1997).

1932

Possible existence of a neutron

J. Chadwick

It has been shown by Bothe and others that beryllium when bombarded by α-particles of polonium emits a radiation of great penetrating power, which has an absorption coefficient in lead of about 0.3 $(\text{cm.})^{-1}$. Recently Mme. Curie-Joliot and M. Joliot found, when measuring the ionisation produced by this beryllium radiation in a vessel with a thin window, that the ionisation increased when matter containing hydrogen was placed in front of the window. The effect appeared to be due to the ejection of protons with velocities up to a maximum of nearly 3×10^9 cm. per sec. They suggested that the transference of energy to the proton was by a process similar to the Compton effect, and estimated that the beryllium radiation had a quantum energy of 50×10^6 electron volts.

I have made some experiments using the valve counter to examine the properties of this radiation excited in beryllium. The valve counter consists of a small ionisation chamber connected to an amplifier, and the sudden production of ions by the entry of a particle, such as a proton or α-particle, is recorded by the deflexion of an oscillograph. These experiments have shown that the radiation ejects particles from hydrogen, helium, lithium, beryllium, carbon, air, and argon. The particles ejected from hydrogen behave, as regards range and ionising power, like protons with speeds up to about 3.2×10^9 cm. per sec. The particles from the other elements have a large ionising power, and appear to be in each case recoil atoms of the elements.

If we ascribe the ejection of the proton to a Compton recoil from a quantum of 52×10^6 electron volts, then the nitrogen recoil atom arising by a similar process should have an energy not greater than about

400,000 volts, should produce not more than about 10,000 ions, and have a range in air at N.T.P. of about 1.3 mm. Actually, some of the recoil atoms in nitrogen produce at least 30,000 ions. In collaboration with Dr. Feather, I have observed the recoil atoms in an expansion chamber, and their range, estimated visually, was sometimes as much as 3 mm. at N.T.P.

These results, and others I have obtained in the course of the work, are very difficult to explain on the assumption that the radiation from beryllium is a quantum radiation, if energy and momentum are to be conserved in the collisions. The difficulties disappear, however, if it be assumed that the radiation consists of particles of mass 1 and charge 0, or neutrons. The capture of the α-particle by the Be^9 nucleus may be supposed to result in the formation of a C^{12} nucleus and the emission of the neutron. From the energy relations of this process the velocity of the neutron emitted in the forward direction may well be about 3×10^9 cm. per sec. The collisions of this neutron with the atoms through which it passes give rise to the recoil atoms, and the observed energies of the recoil atoms are in fair agreement with this view. Moreover, I have observed that the protons ejected from hydrogen by the radiation emitted in the opposite direction to that of the exciting α-particle appear to have a much smaller range than those ejected by the forward radiation. This again receives a simple explanation on the neutron hypothesis.

If it be supposed that the radiation consists of quanta, then the capture of the α-particle by the Be^9 nucleus will form a C^{13} nucleus. The mass defect of C^{13} is known with sufficient accuracy to show that the energy of the quantum emitted in this process cannot be greater than about 14×10^6 volts. It is difficult to make such a quantum responsible for the effects observed.

It is to be expected that many of the effects of a neutron in passing through matter should resemble those of a quantum of high energy, and it is not easy to reach the final decision between the two hypotheses. Up to the present, all the evidence is in favour of the neutron, while the quantum hypothesis can only be upheld if the conservation of energy and momentum be relinquished at some point.

Cavendish Laboratory, Cambridge

Feb. 17 [Published 27 February 1932]

1938

Letters to the Editor

The Editor does not hold himself responsible for opinions expressed by his correspondents. He cannot undertake to return, or to correspond with the writers of, rejected manuscripts intended for this or any other part of NATURE. *No notice is tak[illegible] anonymous communications.*

NOTES ON POINTS IN SOME OF THIS WEEK'S LET[illegible]S A[illegible]AR ON P. 83.

CORRESPONDENTS ARE [illegible] T[illegible] SI[illegible]MMA[illegible] THEIR COMMUNICATIONS.

osity of Liquid Helium b[illegible] th[illegible]

abnormally high heat con[illegible]ivity of [illegible] w the λ-point, as first observed by K[illegible]n, ed to me the possibility of an ex[illegible]n in of convection currents. This explanation require helium II to have an abnormally low y; at present, the only viscosity measure- on liquid helium have been made in Toronto[1], owed that there is a drop in viscosity below oint by a factor of 3 compared with liquid at normal pressure, and by a factor of 8 ed with the value just above the λ-point. In xperiments, however, no check was made to that the motion was laminar, and not tur-

important fact that liquid helium has a density ρ of about 0·15, not very different at of an ordinary fluid, while its viscosity μ small comparable to that of a gas, makes its tic viscosity $\nu = \mu/\rho$ extraordinary small. uently when the liquid is in motion in an y viscosimeter, the Reynolds number may very high, while in order to keep the motion r, especially in the method used in Toronto, , the damping of an oscillating cylinder, the ds number must be kept very low. This ment was not fulfilled in the Toronto experi- and the deduced value of viscosity thus refers ulent motion, and consequently may be higher amount than the real value.

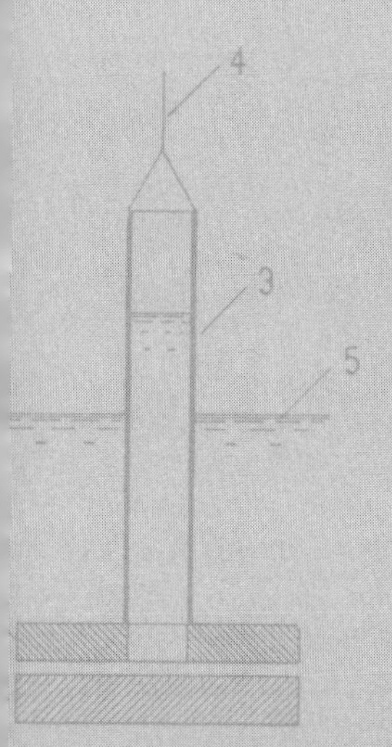

The very small kinematic viscosity of liquid helium II thus makes it diffi- cult to measure the viscosity. In an attempt to get lam- inar motion the following method (shown diagramat- ically in the accom- panying illustra- tion) was devised. The viscosity was measured by the pressure drop when the liquid flows through the gap between the disks 1 and 2; these disks were of glass and were optically e gap between them being adjustable by stance pieces. The upper disk, 1, was 3 cm. in r with a central hole of 1·5 cm. diameter, ich a glass tube (3) was fixed. Lowering and

[illegible] co[illegible]e [illegible]bove or below the level (5 the [illegible]id [illegible]rrounding Dewar flask. amo[illegible] of flow and the pressure were deduced f [illegible]erence of the two levels, which was measu by cathetometer.

The results of the measurements were rat striking. When there were no distance pieces betw the disks, and the plates 1 and 2 were brought i contact (by observation of optical fringes, th separation was estimated to be about half a micro the flow of liquid above the λ-point could be o just detected over several minutes, while below λ-point the liquid helium flowed quite easily, the level in the tube 3 settled down in a few secon From the measurements we can conclude that viscosity of helium II is at least 1,500 times sma than that of helium I at normal pressure.

The experiments also showed that in the case helium II, the pressure drop across the gap proportional to the square of the velocity of flo which means that the flow must have been turbule If, however, we calculate the viscosity, assuming flow to have been laminar, we obtain a value of order 10^{-9} C.G.S., which is evidently still only upper limit to the true value. Using this estima the Reynolds number, even with such a small g comes out higher than 50,000, a value for wh turbulence might indeed be expected.

We are making experiments in the hope of s further reducing the upper limit to the viscosity liquid helium II, but the present upper limit (name 10^{-9} C.G.S.) is already very striking, since it is mo than 10^4 times smaller than that of hydrogen g (previously thought to be the fluid of least viscosit The present limit is perhaps sufficient to suggest, analogy with supraconductors, that the helium bel the λ-point enters a special state which might called a 'superfluid'.

As we have already mentioned, an abnormally l viscosity such as indicated by our experiments mig indeed provide an explanation for the high therm conductivity, and for the other anomalous propert observed by Allen, Peierls, and Uddin[2]. It is eviden possible that the turbulent motion, inevitably set in the technical manipulation required in worki with the liquid helium II, might on account of t great fluidity, not die out, even in the small capilla tubes in which the thermal conductivity w measured; such turbulence would transport he extremely efficiently by convection.

P. KAPITZA.

Institute for Physical Problems,
Academy of Sciences,
Moscow.
Dec. 3.

FOUR

Superfluidity: a new state of matter

Allan Griffin

In January 1938, the Russian physicist Pyotr Kapitza reported that, at temperatures below 2.2 degrees above absolute zero, liquid helium seemed to flow with no loss of energy due to friction. Kapitza suggested the term "superfluid" to describe this new behavior—which was eventually understood as a dramatic illustration of the quantum-mechanical wave nature of matter. The study of such macroscopic quantum effects now lies at the heart of modern physics.

Helium gas becomes a liquid below 4.2 degrees above absolute zero (4.2 K). After decades of research to reach lower and lower temperatures, liquid helium was finally produced in 1908 by the Dutch physicist Heike Kamerlingh Onnes in Leiden, who won the 1913 Nobel Prize in Physics for this achievement. But it was another twenty years before it became clear that the liquid undergoes a transformation at a lower temperature of 2.2 K: the so-called lambda (λ) phase transition, marked by sudden changes in the density, specific heat, and other thermodynamic properties of the liquid. The lower-temperature phase was christened helium II, and the higher-temperature phase helium I.

With hindsight, the first indication that helium II (He II) was a very different kind of liquid was the observation[1] that the rapid bubbling of the liquid suddenly ceased as the λ-point was crossed. But at the time, this strange behavior was just used as a convenient experimental marker for the transition, rather than prompting attempts at explanation. From 1935 to 1937, attempts in Toronto to measure the shear viscosity of He II (the friction when the liquid was subjected to shear), and in Leiden and Cambridge,

U.K., to measure its ability to conduct heat, culminated in Kapitza's classic experiment.

Kapitza observed that He II flowed between two closely spaced parallel plates extremely rapidly compared to He I, for the same pressure difference. This result,[2] published in *Nature* on 8 January 1938, showed unambiguously that here was a new and mysterious kind of liquid—one with almost no viscosity. On the page facing Kapitza's one-page paper was another by the young Canadian physicists Jack Allen and Donald Misener[3] (working at the University of Cambridge), with essentially equivalent results on helium flow in long capillary tubes. It was submitted two weeks after Kapitza's, but both papers are the standard references for the discovery of superfluidity. These two short notes ushered in a golden period of low-temperature physics and the study of what later came to be called "quantum liquids."

In the 1920s and early '30s, Kapitza had been a protégé of Ernest Rutherford at Cambridge, where Kapitza was held in awe for his brilliant mind and commanding presence. But on a visit to the Soviet Union in 1934, he was put under house arrest and asked to set up a new low-temperature laboratory in Moscow. Superfluidity was the first of many significant discoveries made at this laboratory, the Institute for Physical Problems.

Further illustrations of the bizarre fluid behavior of He II, such as the fountain effect (fig. 4.1), quickly followed. Almost all of this work was published as short notes in *Nature* in 1938. Astoundingly, within weeks the strange behavior found a qualitative explanation, reported by Fritz London[4] and Laszlo Tisza.[5] In one of the great conceptual leaps in the history of physics, London suggested that the lambda transition at 2.2 K involved some sort of analogue of Bose-Einstein condensation (BEC), a phenomenon first predicted to occur in dilute gases by Albert Einstein,[6] in a largely forgotten paper published in 1925 (see box 4.1).

Within days of talking to London, Tisza went further and suggested a "two-fluid" model of He II, comprising a normal viscous liquid plus a new component, the Bose condensate, which could move in a coherent, collective way without dissipation of energy. In other words, the Bose-condensed component was acting as a superfluid. The London-Tisza model suggested that the superfluidity exhibited by He II was a dramatic manifestation of the (usually hidden) quantum-mechanical wave nature of matter. These ideas received wide publicity and made superfluidity of liquid helium a hot topic in the years just before the Second World War. Although it would take two decades for a quantitative theory to be fleshed out, London and Tisza's suggestion turned superfluidity from a specific feature of liquid helium into something much more generic about the quantum nature of the physical world.

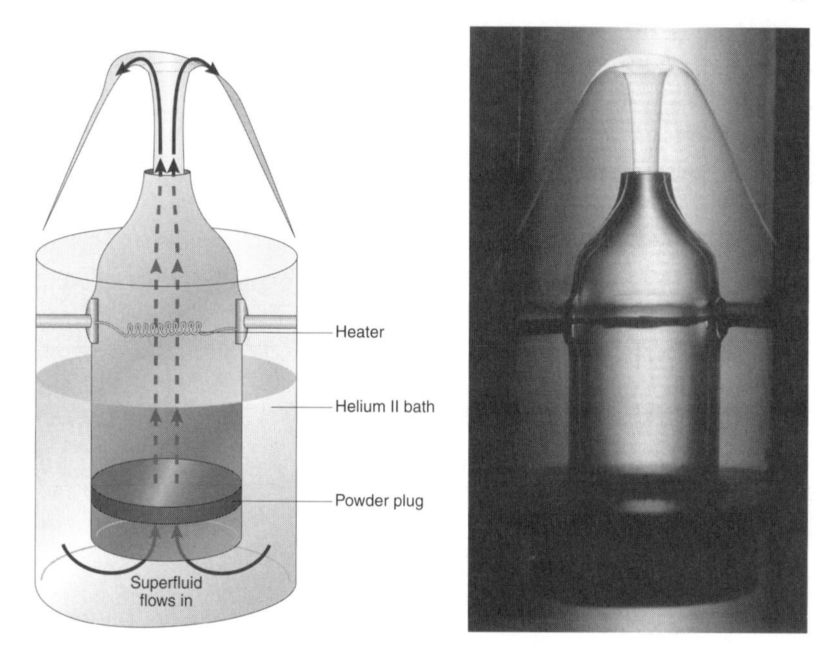

Fig. 4.1. The fountain effect, first reported in 1938 by Jack Allen and Harry Jones,[10] is a manifestation of the two-fluid character of He II (see box 4.1). Below the λ-point, liquid helium is a mixture of a superfluid and a normal fluid. One of the odd properties of the superfluid is that it flows spontaneously from colder to hotter regions. In the experiment shown here (see schematic, left), a glass tube containing He II is suspended in a bath of the same liquid. The tube has a plug at its bottom, made of a tightly packed fine powder, which allows superfluid but not normal fluid to flow through it. When the liquid in the tube is heated, the added energy increases the temperature, and hence the fraction of normal fluid in the tube. To preserve the balance between the normal and superfluid components, superfluid from the bath enters the tube through the powder plug, forcing the contents of the tube out through the top.

In He II the superfluid flows from colder to hotter regions, while the normal fluid flows in the opposite direction; this provides an extremely efficient mechanism for transporting heat. The sudden cessation of bubbling at the λ-point[1] was the first evidence for superefficient heat transport in He II, although this was not understood until the two-fluid model was developed.

One problem with the London-Tisza model was that the only system susceptible to quantitative calculations was a dilute gas of noninteracting atoms, and it was not clear whether BEC would still occur in a liquid, where the closely spaced atoms interact strongly with each other. Nonetheless, London continued to develop his ideas about superfluid helium as a manifestation of a macroscopic quantum state, and emphasized the similarity, first alluded to by Kapitza,[2] of the physics underlying both superconductivity (the ability of certain metals at low temperature to conduct electricity with

Box 4.1 *Bose-Einstein condensation and the two-fluid model of helium II*

Most elementary particles (for example, electrons, protons, and neutrons) are **fermions**—particles with a "spin" of ½. According to Wolfgang Pauli's **exclusion principle**, only one fermion can occupy a particular quantum state at a given time. This fact leads to the characteristic structure of atoms—in which electrons occupy orbitals of different shape and size—and hence to the periodic table of elements. In contrast, composite particles such as atoms, which are made up of neutrons and protons (in the nucleus) and electrons, can be either fermions (net spin ½) or **bosons** (net spin 0, 1, 2, or some other integer). The net spin of a composite particle is the sum of the spins of the constituent fermions, which can contribute positively or negatively, according to whether each spin is "up" or "down." Thus, for example, a pair of electrons with opposite spin, bound to each other (a "Cooper pair"), has zero net spin and thus is a boson.

Unlike fermions, bosons do not obey the Pauli exclusion principle, and so any number of Bose atoms can occupy the same quantum state. The classic examples of the two kinds of atom are the two isotopes of helium: ^{4}He, with two protons, two electrons, and two neutrons (net spin 0, and hence a boson), and ^{3}He, with two protons, two electrons, and one neutron (net spin ½, because of the odd number of neutrons, and hence a fermion). Natural helium is 99.9999 percent ^{4}He, with only a trace of ^{3}He. More generally, atoms with an even number of neutrons, such as the rubidium-87 used recently to demonstrate BEC in an atomic gas,[8] are bosons.

Einstein[6] predicted that a gas of noninteracting Bose atoms could undergo a "phase transition," at which the number of atoms in the lowest-energy quantum state would become very large ("macroscopic"). This would start to occur at a transition temperature where the average de Broglie wavelength of the atoms (which increases as the temperature decreases) becomes comparable to the average distance between atoms in the gas. In other words, at the transition temperature, the "matter waves" corresponding to the atoms would start to overlap. This theoretical work predated the first observation of the de Broglie matter waves of electrons by Davisson and Germer in 1927 (see chapter 2).

This macroscopically occupied quantum state (or "matter wave") is what is now called the **Bose condensate**: all of the atoms are in phase with one another ("coherent"), and behave quite differently from "normal" atoms, which occupy many different incoherent states. It was this Bose-Einstein condensation that London and Tisza in 1938 suggested was the key to what was going on in He II and gave a natural basis for a two-fluid model, in which some of the helium atoms are distributed normally over many different quantum states, while others are "condensed" into the lowest-energy (superfluid) state.

no electrical resistance) and superfluidity. London had a vision of a new world, but the theoretical tools needed to reach it became available only in the decade after his death in 1954.

A giant step forward came in 1941, when the great Russian theoretical physicist L. D. Landau published[7] a derivation of the two-fluid hydrodynamics for superfluidity in liquid ^{4}He (an isotope of helium; see box 4.1). Landau's theory was phenomenological in nature (that is, it was not grounded in a truly microscopic theory), but, unlike Tisza's theory, it could generate quantitative predictions about the properties of superfluid ^{4}He. These included a new kind of collective oscillation, involving out-of-phase motion of the two fluid components (called "second sound"), observed by another Russian physicist, V. P. Peshkov, in 1946. Over the next two decades, Landau's theory and Kapitza's institute dominated the study of superfluid ^{4}He, producing work that was widely viewed as the diamond in the crown of Soviet physics. Both Landau and Kapitza received the Nobel Prize in Physics (in 1962 and 1978, respectively) for their pioneering work on quantum liquids. These prizes were well deserved, but it is unfortunate that the citations made only passing reference to the independent discovery by Allen and Misener, and omitted to mention the important contributions of London and Tisza.

For a long time, Landau argued that his theory had nothing to do with Bose-Einstein condensation. But the evidence for BEC as the root cause of superfluidity was strengthened immensely in 1949, when experiments on liquid ^{3}He (which was thought not to be capable of BEC) showed no superfluidity down to 1 K. By the early 1960s, the theoretical tools were available to give a microscopic basis for Landau's two-fluid hydrodynamics, with the superfluid component being rooted in an underlying Bose condensate. It was realized that superfluidity was always associated with the coherence of the macroscopic wavefunction (all the Bose-condensed particles in the same quantum state, and in phase with each other), and the superfluid velocity could be calculated from the phase of this wavefunction. In 1957, superconductivity in metals was also explained by a kind of BEC of Bose particles (in this case, the so-called Cooper pairs of electrons; see box 4.1). A similar pairing mechanism is now known to occur between ^{3}He atoms, which allows ^{3}He to exhibit superfluidity, but only below 0.002 K.

These developments in superfluid ^{4}He laid the foundation for the much more recent discovery and study of BEC in ultracold atomic gases, for which Eric Cornell, Wolfgang Ketterle, and Carl Wieman were awarded the 2001 Nobel Prize in Physics. In liquid ^{4}He, superfluidity is very visible, but the underlying Bose condensate is much less accessible to experiments.[8] By contrast, in the atomic gases, the macroscopic occupation of the lowest quantum state (see box 4.1) was already apparent in the first experiment.[9] But the specific effects related to "superfluidity" in the gases—such as quantized vortices, first found in 1999—took some time to be detected.

Although the theory underlying superfluidity is the same in the atomic Bose gases and superfluid ^{4}He, the detailed properties of the two systems can be quite different. This is because in a dilute gas the superfluid can be directly identified with the Bose condensate and the normal fluid with the noncondensed atoms. But in a liquid, the strong interactions between atoms complicate the picture, so that in liquid ^{4}He the superfluid component includes some of the noncondensed atoms moving along with the condensate.

We have now come full circle. The complete analogue of the Landau two-fluid hydrodynamic equations, including various transport coefficients associated with the normal fluid, has recently been derived for the trapped atomic Bose gases. In the next few years, experiments should be able to probe this two-fluid domain in Bose gases, a region first explored with such spectacular success in superfluid helium by Kapitza[2] and Allen and Misener.[3] The complexity of analyzing any liquid at a microscopic level (let alone a Bose-condensed liquid) has resulted in many aspects of the theory of superfluid ^{4}He having never been closely checked by experiment. The Bose-condensed atomic gases have finally given us a superfluid system where this can be done. This will complete one of the most exciting chapters in physics, the understanding of the strange behavior of liquid ^{4}He as the macroscopic manifestation of the hidden wave nature of matter.

References

1. McLennan, J. C., Smith, H. D. & Wilhelm, J. O. Scattering of light by liquid helium. *Phil. Mag.* **14,** 161–167 (1932).
2. Kapitza, P. Viscosity of liquid helium below the λ-point. *Nature* **141,** 74 (1938).
3. Allen, J. F. & Misener, A. D. Flow of liquid helium II. *Nature* **141,** 75 (1938).
4. London, F. The λ-phenomenon of liquid helium and the Bose-Einstein degeneracy. *Nature* **141,** 643–644 (1938).
5. Tisza, L. Transport phenomena in helium II. *Nature* **141,** 913 (1938).
6. Einstein, A. Quantum theory of monatomic ideal gases. *Sitzungber. Berlin Preuss. Akad. Wiss.,* 3–14 (1925).
7. Landau, L. D. The theory of superfluidity of helium II. *J. Phys. (U.S.S.R.)* **5,** 71–90 (1941).
8. Griffin, A. Condensate forced out of hiding. *Nature* **391,** 25–26 (1998).
9. Anderson, M. H. *et al.* Observation of Bose-Einstein condensation in a dilute atomic vapor. *Science* **269,** 198–201 (1995).
10. Allen, J. F. & Jones, H. New phenomena connected with heat flow in helium II. *Nature* **141,** 243–244 (1938).

Further reading

Allen, J. The beginning of superfluidity. *Physics World* 29–31 (November 1988).

Boag, J. W., Rubinin, P. E. & Shoenberg, D. (eds) *Kapitza in Cambridge and Moscow: Life and Letters of a Russian Physicist* (North-Holland, New York, 1990).

Collins, G. P. The coolest gas in the universe. *Scientific American* 92–99 (December 2000).

Donnelly, R. J. The discovery of superfluidity. *Physics Today* 30–36 (July 1995).

Nobel e-Museum. *The Nobel Prize in Physics 1978* (http://www.nobel.se/physics/laureates/1978/). (Also physics prizes in 1962, 1913, and 2001.)

1938

Viscosity of liquid helium below the λ-point

P. Kapitza

The abnormally high heat conductivity of helium II below the λ-point, as first observed by Keesom, suggested to me the possibility of an explanation in terms of convection currents. This explanation would require helium II to have an abnormally low viscosity; at present, the only viscosity measurements on liquid helium have been made in Toronto[1], and showed that there is a drop in viscosity below the λ-point by a factor of 3 compared with liquid helium at normal pressure, and by a factor of 8 compared with the value just above the λ-point. In these experiments, however, no check was made to ensure that the motion was laminar, and not turbulent.

The important fact that liquid helium has a specific density ρ of about 0.15, not very different from that of an ordinary fluid, while its viscosity μ is very small comparable to that of a gas, makes its kinematic viscosity $\nu = \mu/\rho$ extraordinarily small. Consequently when the liquid is in motion in an ordinary viscosimeter, the Reynolds number may become very high, while in order to keep the motion laminar, especially in the method used in Toronto, namely, the damping of an oscillating cylinder, the Reynolds number must be kept very low. This requirement was not fulfilled in the Toronto experiments, and the deduced value of viscosity thus refers to turbulent motion, and consequently may be higher by any amount than the real value.

The very small kinematic viscosity of liquid helium II thus makes it difficult to measure the viscosity. In an attempt to get laminar motion the following method (shown diagramatically in the accompanying illustration) was devised. The viscosity was measured by the pressure drop

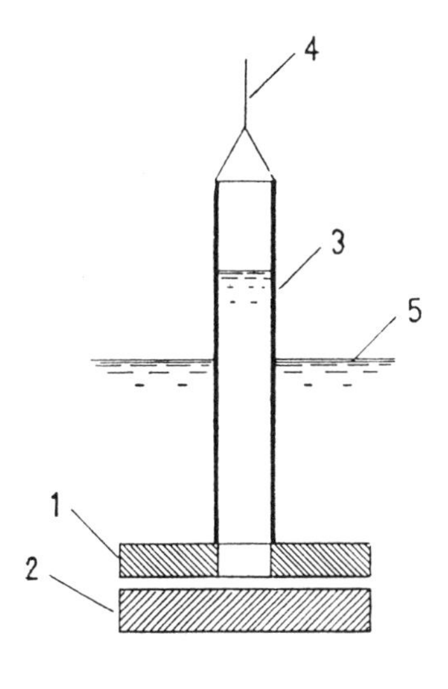

when the liquid flows through the gap between the disks 1 and 2; these disks were of glass and were optically flat, the gap between them being adjustable by mica distance pieces. The upper disk, 1, was 3 cm. in diameter with a central hole of 1.5 cm. diameter, over which a glass tube (3) was fixed. Lowering and raising this plunger in the liquid helium by means of the thread (4), the level of the liquid column in the tube 3 could be set above or below the level (5) of the liquid in the surrounding Dewar flask.
The amount of flow and the pressure were deduced from the difference of the two levels, which was measured by cathetometer.

The results of the measurements were rather striking. When there were no distance pieces between the disks, and the plates 1 and 2 were brought into contact (by observation of optical fringes, their separation was estimated to be about half a micron), the flow of liquid above the λ-point could be only just detected over several minutes, while below the λ-point the liquid helium flowed quite easily, and the level in the tube 3 settled down in a few seconds. From the measurements we can conclude that the viscosity of helium II is at least 1,500 times smaller than that of helium I at normal pressure.

The experiments also showed that in the case of helium II, the pressure drop across the gap was proportional to the square of the velocity of flow, which means that the flow must have been turbulent. If, however, we calculate the viscosity, assuming the flow to have been laminar, we obtain a value of the order 10^{-9} C.G.S., which is evidently still only an upper limit to the true value. Using this estimate, the Reynolds number, even with such a small gap, comes out higher than 50,000, a value for which turbulence might indeed be expected.

We are making experiments in the hope of still further reducing the upper limit to the viscosity of liquid helium II, but the present upper limit (namely, 10^{-9} C.G.S.) is already very striking, since it is more than 10^4 times smaller than that of hydrogen gas (previously thought to be the fluid of least viscosity). The present limit is perhaps sufficient to suggest, by analogy with supraconductors, that the helium below the λ-point enters a special state which might be called a 'superfluid'.

As we have already mentioned, an abnormally low viscosity such as indicated by our experiments might indeed provide an explanation for the high thermal conductivity, and for the other anomalous properties observed by Allen, Peierls, and Uddin[2]. It is evidently possible that the turbulent motion, inevitably set up in the technical manipulation required in working with the liquid helium II, might on account of the great fluidity, not die out, even in the small capillary tubes in which the thermal conductivity was measured; such turbulence would transport heat extremely efficiently by convection.

Institute for Physical Problems, Academy of Sciences, Moscow

Dec. 3 [1937] [Published 8 January 1938]

References

1. Burton, Nature, 135, 265 (1935); Wilhelm, Misener and Clark, *Proc. Roy. Soc., A*, 151, 342 (1935).
2. Nature, 140, 62 (1937).

Letters to the Editor

The Editor does not hold himself responsible for opinions expressed by his correspondents. He cannot undertake to return, or to correspond with the writers of, rejected manuscripts intended for this or any other part of NATURE. *No notice is taken of anonymous communications.*

NOTES ON POINTS IN SOME OF THIS WEEK'S LETTERS APPEAR ON P. 247.

CORRESPONDENTS ARE ... D ... S ... MM ... THEIR COMMUNICATIONS.

ntegration of Uranium by ... trons
Type of Nuclear Reaction

bombarding uranium with neutron ... and
orators[1] found that at least four radioactive
ances were produced, to two of which atomic
ers larger than 92 were ascribed. Further
igations[2] demonstrated the existence of at least
radioactive periods, six of which were assigned
ments beyond uranium, and nuclear isomerism
to be assumed in order to account for their
ical behaviour together with their genetic
ons.

making chemical assignments, it was always
ned that these radioactive bodies had atomic
ers near that of the element bombarded, since
particles with one or two charges were known
emitted from nuclei. A body, for example,
similar properties to those of osmium was
ned to be eka-osmium ($Z = 94$) rather than
um ($Z = 76$) or ruthenium ($Z = 44$).

llowing up an observation of Curie and Savitch[3],
and Strassmann[4] found that a group of at
three radioactive bodies, formed from uranium
neutron bombardment, were chemically similar
arium and, therefore, presumably isotopic with
m. Further investigation[5], however, showed
it was impossible to separate these bodies from
m (although mesothorium, an isotope of radium,
readily separated in the same experiment), so
Hahn and Strassmann were forced to conclude
isotopes of barium ($Z = 56$) *are formed as a*
quence of the bombardment of uranium ($Z = 92$)
neutrons.

first sight, this result seems very hard to under-
l. The formation of elements much below
ium has been considered before, but was always
ted for physical reasons, so long as the chemical
nce was not entirely clear cut. The emission,
n a short time, of a large number of charged
cles may be regarded as excluded by the small
trability of the 'Coulomb barrier', indicated by
ov's theory of alpha decay.

the basis, however, of present ideas about the
viour of heavy nuclei[6], an entirely different and
tially classical picture of these new disintegration
esses suggests itself. On account of their close
ing and strong energy exchange, the particles
heavy nucleus would be expected to move in a
ctive way which has some resemblance to the
ement of a liquid drop. If the movement is made
ciently violent by adding energy, such a drop
divide itself into two smaller drops.

the discussion of the energies involved in the
mation of nuclei, the concept of surface tension

th... s... ion of a charged drople
dim... ed by i... harge, and a rough esti
... hat ... ce tension of nuclei, decrea
... increa... nuclear charge, may become zer
atomic numbers of the order of 100.

It seems therefore possible that the uran
nucleus has only small stability of form, and m
after neutron capture, divide itself into two n
of roughly equal size (the precise ratio of sizes dep
ing on finer structural features and perhaps partl
chance). These two nuclei will repel each other
should gain a total kinetic energy of c. 200 Mev
calculated from nuclear radius and charge.
amount of energy may actually be expected t
available from the difference in packing frac
between uranium and the elements in the midd
the periodic system. The whole 'fission' process
thus be described in an essentially classical
without having to consider quantum-mechan
'tunnel effects', which would actually be extrem
small, on account of the large masses involved.

After division, the high neutron/proton rati
uranium will tend to readjust itself by beta d
to the lower value suitable for lighter eleme
Probably each part will thus give rise to a chai
disintegrations. If one of the parts is an iso
of barium[5], the other will be krypton ($Z = 92 -$
which might decay through rubidium, stront
and yttrium to zirconium. Perhaps one
two of the supposed barium-lanthanum-cer
chains are then actually strontium-yttrium-zircon
chains.

It is possible[5], and seems to us rather proba
that the periods which have been ascribed to elem
beyond uranium are also due to light eleme
From the chemical evidence, the two short per
(10 sec. and 40 sec.) so far ascribed to ^{239}U migh
masurium isotopes ($Z = 43$) decaying through r
enium, rhodium, palladium and silver into cadm

In all these cases it might not be necessar
assume nuclear isomerism; but the different radioac
periods belonging to the same chemical element
then be attributed to different isotopes of this elem
since varying proportions of neutrons may be g
to the two parts of the uranium nucleus.

By bombarding thorium with neutrons, activ
are obtained which have been ascribed to rad
and actinium isotopes[8]. Some of these periods
approximately equal to periods of barium
lanthanum isotopes[5] resulting from the bomb
ment of uranium. We should therefore like
suggest that these periods are due to a 'fission'
thorium which is like that of uranium and res
partly in the same products. Of course it wo

FIVE

From nuclear physics to nuclear weapons

Joseph Rotblat

The discovery of nuclear fission, reported in 1939 by Lise Meitner and Robert Frisch, would have attracted little public attention as a purely scientific advance. But by opening the way to the use of nuclear energy for peaceful and military purposes, it had an impact on the course of human affairs out of all proportion to its contribution to fundamental physics.

The story of nuclear fission begins in 1932, the annus mirabilis of physics. Among the events of that year, the most momentous was the discovery of the neutron by James Chadwick[1] (see chapter 3). Not only did this discovery help to establish the constitution of all atoms (see box 5.1), but neutrons became an important tool for the study of nuclear reactions. Being electrically neutral, neutrons can penetrate into the atomic nucleus even at low energies; indeed—as was soon discovered—slow neutrons (traveling with the velocities of light gas molecules at a given temperature) are more efficient in causing nuclear reactions.

The subsequent flurry of research in many laboratories led to the production of radioactive forms of many elements (see box 5.2), with many practical applications, particularly in medicine. Some of the natural radioactive elements, such as radium, had been used in medicine since the beginning of the twentieth century, but the scope of medical applications, for both therapy and diagnosis, increased enormously when radioactive isotopes of all elements became available.

Neutron research also led to a better understanding of the interactions between protons and neutrons in atomic nuclei. In particular, the Danish physicist Niels Bohr and others developed the model of the "compound

Box 5.1 The building blocks of matter

For practical purposes, all matter can be considered to be composed of three elementary particles: protons and neutrons in the nucleus of the atom, and electrons orbiting the nucleus. The number of electrons is equal to the number of protons, so two numbers are sufficient to characterize the atom. The **atomic number** is the number of protons; it specifies an element's place in the periodic table and its chemical behavior. The **mass number** gives the total number of particles in the nucleus: protons plus neutrons. For example, uranium is element 92, and its most common form is uranium-238, which has 92 protons and 146 (238 minus 92) neutrons.

A given element may exist in forms with different numbers of neutrons: these are **isotopes** of the element. Thus, another form of uranium is uranium-235, with 92 protons and 143 neutrons. Having the same number of electrons, all isotopes of an element have the same chemical properties, but different nuclear properties.

nucleus." According to this model, when a neutron hits a nucleus it does not bounce off, as if it were one billiard ball hitting another. Instead, the neutron is captured by the nucleus, forming a transient compound nucleus, with the energy of the colliding neutron distributed among all protons and neutrons.

The most intensive experimental studies with neutrons were carried out in Rome, where Enrico Fermi and his team systematically bombarded the nuclei of nearly all the elements with slow neutrons. The results from uranium proved to be very puzzling because of the large number of radioactive species produced, each characterized by its half-life (see box 5.2). Some of these were identified as elements beyond uranium in the periodic table, but the origin and nature of the majority remained unexplained. It became the job of chemists to figure out what elements had been produced by investigating their chemical (rather than physical) properties. This was a demanding task, as the radioactive products were present in minute quantities, distributed throughout the uranium from which they had formed.

Much of this research was carried out in Berlin, where Otto Hahn and Fritz Strassmann, advised by the Austrian physicist Lise Meitner, investigated the nature of the products of the bombardment of uranium with neutrons. Try as they might, they were unable to identify most of these products as any of the elements near uranium (atomic number 92) in the periodic table. Eventually, at the end of 1938, Hahn and Strassmann concluded that at least one of the radioactive products was an isotope of barium—an element with atomic number 56. Until then, in all nuclear transformations the product nucleus had differed from the target nucleus by only one or two atomic numbers. So this was an astonishing finding.

By this time, Meitner was living in Sweden, having been forced to emigrate, like most of the Jewish scientists in Germany. Her nephew, Otto Robert Frisch, then working in Bohr's institute in Copenhagen, decided to spend Christmas with his aunt in the small Swedish town of Kungälv. There Meitner showed him the letter she had received from Hahn about his seemingly inexplicable findings.

The explanation dawned on Meitner and Frisch as they walked through the woods in the snow. It was based on Bohr's concept of the compound nucleus. A heavy nucleus, such as that of uranium, is already close to breaking apart, owing to the repulsive electrical forces between its many positively charged protons. When such a nucleus captures a neutron, the neutron's

Box 5.2 Radioactivity

Only certain combinations of protons and neutrons can form a stable nucleus; all other forms decay spontaneously into stable nuclei by the emission of radiation—hence the name "radioactive." The two main types of transformation are **α-decay,** in which the radioactive nucleus ejects an α-particle (two protons and two neutrons), and **β-decay,** in which a neutron in the nucleus changes into a proton, and a β-ray (a fast electron) is emitted. After the transformation the new nucleus may still be unstable and decay further, until a stable form is reached.

Each nucleus has a characteristic decay rate, described by the **half-life**—the time it takes for half of the nuclei in any sample to decay. The half-lives of radioactive nuclei range from a tiny fraction of a second to billions of years.

Many of the nuclei created in the Big Bang and in stars have decayed into stable forms. Of the elements that were incorporated into the Earth 4.5 billion years ago, only those with half-lives of hundreds of millions of years or more have survived; they, and their products, are the naturally occurring radioactive elements. Nowadays, however, it is possible to re-create many of the decayed forms by bombardment in accelerators or through neutron-induced fission. For example, uranium-238 bombarded with neutrons becomes uranium-239, a radioactive isotope that undergoes β-decay. It is transformed into neptunium-239, which, by a further β-decay, becomes plutonium-239. This is how the plutonium used in weapons is produced.

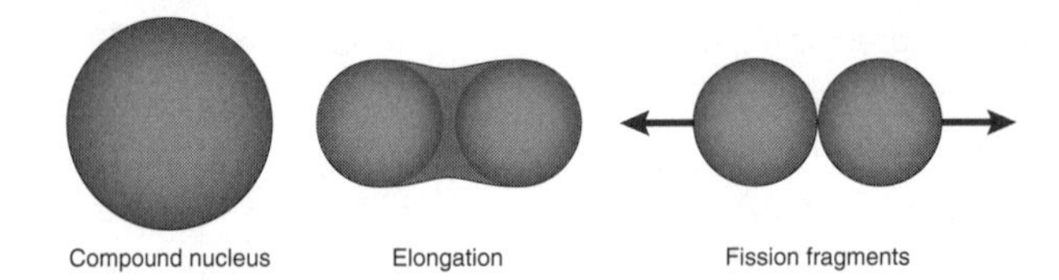

Fig. 5.1. The fission process, as described in the text.

energy is distributed among the protons and neutrons, causing the nucleus to vibrate and become elongated (fig. 5.1). The repulsive electrical force then tears the nucleus apart, and the two fragments fly off at great speed. Frisch named this process "fission"—a term borrowed from biology, where it describes the division of living cells.

When, a few days later, Frisch described the fission concept to Bohr, the latter (in Frisch's words) "smote his forehead with his hand and exclaimed: 'Oh what idiots we all have been! Oh but this is wonderful! This is just as it must be.' "

Meitner and Frisch's report of fission[2] was published in *Nature* on 11 February 1939. Before sending the paper off, Frisch carried out a simple experiment confirming the high speed of the fission fragments. He described this in a separate paper,[3] which appeared a week later.

Even before the papers appeared, the news leaked out during a visit by Bohr to the United States in January 1939. Further experimental and theoretical findings quickly emerged. For example, Bohr and the American physicist John Wheeler worked out that the effect observed with slow neutrons was due to uranium-235, an isotope that forms 0.7 percent of natural uranium.

Of far greater significance, however, was the simultaneous finding by several scientists (including myself) that new neutrons are emitted at each fission event (fig. 5.2). If more than one neutron is available after each fission

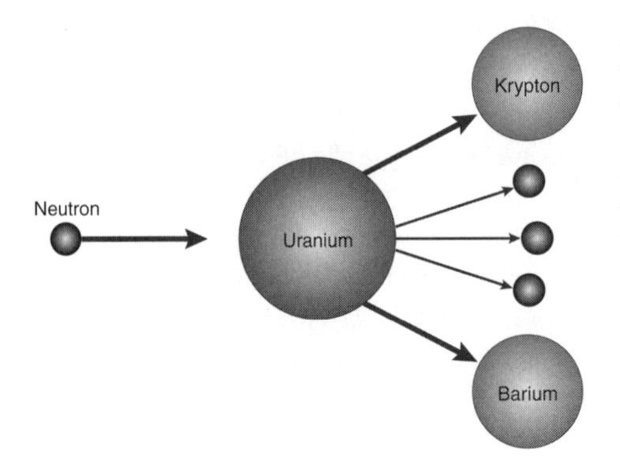

Fig. 5.2. Schematic diagram of the fission of uranium, induced by neutron bombardment. The uranium nucleus splits to yield two new nuclei and several neutrons, each of which can go on to induce another fission event. This is the basis of the fission chain reaction.

to produce further fissions, a chain reaction will ensue, with an exponential increase in the number of fissions and the energy released. Compared with chemical reactions, a fission event releases enormous amounts of energy—about forty million times more than comes from burning one atom of carbon. But on an absolute scale this is infinitesimal: more than one hundred billion billion (10^{20}) fissions per second are needed to make this process a large source of energy. This became possible, thanks to the emission of new neutrons with each fission.

One practical outcome of this discovery was the generation of electricity in nuclear reactors. When fission fragments are stopped in passing through matter, their kinetic energy is converted into heat, which can be used to make steam to drive a generator. Most reactors are based on the fission of uranium by slow neutrons, the high energy of the neutrons emitted at fission being reduced by collisions with the atoms of hydrogen in the water surrounding the uranium core. The chain reaction is kept at a steady rate by control rods, usually made of boron, which is an avid absorber of neutrons. More than 440 nuclear reactors are now in operation in thirty-one countries, with a total electric power output of 360 million kilowatts, providing 16 percent of the total electricity generated in those countries.

But there was another, frightful, consequence of the observation of the chain reaction. Physicists soon realized that if the fission process were propagated by high-energy ("fast") neutrons, a huge energy release could occur in less than a microsecond—making it equivalent to the explosion of many thousands of tons of high-explosive. In other words, nuclear fission could lead to a weapon of unprecedented destructive power.

When the idea of an atomic bomb occurred to me (as it did to several other physicists) early in 1939, I gave it only a passing thought: working on a weapon of mass destruction was not my concept of science. But in the back of my mind a fear kept gnawing away, that other scientists might not have such scruples. The particular worry was that German scientists might work on such a bomb and provide Hitler with a tool to win the war that was imminent. When the Second World War broke out with Hitler's invasion of Poland, revealing Germany's military might, this fear prompted me to suggest to James Chadwick that we should investigate the feasibility of the atomic bomb.

A few months later, in March 1940, Frisch (then in Birmingham, U.K.) and Rudolf Peierls calculated that the amount of material needed to start an explosive chain reaction in uranium-235—the "critical mass"—was only a few kilograms. Frisch and Peierls wrote a memorandum describing their findings, which was sent to the U.K. Committee on Air Defence.[4] This led to the creation of a committee, code-named MAUD, charged to work on the development of the atomic bomb.

Most of the experimental research was done in Liverpool, where Frisch joined Chadwick's team. By 1941 these experiments established the scientific basis of the bomb, but the problem was the separation of the 235 isotope from natural uranium; this required an industrial effort beyond Britain's capability in wartime.

Meanwhile, in the United States the emphasis in nuclear research was on building a reactor. Hardly any attention was paid to the military aspect, even though the report of the MAUD committee had been made available to U.S. scientists. A visit to the United States by the Australian physicist Mark Oliphant in autumn 1941 stirred the U.S. government to start the Manhattan Project. It was mainly the work of scientists in Los Alamos, New Mexico, led by J. Robert Oppenheimer, that led to the making of bombs based on the fission of uranium-235 and plutonium-239, and the testing of the latter at Alamogordo in July 1945. Soon afterward, such bombs were used to destroy Hiroshima and Nagasaki.

The motivation of the scientists who worked on the atomic bomb was that the Allies needed the bomb not to use it, but to prevent, by the threat of retaliation, its use by Hitler. Whether this would have worked with a psychopath like Hitler will never be known, as the German scientists gave up their atomic bomb project in 1942.

The bombs dropped on the Japanese cities brought the war to a dramatic end. They were also the start of an arms race, which included the development of the hydrogen bomb—based on the fusion of isotopes of hydrogen, with an energy release thousands of times greater than that of the atomic bomb. During the four decades of the Cold War the arms race resulted in the accumulation of huge nuclear arsenals, at one time amounting to seventy thousand warheads (reduced by about half after the end of the Cold War)—enough, if detonated, to destroy not only civilization but perhaps even the human species.

All this came from the discovery of nuclear fission in 1939. The world was never the same again.

References

1. Chadwick, J. Possible existence of a neutron. *Nature* **129,** 312 (1932).
2. Meitner, L. & Frisch, O. R. Disintegration of uranium by neutrons: a new type of nuclear reaction. *Nature* **143,** 239–240 (1939).
3. Frisch, O. R. Physical evidence for the division of heavy nuclei under nuclear bombardment. *Nature* **143,** 276 (1939).
4. Gowing, M. *Britain and Atomic Energy 1939–1945,* 28–46 (Macmillan, London, 1964).

Further reading

Frisch, O. R. *What Little I Remember* (Cambridge University Press, 1979).
Garwin, R. L. & Charpak, G. *Megawatts and Megatons* (University of Chicago Press, 2003).
Peierls, R. *Bird of Passage* (Princeton University Press, 1985).
Rhodes, R. *The Making of the Atomic Bomb* (Simon & Schuster, New York, 1986).
Rotblat, J. Leaving the bomb project. *Bulletin of the Atomic Scientists,* August 1985, 16–19.

1939

Disintegration of uranium by neutrons: a new type of nuclear reaction

Lise Meitner and O. R. Frisch

On bombarding uranium with neutrons, Fermi and collaborators[1] found that at least four radioactive substances were produced, to two of which atomic numbers larger than 92 were ascribed. Further investigations[2] demonstrated the existence of at least nine radioactive periods, six of which were assigned to elements beyond uranium, and nuclear isomerism had to be assumed in order to account for their chemical behaviour together with their genetic relations.

In making chemical assignments, it was always assumed that these radioactive bodies had atomic numbers near that of the element bombarded, since only particles with one or two charges were known to be emitted from nuclei. A body, for example, with similar properties to those of osmium was assumed to be eka-osmium ($Z = 94$) rather than osmium ($Z = 76$) or ruthenium ($Z = 44$).

Following up an observation of Curie and Savitch[3], Hahn and Strassmann[4] found that a group of at least three radioactive bodies, formed from uranium under neutron bombardment, were chemically similar to barium and, therefore, presumably isotopic with radium. Further investigation[5], however, showed that it was impossible to separate these bodies from barium (although mesothorium, an isotope of radium, was readily separated in the same experiment), so that Hahn and Strassmann were forced to conclude that *isotopes of barium* ($Z = 56$) *are formed as a consequence of the bombardment of uranium* ($Z = 92$) *with neutrons.*

At first sight, this result seems very hard to understand. The formation of elements much below uranium has been considered before, but was always rejected for physical reasons, so long as the chemical evidence was not entirely clear cut. The emission, within a short time, of a large number of charged particles may be regarded as excluded by the small penetrability of the 'Coulomb barrier', indicated by Gamov's theory of alpha decay.

On the basis, however, of present ideas about the behaviour of heavy nuclei[6], an entirely different and essentially classical picture of these new disintegration processes suggests itself. On account of their close packing and strong energy exchange, the particles in a heavy nucleus would be expected to move in a collective way which has some resemblance to the movement of a liquid drop. If the movement is made sufficiently violent by adding energy, such a drop may divide itself into two smaller drops.

In the discussion of the energies involved in the deformation of nuclei, the concept of surface tension of nuclear matter has been used[7] and its value has been estimated from simple considerations regarding nuclear forces. It must be remembered, however, that the surface tension of a charged droplet is diminished by its charge, and a rough estimate shows that the surface tension of nuclei, decreasing with increasing nuclear charge, may become zero for atomic numbers of the order of 100.

It seems therefore possible that the uranium nucleus has only small stability of form, and may, after neutron capture, divide itself into two nuclei of roughly equal size (the precise ratio of sizes depending on finer structural features and perhaps partly on chance). These two nuclei will repel each other and should gain a total kinetic energy of *c.* 200 Mev., as calculated from nuclear radius and charge. This amount of energy may actually be expected to be available from the difference in packing fraction between uranium and the elements in the middle of the periodic system. The whole 'fission' process can thus be described in an essentially classical way, without having to consider quantum-mechanical 'tunnel effects', which would actually be extremely small, on account of the large masses involved.

After division, the high neutron/proton ratio of uranium will tend to readjust itself by beta decay to the lower value suitable for lighter elements. Probably each part will thus give rise to a chain of disintegrations. If one of the parts is an isotope of barium[5], the other will be krypton ($Z = 92 - 56$), which might decay through rubidium, strontium and yttrium to zirconium. Perhaps one or two of the supposed barium-lanthanum-cerium chains are then actually strontium-yttrium-zirconium chains.

It is possible[5], and seems to us rather probable, that the periods which have been ascribed to elements beyond uranium are also due to light elements. From the chemical evidence, the two short periods (10 sec. and

40 sec.) so far ascribed to ^{239}U might be masurium isotopes ($Z = 43$) decaying through ruthenium, rhodium, palladium and silver into cadmium.

In all these cases it might not be necessary to assume nuclear isomerism; but the different radioactive periods belonging to the same chemical element may then be attributed to different isotopes of this element, since varying proportions of neutrons may be given to the two parts of the uranium nucleus.

By bombarding thorium with neutrons, activities are obtained which have been ascribed to radium and actinium isotopes[8]. Some of these periods are approximately equal to periods of barium and lanthanum isotopes[5] resulting from the bombardment of uranium. We should therefore like to suggest that these periods are due to a 'fission' of thorium which is like that of uranium and results partly in the same products. Of course, it would be especially interesting if one could obtain one of these products from a light element, for example, by means of neutron capture.

It might be mentioned that the body with half-life 24 min.[2] which was chemically identified with uranium is probably really ^{239}U, and goes over into an eka-rhenium which appears inactive but may decay slowly, probably with emission of alpha particles. (From inspection of the natural radioactive elements, ^{239}U cannot be expected to give more than one or two beta decays; the long chain of observed decays has always puzzled us.) The formation of this body is a typical resonance process[9]; the compound state must have a life-time a million times longer than the time it would take the nucleus to divide itself. Perhaps this state corresponds to some highly symmetrical type of motion of nuclear matter which does not favour 'fission' of the nucleus.

L.M.: Physical Institute, Academy of Sciences, Stockholm; O.R.F.: Institute of Theoretical Physics, University, Copenhagen

Jan. 16 [Published 11 February 1939]

References

1. Fermi, E., Amaldi F., d'Agostino, O., Rasetti, F., and Segrè, E. *Proc. Roy. Soc.*, A, **146**, 483 (1934).
2. See Meitner, L., Hahn, O., and Strassmann, F., *Z. Phys.*, **106**, 249 (1937).
3. Curie, I., and Savitch, P., *C.R.*, **206**, 906, 1643 (1938).
4. Hahn, O., and Strassmann, F., *Naturwiss.*, **26**, 756 (1938).
5. Hahn, O., and Strassmann, F., *Naturwiss.*, **27**, 11 (1939).
6. Bohr, N., NATURE, **137**, 344, 351 (1936).
7. Bohr, N., and Kalckar, F., *Kgl. Danske Vid. Selskab, Math. Phys. Medd.*, **14**, Nr. 10 (1937).
8. See Meitner, L., Strassmann, F., and Hahn, O., *Z. Phys.*, **109**, 538 (1938).
9. Bethe, H. A., and Placzek, G., *Phys. Rev.*, **51**, 450 (1937).

ment, and to Dr. G. E. R. Deacon and the in and officers of R.R.S. *Discovery II* for their n making the observations.

., F. B., Gerrard, H., and Jevons, W., *Phil. Mag.*, 40, 149 920).
het-Higgins, M. S., *Mon. Not. Roy. Astro. Soc., Geophys. Supp.*, 285 (1949).
Arx, W. S., Woods Hole Papers in Phys. Oceanog. Meteor., 11 (1950).
n, V. W., *Arkiv. Mat. Astron. Fysik.* (Stockholm), 2 (11) (1905).

MOLECULAR STRU… NUCLEIC AC…

tructure for Deoxyribose Nuc… cid

E wish to suggest a structure for the salt of deoxyribose nucleic acid (D.N.A.). This ture has novel features which are of considerable gical interest.

structure for nucleic acid has already been osed by Pauling and Corey[1]. They kindly made manuscript available to us in advance of cation. Their model consists of three inter- d chains, with the phosphates near the fibre and the bases on the outside. In our opinion, structure is unsatisfactory for two reasons: We believe that the material which gives the y diagrams is the salt, not the free acid. Without cidic hydrogen atoms it is not clear what forces d hold the structure together, especially as the tively charged phosphates near the axis will each other. (2) Some of the van der Waals nces appear to be too small.

nother three-chain structure has also been sug- d by Fraser (in the press). In his model the phates are on the outside and the bases on the e, linked together by hydrogen bonds. This ture as described is rather ill-defined, and for this reason we shall not comment on it.

We wish to put forward a radically different structure for the salt of deoxyribose nucleic acid. This structure has two helical chains each coiled round the same axis (see diagram). We have made the usual chemical assumptions, namely, that each chain consists of phosphate di-ester groups joining β-D-deoxy-ribofuranose residues with 3′,5′ linkages. The two chains (but not their bases) are related by a dyad perpendicular to the fibre axis. Both chains follow right-handed helices, but owing to the dyad the sequences of the atoms in the two chains run in opposite directions. Each chain loosely resembles Furberg's[2] model No. 1; that is, the bases are on the inside of the helix and the phosphates on the outside. The configuration of the sugar and the atoms near it is close to Furberg's 'standard configuration', the

figure is purely ammatic. The two ns symbolize the phosphate—sugar s, and the hori- l rods the pairs of

is a residue on each chain every 3·4 A. in the z-di tion. We have assumed an angle of 36° betw adjacent residues in the same chain, so that structure repeats after 10 residues on each chain, t is, after 34 A. The distance of a phosphorus a from the fibre axis is 10 A. As the phosphates are the outside, cations have easy access to them.

The structure is an open one, and its water con is rather high. At lower water contents we wo …e … tilt so that the structure c …me mor… …ct.

…e novel fe… of the structure is the man … th… chains are held together by puri…d pyri…e bases. The planes of the b are …ndicula… the fibre axis. They are jo … in… single base from one chain b hydrogen-b… to a single base from the o chain, so that the two lie side by side with ident z-co-ordinates. One of the pair must be a purine the other a pyrimidine for bonding to occur. hydrogen bonds are made as follows: purine posi 1 to pyrimidine position 1; purine position 6 pyrimidine position 6.

If it is assumed that the bases only occur in structure in the most plausible tautomeric fo (that is, with the keto rather than the enol c figurations) it is found that only specific pairs bases can bond together. These pairs are: ade (purine) with thymine (pyrimidine), and gua (purine) with cytosine (pyrimidine).

In other words, if an adenine forms one membe a pair, on either chain, then on these assumpt the other member must be thymine; similarly guanine and cytosine. The sequence of bases o single chain does not appear to be restricted in way. However, if only specific pairs of bases ca formed, it follows that if the sequence of bases one chain is given, then the sequence on the o chain is automatically determined.

It has been found experimentally[3,4] that the r of the amounts of adenine to thymine, and the r of guanine to cytosine, are always very close to u for deoxyribose nucleic acid.

It is probably impossible to build this struc with a ribose sugar in place of the deoxyribose the extra oxygen atom would make too close a der Waals contact.

The previously published X-ray data[5,6] on de ribose nucleic acid are insufficient for a rigorous of our structure. So far as we can tell, it is rou compatible with the experimental data, but it m be regarded as unproved until it has been chec against more exact results. Some of these are g in the following communications. We were not a of the details of the results presented there whe devised our structure, which rests mainly though entirely on published experimental data and st chemical arguments.

It has not escaped our notice that the spe pairing we have postulated immediately sugges possible copying mechanism for the genetic mate

Full details of the structure, including the ditions assumed in building it, together with a of co-ordinates for the atoms, will be publi elsewhere.

We are much indebted to Dr. Jerry Donohue constant advice and criticism, especially on i atomic distances. We have also been stimulate a knowledge of the general nature of the unpubli

SIX

The double helix

Sydney Brenner

A brief note published in the middle of the last century described a possible structure of DNA: it included a throwaway comment that the structure suggested how DNA could be copied, and so could act as heritable genetic material. The proposal did not meet with instant acceptance, except among a small group of enthusiasts. But its explanatory power was quickly tested and extended, and the structure has assumed iconic status in both biology and the wider world.

J. D. Watson and F. H. C. Crick's paper,[1] "A structure for deoxyribose nucleic acid," published on 25 April 1953, ushered in the modern era of biology. The little figure it included, showing the double-helical structure, has become a symbol of that era and an icon of our time (Fig. 6.1). Today, when DNA, genes, and genetics are on everybody's lips, it is hard to believe that there was a time when many viewed DNA as an inert polymer that was unlikely to have anything to do with heredity.

In the early 1950s, Watson and Crick were at a Medical Research Council unit in the Cavendish Laboratory of Cambridge University. I was working at Oxford University then, and vividly remember the day in early April 1953 when I and others went over to Cambridge to see the Watson-Crick model of DNA. I found it a revelation, and immediately knew what direction I wanted to follow in science. But when the paper appeared later that month, it by no means received universal acclaim. The scientific establishment of the time consisted largely of biochemists who were preoccupied with the transformation of energy and matter in biological systems, and they failed to see how the structure would revolutionize the study of biological information. But that it did. It revealed the basis on which genetic material is passed

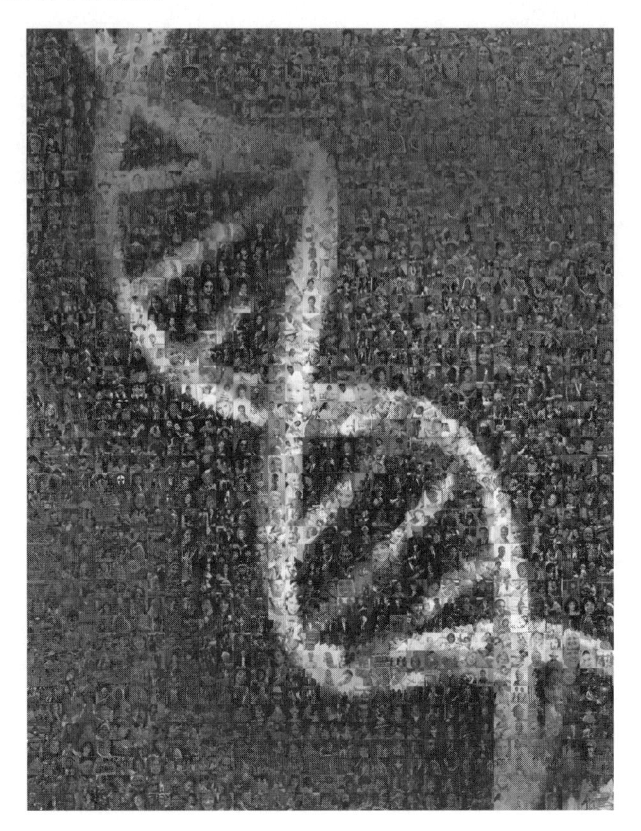

Fig. 6.1. An icon of our time: the double-helical structure of DNA as portrayed on the cover of *Nature* of 15 February 2001. This issue[4] contained milestone papers from the Human Genome Project announcing a draft DNA sequence of the entire human genome. Simultaneously, a draft sequence produced by a company, Celera Genomics, appeared in *Science*.[5] The "rungs" on the double-helical ladder are pairs of the four molecules (adenine, thymine, guanine, and cytosine) known as bases, while the strands that constitute the outside of the ladder are chains composed of sugar and phosphate. (The image, by Eric Lander, was created by Runaway Technology Inc. using PhotoMosaic by Robert Silvers from artwork by Darryl Leja. Reproduced courtesy of the Whitehead Institute)

from one generation to the next, and it showed the principle behind the genetic code through which proteins are built from a DNA blueprint.

Although most of modern biology stems from this brief paper, there is also a long prehistory of thought in the subject that we now know as genetics. It begins in the 1850s and 1860s with the famous pea-breeding experiments of Gregor Mendel, a monk living in what is now the Czech Republic. His results were published in an obscure journal, and only in 1900 were they rediscovered and their significance recognized. Put simply, Mendel postulated that there were internal factors that specified external characters; and that to identify a factor specifying a normal or "wild-type" character in an organism, one had to find an example of that organism displaying a variant character. Today the factors are called genes. The characters are the phenotype, an organism's observable characteristics. And a variant character is a mutant phenotype, which results from the action of a mutant rather than wild-type form of a gene.

Following the rediscovery of Mendel's work, genetics flourished largely through work on the fruit fly, *Drosophila melanogaster.* This tiny insect is easily bred in large numbers and has large numbers of mutants (in eye color, for instance, among many other characteristics). The pioneer was Thomas Hunt Morgan, who spent most of his career at Columbia University. As a frustrated embryologist, he had entered the new field of genetics around the turn of the century in the hope that knowledge about genes would throw light on the then intractable problems of developmental biology.

Morgan's research was largely responsible for revealing the relationship between Mendel's findings and genes and chromosomes (of which the fruit fly has four pairs; we have twenty-three pairs). Most phenotypic characters are inherited independently of each other, but some are "linked": that is, they tend to be inherited together, implying that they occur on the same chromosome. Linkage can be broken, because pairs of chromosomes can swap segments in a process known as recombination. The closer genes are to each other on a chromosome, the stronger the linkage, and the more likely they or their mutant forms are to be inherited together. In consequence, from the breeding and "gene-mapping" work in *Drosophila,* analyzing patterns of inheritance of mutant phenotypes, Morgan could assign a physical reality to the gene as part of a chromosome and with a defined location on that chromosome.

Meanwhile, research on enzymes was developing in the new field of biochemistry. Enzymes are proteins that catalyze chemical transformations in a cell. Their chemical and physical constitution were to remain enigmas for some time. But in 1909 a British physician, Archibald Garrod, proposed a connection between genes and enzymes: he observed that certain inherited diseases reflected a patient's production of malfunctioning enzymes (he called these "inborn errors of metabolism"). In the 1940s this line of thinking culminated in the research of G. W. Beadle and E. L. Tatum in the United States, who investigated nutritional deficiencies due to mutant enzymes in the bread mold *Neurospora.* From the results Beadle formulated the "one gene, one enzyme" principle, which holds that a single gene specifies a single enzyme.

As early as the last years of the nineteenth century, E. B. Wilson had stated in a book that DNA (which he called "chromonucleic acid") was the physical carrier of heredity. And in 1944, in what at least in retrospect were highly convincing experiments, Oswald Avery and colleagues showed that a heritable alteration to one strain of the bacterium *Pneumococcus* could be caused by adding an extract of a second strain to the growth medium of the first. They identified the active material in the extracts as DNA. Nonetheless, as the 1950s dawned, most people still believed that proteins would have to have the leading role in copying genes: only proteins seemed to have the required chemical specificity.

Thus was the stage set. Genetics had arrived at the conclusion that the gene was the fundamental unit of heredity, indivisible like the atom of early physics. Biochemistry had shown that the work of cells was performed by enzymes, specialized proteins of elaborate but unknown structures, performing all the various steps of biosynthesis and energy metabolism. Each enzyme was specified by a gene, but no enzymes seemed to be concerned with specifying genes. DNA was a candidate as the genetic material, but by no means was this widely accepted.

When Watson and Crick's paper[1] appeared, it was not at all immediately evident that it provided an answer to any of the problems of genetics confronting us at that time. The structure was described and the notion of the specificity of the base pairs stated: the authors proposed that the four types of molecule (bases) of which DNA was known to be composed—adenine, thymine, guanine, and cytosine—could pair off only as A–T and G–C. However, the only comment about the structure's potential biological significance was contained in the famous sentence: "It has not escaped our notice that the specific pairing we have postulated immediately suggests a possible copying mechanism for the genetic material."

In an article[2] marking the twenty-first anniversary of the paper's appearance, we learn from Francis Crick that the sentence was inserted as a compromise between the two authors. Crick was keen to discuss the genetic implications.

> Watson [writes Crick][2] was against it. He suffered from periodic fears that the structure might be wrong and that he had made an ass of himself. I yielded to his point of view but insisted that something be put in the paper, otherwise someone else would certainly write to make the suggestion, assuming we had been too blind to see it.

I was an early convert: my own memory of having seen the structure earlier in Cambridge is that it was certain to be right, explaining as it did instantly to me the central puzzle of genetics. But it didn't take the authors themselves long to present their argument in more detail, for a second paper[3] appeared on 30 May 1953 which spelled out the biological implications. In it Watson and Crick give more detail about "complementary replication"—how the invariable pairing of A with T and C with G (fig. 6.2) could lead to reconstruction of a double helix from the information contained in one strand only. They explain how mutations could occur through one base being substituted for another. And they mention that "it . . . seems likely that the precise sequence of the bases is the code which carries the genetical information." This generated the whole field of study of the genetic code.

Before these possibilities registered with most scientists, the proposed structure of DNA had to be combined with two other pieces of research to result in the birth of modern biology. The first, which preceded the double helix, was Frederick Sanger's work in Cambridge on the structure of the protein insulin. Sanger showed that proteins have a precise chemical structure as defined by the sequence of amino acids linked in a chain. As became evident in later work on the genetic code in which I was involved, the sequence of bases in the DNA specifies the sequence of amino acids in a protein via an intermediate molecule, messenger RNA.

The second piece of research was Seymour Benzer's subsequent investigations on the fine structure of the gene. Benzer was at Purdue University, working with bacteriophages—viruses that infect bacteria. He isolated hundreds of mutants of a gene; by using techniques to detect viruses that were the result of genetic recombination, he was able to resolve and map mutant phenotypes onto many different sites on a DNA sequence and show that they corresponded to distances between the base pairs in the DNA structure. This destroyed the classical, bead-on-a-string notion of genes as indivisible units of function, mutation, and recombination. It was replaced by the concept of the gene as a stretch of DNA, with hundreds of possible mutational states caused by alterations of single base pairs.

Today we no longer have to study genes indirectly through their effects on the phenotypes of organisms, but can use the technologies of DNA cloning and sequencing to characterize genes directly. Today the genetic code is no longer a daring speculation, but a working tool that allows us to derive the sequences of amino acids in a protein from the DNA sequence of the gene encoding it. Today we know in great detail the molecular machinery that copies DNA, repairs it and expresses it.

It all started with this little paper of Watson and Crick.[1] The fusion of

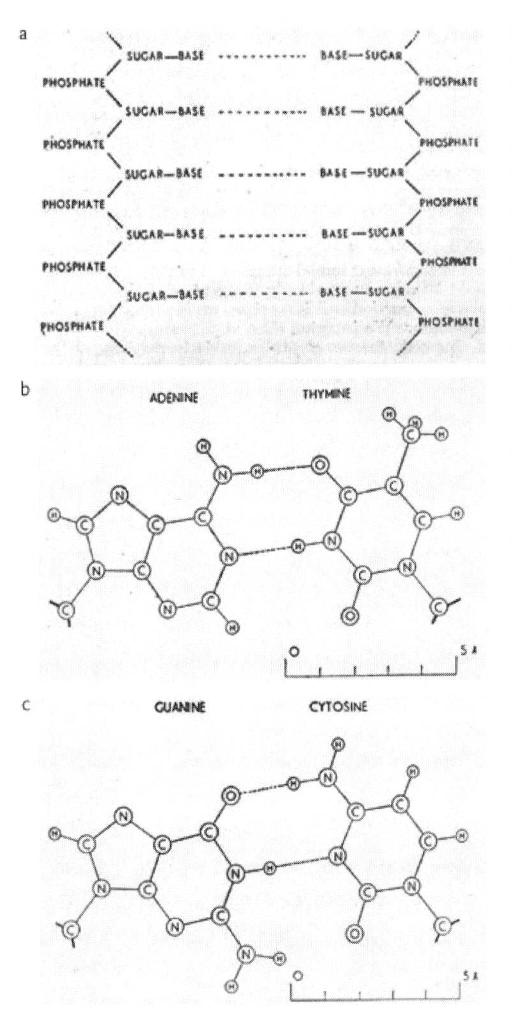

Fig. 6.2. The greater detail about DNA structure and base-pairing that followed in Watson and Crick's second paper,[3] of 30 May 1953. These diagrams are facsimiles of the originals. a, The chemical formula of a pair of DNA strands, with sugar-phosphate chains and bases attached to the chains. Forces known as hydrogen bonds (dotted lines) connect each pair of bases and so the two strands of the double helix. b, The pairing of adenine with thymine, and c, that of guanine with cytosine. Watson and Crick's most telling insight[1] was how, because A always pairs with T and G with C, the double helix can be re-created from a single strand—that is, how the genetic material can be copied.

genetics and biochemistry was then already under way. But the structure of DNA greatly intensified the process, which resulted in the creation of molecular biology and the tools to study the chemistry of information in biological systems. The route to the double helix was strewn with Nobel Prizes (see Further reading). And in 1962 Watson and Crick, along with Maurice Wilkins for his analysis of DNA structure using the X-ray technique, received the prize in physiology or medicine. The fiftieth anniversary of the structure's publication was widely celebrated in 2003 with various events and publications.[6]

Watson and Crick, of course, are among those rare scientists whose names are known to the general public. Watson has been director, then president, of Cold Spring Harbor Laboratory, New York, and was a prime mover behind the Human Genome Project. In recent years Crick has been at the Salk Institute in California, his interests centering on neurobiology and the problem of consciousness.

Let Crick have the last word on the double helix. Here he is again, in his 1974 article,[2] responding to the argument that a scientific discovery is akin to a work of art, and style is as important as content:

> Rather than believe that Watson and Crick made the DNA structure, I would rather stress that the structure made Watson and Crick. After all, I was almost totally unknown at the time and Watson was regarded, in most circles, as too bright to be really sound. But what I think is overlooked in such arguments is the intrinsic beauty of the DNA double helix. It is the molecule which has style, quite as much as the scientists.

References

1. Watson, J. D. & Crick, F. H. C. A structure for deoxyribose nucleic acid. *Nature* **171**, 737–738 (1953).
2. Crick, F. The double helix: a personal view. *Nature* **248**, 766–769 (1974).
3. Watson, J. D. & Crick, F. H. C. Genetical implications of the structure of deoxyribonucleic acid. *Nature* **171**, 964–967 (1953).
4. International Human Genome Seqencing Consortium *Nature* **409**, 860–921 (2001).
5. Venter, J. C. *et al. Science* **291**, 1304–1351 (2001).
6. Dennis, C. (ed.) The double helix: 50 years. *Nature* **421**, 395–453 (2003).

Further reading

Judson, H. F. *The Eighth Day of Creation: Makers of the Revolution in Biology* (Cold Spring Harbor Laboratory Press, Cold Spring Harbor, NY, 1996; expanded edition).

Nobel e-Museum. *The Nobel Prize in Physiology or Medicine 1962* (http://www.nobel.se/medicine/laureates/1962/). (Also physiology or medicine prizes in 1933 and 1958, and chemistry prize in 1958.)

Olby, R. C. *The Path to the Double Helix: The Discovery of DNA* (Dover, Mineola, NY, 1994; reprint of the edition of 1974).

Watson, J. D. *The Double Helix* (Penguin, London, 1999; new edition with introduction by S. Jones).

Watson, J. D. with Berry, A. *DNA: The Secret of Life* (Knopf, New York, 2003).

1953

A structure for deoxyribose nucleic acid

J. D. Watson and F. H. C. Crick

We wish to suggest a structure for the salt of deoxyribose nucleic acid (D.N.A.). This structure has novel features which are of considerable biological interest.

A structure for nucleic acid has already been proposed by Pauling and Corey[1]. They kindly made their manuscript available to us in advance of publication. Their model consists of three intertwined chains, with the phosphates near the fibre axis, and the bases on the outside. In our opinion, this structure is unsatisfactory for two reasons: (1) We believe that the material which gives the X-ray diagrams is the salt, not the free acid. Without the acidic hydrogen atoms it is not clear what forces would hold the structure together, especially as the negatively charged phosphates near the axis will repel each other. (2) Some of the van der Waals distances appear to be too small.

Another three-chain structure has also been suggested by Fraser (in the press). In his model the phosphates are on the outside and the bases on the inside, linked together by hydrogen bonds. This structure as described is rather ill-defined, and for this reason we shall not comment on it.

We wish to put forward a radically different structure for the salt of deoxyribose nucleic acid. This structure has two helical chains each coiled round the same axis (see diagram). We have made the usual chemical assumptions, namely, that each chain consists of phosphate diester groups joining β-D-deoxyribofuranose residues with 3′,5′ linkages. The two chains (but not their bases) are related by a dyad perpendicular to the fibre axis. Both chains follow right-handed helices, but owing to the dyad the sequences of the atoms in the two chains run in opposite di-

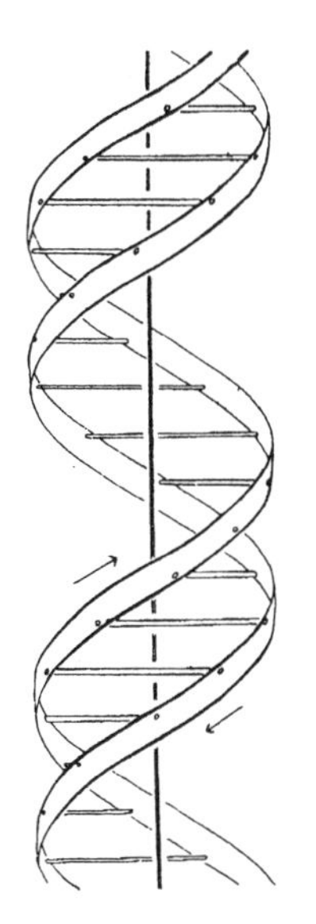

This figure is purely diagrammatic. The two ribbons symbolize the two phosphate—sugar chains, and the horizontal rods the pairs of bases holding the chains together. The vertical line marks the fibre axis.

rections. Each chain loosely resembles Furberg's[2] model No. 1; that is, the bases are on the inside of the helix and the phosphates on the outside. The configuration of the sugar and the atoms near it is close to Furberg's 'standard configuration', the sugar being roughly perpendicular to the attached base. There is a residue on each chain every 3.4 A. in the *z*-direction. We have assumed an angle of 36° between adjacent residues in the same chain, so that the structure repeats after 10 residues on each chain, that is, after 34 A. The distance of a phosphorus atom from the fibre axis is 10 A. As the phosphates are on the outside, cations have easy access to them.

The structure is an open one, and its water content is rather high. At lower water contents we would expect the bases to tilt so that the structure could become more compact.

The novel feature of the structure is the manner in which the two chains are held together by the purine and pyrimidine bases. The planes of the bases are perpendicular to the fibre axis. They are joined together in pairs, a single base from one chain being hydrogen-bonded to a single base from the other chain, so that the two lie side by side with identical *z*-co-ordinates. One of the pair must be a purine and the other a pyrimidine for bonding to occur. The hydrogen bonds are made as follows: purine position 1 to pyrimidine position 1; purine position 6 to pyrimidine position 6.

If it is assumed that the bases only occur in the structure in the most plausible tautomeric forms (that is, with the keto rather than the enol configurations) it is found that only specific pairs of bases can bond together. These pairs are: adenine (purine) with thymine (pyrimidine), and guanine (purine) with cytosine (pyrimidine).

In other words, if an adenine forms one member of a pair, on either chain, then on these assumptions the other member must be thymine; similarly for guanine and cytosine. The sequence of bases on a single chain does not appear to be restricted in any way. However, if only specific pairs of bases can be formed, it follows that if the sequence of bases on one chain is given, then the sequence on the other chain is automatically determined.

It has been found experimentally[3,4] that the ratio of the amounts of adenine to thymine, and the ratio of guanine to cytosine, are always very close to unity for deoxyribose nucleic acid.

It is probably impossible to build this structure with a ribose sugar in place of the deoxyribose, as the extra oxygen atom would make too close a van der Waals contact.

The previously published X-ray data[5,6] on deoxyribose nucleic acid are insufficient for a rigorous test of our structure. So far as we can tell, it is roughly compatible with the experimental data, but it must be regarded as unproved until it has been checked against more exact results. Some of these are given in the following communications. We were not aware of the details of the results presented there when we devised our structure, which rests mainly though not entirely on published experimental data and stereochemical arguments.

It has not escaped our notice that the specific pairing we have postulated immediately suggests a possible copying mechanism for the genetic material.

Full details of the structure, including the conditions assumed in building it, together with a set of co-ordinates for the atoms, will be published elsewhere.

We are much indebted to Dr. Jerry Donohue for constant advice and criticism, especially on interatomic distances. We have also been stimulated by a knowledge of the general nature of the unpublished experimental results and ideas of Dr. M. H. F. Wilkins, Dr. R. E. Franklin and their co-workers at King's College, London. One of us (J. D. W.) has been aided by a fellowship from the National Foundation for Infantile Paralysis.

Medical Research Council Unit for the Study of the Molecular Structure of Biological Systems, Cavendish Laboratory, Cambridge

April 2 [Published 25 April 1953]

References

1. Pauling, L., and Corey, R. B., *Nature*, 171, 346 (1953); *Proc. U.S. Nat. Acad. Sci.*, 39, 84 (1953).
2. Furberg, S., *Acta Chem. Scand.*, 6, 634 (1952).
3. Chargaff, E., for references see Zamenhof, S., Brawerman, G., and Chargaff, E., *Biochim. et Biophys. Acta*, 9, 402 (1952).
4. Wyatt, G. R., *J. Gen. Physiol.*, 86, 201 (1952).
5. Astbury, W. T., Symp. Soc. Exp. Biol. 1, Nucleic Acid, 66 (Camb. Univ. Press, 1947).
6. Wilkins, M. H. F., and Randall, J. T., *Biochim. et Biophys. Acta*, 10, 192 (1953).

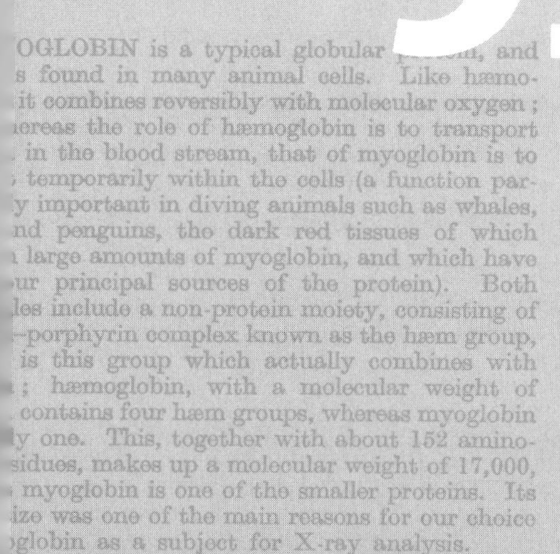

1958

THREE-DIMENSIONAL MODEL OF THE MYOGLOBIN MOLECULE OBTAINED BY X-RAY ANALYSIS

y Drs. J. C. KENDREW, G. BODO, H. M. DINTZIS, R. . RISH and H. WYCKOFF
Medical Research Unit ... ecular ... Cav ... h ... ratory, Cambridge
A
F
Da ... raday ... , The Ro ... stit ... on

OGLOBIN is a typical globular p..., and s found in many animal cells. Like hæmo- it combines reversibly with molecular oxygen; ereas the role of hæmoglobin is to transport in the blood stream, that of myoglobin is to temporarily within the cells (a function par- y important in diving animals such as whales, nd penguins, the dark red tissues of which large amounts of myoglobin, and which have ur principal sources of the protein). Both les include a non-protein moiety, consisting of –porphyrin complex known as the hæm group, is this group which actually combines with ; hæmoglobin, with a molecular weight of contains four hæm groups, whereas myoglobin y one. This, together with about 152 amino- sidues, makes up a molecular weight of 17,000, myoglobin is one of the smaller proteins. Its ize was one of the main reasons for our choice globin as a subject for X-ray analysis.

escribing a protein it is now common to dis- h the primary, secondary and tertiary struc- The *primary structure* is simply the order, or ce, of the amino-acid residues along the ptide chains. This was first determined by using chemical techniques for the protein [1], and has since been elucidated for a number ides and, in part, for one or two other small s. The *secondary structure* is the type of , coiling or puckering adopted by the poly- chain: the α-helix and the pleated sheet are les. Secondary structure has been assigned in outline to a number of fibrous proteins such as ratin and collagen; but we are ignorant of ure of the secondary structure of any globular . True, there is suggestive evidence, though no proof, that α-helices occur in globular s, to an extent which is difficult to gauge tatively in any particular case. The *tertiary* e is the way in which the folded or coiled ptide chains are disposed to form the protein le as a three-dimensional object, in space. emical and physical properties of a protein be fully interpreted until all three levels of re are understood, for these properties depend spatial relationships between the amino-acids, ese in turn depend on the tertiary and ary structures as much as on the primary.

X-ray diffraction methods seem capable, even ciple, of unravelling the tertiary and secondary res. But the great efforts which have been l to the study of proteins by X-rays, while ng successes in clarifying the secondary

the metabolically more important globular, or cr talline, proteins. Progress here has been s because globular proteins are much more complica then the organic molecules which are the nor objects of X-ray analysis (not counting hydroge myoglobin contains 1,200 atoms, whereas the m complicated molecule the structure of which been completely determined by X-rays, vitamin B contains 93). Until five years ago, no one knew h in practice, the complete structure of a crystall protein might be found by X-rays, and it was reali that the methods then in vogue among prot crystallographers could at best give the most sketc indications about the structure of the molecule. T situation was transformed by the discovery, m by Perutz and his colleagues[2], that heavy ato could be attached to protein molecules in spec sites and that the resulting complexes gave diffract patterns sufficiently different from normal to ena a classical method of structure analysis, the so-cal 'method of isomorphous replacement', to be used determine the relative phases of the reflexions. T method can most easily be applied in two dimensio giving a projection of the contents of the unit c along one of its axes. Perutz attached a *p*-chlo mercuri-benzoate molecule to each of two f sulphydryl groups in hæmoglobin and used t resulting changes in certain of the reflexions prepare a projection along the *y*-axis of the u cell[3]. Disappointingly, the projection was larg uninterpretable. This was because the thickness the molecule along the axis of projection was 63 (corresponding to some 40 atomic diameters), so th the various features of the molecule were superpos in inextricable confusion, and even at the increas resolution of 2·7 A. it has proved impossible disentangle them[4]. It was clear that further progre could only be made if the analysis were extended three dimensions. As we shall see, this involves t collection of many more observations and the pr duction of three or four different isomorpho replacements of the same unit cell, a requireme which presents great technical difficulties in m proteins.

The present article describes the application, low resolution, of the isomorphous replaceme method in three dimensions to type *A* crystals sperm whale myoglobin[5]. The result is a thr dimensional Fourier, or electron-density, map of t unit cell, which for the first time reveals the gener nature of the tertiary structure of a protein molecu

Isomorphous Replacement in Myoglobin

SEVEN

Dawn of structural biology

Gregory A. Petsko

To understand how a biological molecule such as a protein works, you have to know how it is built. Achieving such knowledge is the province of what is now called structural biology: because all living things are largely constructed from proteins, this discipline is central to biology. The first dramatic but hard-won success of the approach, the determination of the three-dimensional structure of a protein called myoglobin, was announced in 1958.

The twentieth century was a century of revelation in biology. It began with the alliance of genetics with Darwin's theory of evolution by natural selection, and creation of the modern subject. There followed, in midcentury, the birth of the twin disciplines of molecular and structural biology, which turned the subject into a science that concentrated on the structures and functions of some of the large molecules that are the most important constituents of living cells. The century ended with the advent of genomics, a return to study of the behavior of whole cells or organisms, but now interpreted through knowledge of their genes and the genes' protein products (see p. 238).

Two findings laid the foundation for the age of structure in biology: first, that the positions of atoms in a crystallized substance could be determined from X-rays passing through, and scattered by, the crystal; second, that the approach could be applied to very simple biological molecules. Arguably, however, the age of structure truly dawned only in the late 1950s, at the University of Cambridge. There, a team mostly composed of physicists and led by John Kendrew, "solved" the first detailed three-dimensional structure of a protein. The protein was sperm whale myoglobin, and its structure was described in *Nature*[1] on 8 March 1958.

The paper was the outcome of a truly Herculean task. When the project began, there were no automated instruments or digital computers for generating or analyzing the huge amounts of data necessary. Every step had to be carried out by hand, and there were many steps, each repeated thousands of times. Kendrew was to share the 1962 Nobel Prize in Chemistry for his achievement (with Max Perutz, who determined the second protein structure to be solved, that of hemoglobin).[2] But Kendrew never tried to solve another protein crystal structure, and it is easy to understand why.

Proteins are the products of genetic information and are large polymer molecules made up of building blocks called amino acids. They can be thought of as being like a charm bracelet, with a backbone (the links) that is the same for every amino acid building block, and side-chains (the charms) that differ for each different amino acid. Twenty naturally occurring amino acids are all that are needed to build all of the thousands of different proteins in a living cell. Each protein contains a specific number of each type of amino acid, arranged in a specific sequence along the chain.

Proteins carry out most of the main functions of living cells—from catalysis, to communication within and between cells, to specific molecular recognition (in, for instance, the immune system). Unlike most man-made polymers, which tend to behave as random coils like the strands of spaghetti on a plate, proteins fold up into precise three-dimensional structures that are essential for their proper function. There are three levels of protein structure: "primary," which is the sequence of the amino acid building blocks in the polymer chain; "secondary," which refers to regular conformations of the backbone of the folded protein chain such as the alpha helix (about which more later); and "tertiary," the detailed three-dimensional structure, specified by the relative positions of all of the atoms in the molecule, backbone as well as side-chains.

Knowledge of protein structure is central to understanding what proteins do and how they do it. Structure, for instance, can show how the working parts of biological motors move, can reveal the catalytic sites of enzymes, and can pinpoint interaction areas between molecules—for, say, developing drugs that can block or enhance such interactions.

But proteins are very small (10^{18}—one million trillion—of them would fit on the head of a pin), so their structures can be seen in atomic detail only through the use of light that has a wavelength comparable to the size of an atom. X-rays are of the right wavelength, but a single protein molecule would not scatter X-rays strongly enough for the scattered beams to be measurable. Instead, one needs trillions of the same molecule, all arranged precisely in space so that their individual scattering adds up to give scattered intensities of sufficient strength. Only if the protein molecules can be induced to form a crystal will this condition be met.

The process Kendrew and his team went through is much the same as that practiced today. But at that time few proteins had even been purified in large quantities, much less crystallized. So Kendrew had to find one that was abundant in the tissue or organism in which it functioned. Myoglobin, an oxygen-binding protein from muscle, seemed a good candidate. It is comparatively small (about 150 amino acids long). It is a bright red color, because of the iron-containing chemical entity called a heme group bound to it, and is therefore easy to follow during purification. And it is as stable as a brick. Horse meat was tried first as a myoglobin source. But that protein gave crystals which didn't behave as well as hoped for. Diving mammals promised to be a good source of large quantities of material because they need to store large quantities of oxygen, and sperm whale meat eventually provided a protein that gave the necessary huge crystals.

The next step for Kendrew's group was to measure the amplitudes of the "reflections"—the scattered X-ray beams—produced when the crystals were exposed to X-radiation. And the step after that was to add up the measured reflections, in a process called Fourier summation, to produce an image of the molecules in the crystal lattice.

Photographic film was used to record the reflection amplitudes, thousands of them, and then teams of young women were trained to estimate their values by comparison with a known scale. The tedium of the work can only be imagined, and the task by no means ended there. The correct relative registration of each scattered wave (its phase) had to be determined before the Fourier summation could be carried out. Perutz and his associates had already shown how this could be done by labeling the protein with heavy elements, such as mercury and platinum, which bound to specific sites on each molecule in the crystal lattice. Their binding caused small but measurable changes in the magnitudes of each reflection; whether the scattered intensity of a given wave went up or down, and by how much, could be used to estimate the phase of that wave by a graphical procedure. Ambiguity in the phase determination was resolved by combining phase information from a number of different heavy-atom derivatives. But for each of them, a whole new dataset had to be measured. No wonder the process took years.

The final step was the Fourier summation itself. Kendrew and his coworkers had been able to carry out an earlier trial at lower resolution by doing the summation by hand. Resolution is a measure of the distance that two objects can be apart before they appear as separate or "resolved" images in the Fourier summation: the lower the resolution, the less detail in the image of the protein, but the less data needed for the summation. For the more than 10,000 reflections that had to be added up at 2 Å (angstrom) resolution, however, hand summation was impossible (roughly speaking 2 Å is the level at which individual atoms can be seen as resolved features; an angstrom is one ten-millionth of a millimeter). For myoglobin, technology came to the rescue: the first high-resolution protein structure ever determined was calculated on the first electronic computer ever built in Britain.

At first, the resulting electron-density map looked impossible to interpret (X-rays being electromagnetic radiation, the image obtained is of the electron cloud around each atomic nucleus, but one simply places an atom at the center of each such distribution). The iron atom in the heme provided a convenient place to start, because its large number of electrons made it the strongest feature in the map. An ingenious system was devised for building the atomic model of the protein: the map was produced as a set of contoured topographic plots on sections through the crystal lattice; a three-dimensional grid was overlaid onto the map so that atomic coordinates could be measured with respect to three axes, *x*, *y*, and *z*, at right angles to each other; and wire rods were constructed whose height was proportional to the *z*-coordinate of each feature of density. Brass models were fabricated for each amino acid in the protein, and these were fitted by hand into the "forest" of wires and secured by clamps (fig. 7.1). David Phillips (a member of the team, who later headed the first group to determine the three-dimensional structure of an enzyme)[3] once remarked to me that lacerations from the forest were a daily hazard of this model-building process. To those of us raised in an era of computer graphics and semiautomatedprotein modeling, it sounds laborious—and fun.

I won't try to summarize Kendrew's description of the structure: suffice it to say that no one was prepared for the seeming irregularity of the overall polymer fold and the remarkable fact that the heme iron atom where oxygen bound was buried completely inside the protein. And yet there were striking patterns: the polymer chain was folded into a series of eight segments, in each of which the backbone followed a helical path with 3.6 amino acids per turn and a "translation" of 1.5 angstroms per amino acid. Remember that no one had ever seen this, the alpha helix, before; it had been nothing but a gleam in Linus Pauling's eye. (Pauling, the father of so much of modern structural science, had speculated that protein chains would fold into regular patterns. Cooped up in a sickbed in Oxford, he had played with

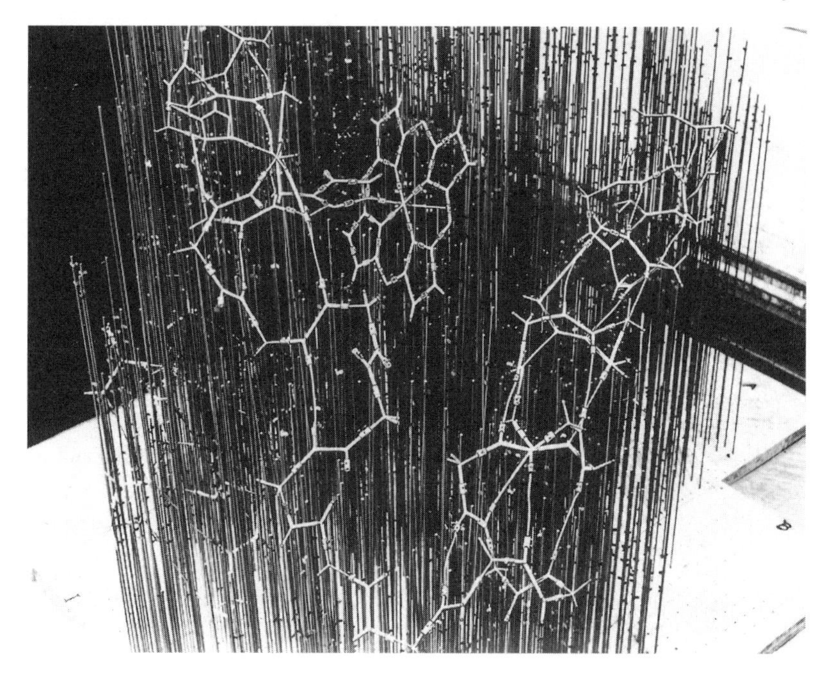

Fig. 7.1. Close-up of part of the "forest" of wire rods. This was the approach used by Kendrew and colleagues[1] to build up the three-dimensional model of myoglobin from the electron-density map produced from the X-ray measurements. (Photo courtesy of Laboratory of Molecular Biology, Cambridge.)

drawings of amino acid polymers on paper. He came up with exactly the helical parameters found in the myoglobin crystal structure, as he and colleagues described[4] in 1951.)

Word of the myoglobin structure spread like wildfire through the biological community. After almost one hundred years of speculation, one could now see what a protein looked like in atomic detail. Disappointment at the "ugliness" of the overall fold was quickly replaced by delight at the beauty of the helices and the excitement that the structure revealed so much about the oxygen-binding properties of the molecule. Almost overnight, biologists and biochemists realized that structure could be a key to understanding protein function. The hemoglobin structure, which came hot on the heels of myoglobin, was even more informative. Yet structure determination was perceived to be so difficult that one leading scientist predicted in 1960 that it would be at least a decade before another protein structure was solved. David Phillips and his coworkers proved him wrong by almost eight years when they solved the structure of the first enzyme, lysozyme,[3] in 1962, the year that Kendrew and Perutz shared the Nobel Prize in Chemistry. The lysozyme structure proved to be enormously informative about how that enzyme—and by implication, all enzymes—worked, and from that point there was no turning back. The Age of Structure had arrived.

Kendrew and Perutz reacted to success in different ways. Having climbed his mountain, Kendrew left research behind and spent the rest of his life as a scientific administrator and adviser to governments. He served with great distinction at posts as varied as head of an Oxford college (St. John's) and founding director of the European Molecular Biology Organization Laboratory in Heidelberg. He died in Cambridge on 23 August 1997, aged eighty. Max Perutz had his own brilliant administrative stint, as founding director of the Medical Research Council Laboratory of Molecular Biology in Cambridge, and had a successful moonlighting career as a popular science writer in publications like the *New Yorker.* But his first love was research and he never forsook it. He continued to publish outstanding papers on subjects ranging from hemoglobin to diseases of protein folding for the rest of his life. He died on 6 February 2002, at eighty-seven years of age, but only the day before had sent off two papers to the *Proceedings of the National Academy of Sciences.*

We are now entering still another era, that of "structural genomics," in which structural information should help us to understand the functions of the thousands of genes in our own genome and those of other organisms. The future seems clear: an exponential growth of structure information, illuminating all areas of biology. In the year 2002 more than 3,300 sets of atomic coordinates depicting the three-dimensional structures of proteins were deposited in the Protein Data Bank. This is the international repository for such information, which now contains almost 20,000 entries And the biological literature is replete with articles that either refer to such information or are illustrated with beautiful ribbon drawings (fig. 7.2) that show the path taken by the folded chain that comprises the protein backbone. In every pharmaceutical company, and many biotech companies, there is a team of crystallographers working on protein structures to aid in drug design or in engineering proteins for medical or industrial applications.

It wasn't always like this. In 1971 there were fewer than a dozen known protein structures, and less than ten new ones were determined per year. A single protein structure, such as that of insulin,[5] took years of work on the part of a team of people to solve. It was widely felt that one protein crystallographer—the term "structural biologist" didn't exist then (it was coined later that year at Brandeis University, my own institution, by Donald

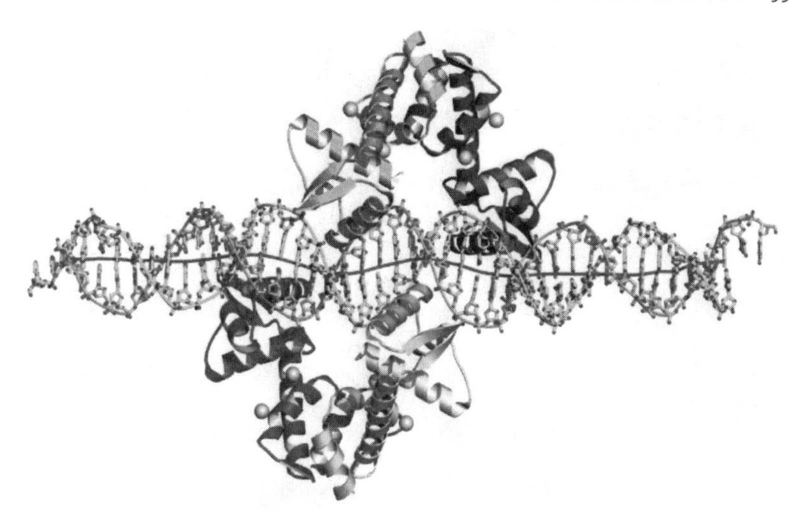

Fig. 7.2. Protein structure, twenty-first-century style. This diagram shows the backbone structure of a complex of four molecules of the diphtheria tox repressor protein bound to its "operator sequence," which forms four turns of double helical B-DNA. Side chains in the protein have been omitted for clarity, and the fold of the polymer backbone is drawn as a ribbon to enhance the sense of the way the chain folds up. In each copy of the repressor, a number of alpha helices can be seen as a continuous coil of ribbon; in the complete structure each coil would have individual amino acid side chains bristling from it at regular intervals. (Figure courtesy of Dagmar Ringe, Brandeis University; see ref. 12.) The myoglobin structure solved by Kendrew and coworkers in 1958 is a single polypeptide chain of 153 amino acids surrounding a single heme cofactor. By contrast, this structure has four molecules of repressor each containing over 200 amino acids plus 43 base pairs of DNA. And the structure of photosystem I (ref. 9) has 12 polypeptide chains containing 2,334 amino acids and 250 cofactors, including 96 light-capturing chlorophyll molecules.

Caspar, Carolyn Cohen, and Susan Lowey)—was sufficient for any large university. And industry had no interest in a technique that seemingly had no practical use. But then we had no idea what was out there.

Now we do: there are probably over a thousand different families of protein fold, and many of them have yet to be determined. Every week brings the announcement of at least one more new and exciting structure. Among the most recent triumphs are the three-dimensional structures of the large and small subunits of the ribosome,[6–8] which is the protein factory of all living cells; of a major component[9] of the machinery of photosynthesis (the process by which plants convert the Sun's light energy into chemical energy); and of various membrane proteins, including the calcium pump essential to the action of muscle cells[10] and the transporter that pumps multiple drugs out of drug-resistant bacteria.[11]

As a structural biologist, I love sitting with a student who has just solved the crystal structure of a new protein and saying, "Do you realize that you are the first person since the beginning of time to see what this molecule looks like?" Few other scientific fields offer this kind of sublime pleasure on a regular basis—this sense of coming over a peak in Darien and staring at the Pacific for the first time. And the first people ever to have that feeling were John Kendrew and his associates.

References

1. Kendrew, J. C. *et al.* A three-dimensional model of the myoglobin molecule obtained by X-ray analysis. *Nature* **181,** 662–666 (1958).
2. Perutz, M. F. *et al.* Structure of haemoglobin: a three-dimensional Fourier synthesis at 5.5 Å resolution, obtained by X-ray analysis. *Nature* **185,** 416–422 (1960).
3. Blake, C. C. F. *et al.* Structure of lysozyme: a Fourier map of the electron density at 6 Å resolution obtained by X-ray diffraction. *Nature* **196,** 1173–1176 (1962).
4. Pauling, L., Corey, R. B. & Branson, H. R. The structure of proteins: Two hydrogen-bonded helical configurations of the peptide chains. *Proc. Natl. Acad. Sci. USA* **37,** 205–211 (1951).
5. Blundell, T. L. *et al.* Atomic positions in rhombohedral 2-zinc insulin crystals. *Nature* **231,** 506–511 (1971).
6. Wimberly, B. T. *et al.* Structure of the 30S ribosomal subunit. *Nature* **407,** 327–339 (2000).
7. Ban, N. *et al.* The complete atomic structure of the large ribosomal subunit at 2.4 Å resolution. *Science* **289,** 905–920 (2000).
8. Pioletti, M. *et al.* Crystal structures of complexes of the small ribosomal subunit with tetracycline, edeine and IF3. *EMBO J.* **20,** 1829–1839 (2001).
9. Jordan, P. *et al.* Three dimensional structure of cyanobacterial photosystem I at 2.5 Å resolution. *Nature* **411,** 909–917 (2001).
10. Toyoshima, C. *et al.* Crystal structure of the calcium pump of sarcoplasmic reticulum at 2.6 Å resolution. *Nature* **405,** 647–655 (2001).
11. Murakami, S. *et al.* Crystal structure of bacterial multidrug efflux transporter AcrB. *Nature* **419,** 587–593 (2002).
12. Chen, C. S. *et al.* Methyl groups of thymine bases are important for nucleic acid recognition by DtxR. *Biochemistry* **39,** 10397–10407 (2000).

Further reading

Kendrew, J. C. The three-dimensional structure of a protein molecule. *Scientific American* 96–110 (June 1961).

Nobel e-Museum. *The Nobel Prize in Chemistry 1962* (http://www.nobel.se/chemistry/laureates/1962/).

1958

A three-dimensional model of the myoglobin molecule obtained by X-ray analysis

J. C. Kendrew, G. Bodo, H. M. Dintzis, R. G. Parrish, H. Wyckoff, and D. C. Phillips

Myoglobin is a typical globular protein, and is found in many animal cells. Like hæmoglobin, it combines reversibly with molecular oxygen; but whereas the role of hæmoglobin is to transport oxygen in the blood stream, that of myoglobin is to store it temporarily within the cells (a function particularly important in diving animals such as whales, seals and penguins, the dark red tissues of which contain large amounts of myoglobin, and which have been our principal sources of the protein). Both molecules include a non-protein moiety, consisting of an iron–porphyrin complex known as the hæm group, and it is this group which actually combines with oxygen; hæmoglobin, with a molecular weight of 67,000, contains four hæm groups, whereas myoglobin has only one. This, together with about 152 amino-acid residues, makes up a molecular weight of 17,000, so that myoglobin is one of the smaller proteins. Its small size was one of the main reasons for our choice of myoglobin as a subject for X-ray analysis.

In describing a protein it is now common to distinguish the primary, secondary and tertiary structures. The *primary structure* is simply the order, or sequence, of the amino-acid residues along the polypeptide chains. This was first determined by Sanger using chemical techniques for the protein insulin[1], and has since been elucidated for a number of peptides and, in part, for one or two other small proteins. The *secondary structure* is the type of folding, coiling or puckering adopted by the polypeptide chain: the α-helix and the pleated sheet are examples. Secondary structure has been assigned in broad outline to a number of fibrous proteins such as silk, keratin and collagen; but we are ignorant of the nature of the secondary structure of any globular protein. True, there is suggestive evidence, though as yet no proof, that α-helices occur in globular proteins, to an extent which is difficult to gauge quantitatively in any particular case. The *tertiary structure* is the way in which the folded or coiled polypeptide chains are disposed to form the protein molecule as a three-dimensional object, in space. The chemical and physical properties of a protein cannot be fully interpreted until all three levels of structure are understood, for these properties depend on the spatial relationships between the amino-acids, and these in turn depend on the tertiary and secondary structures as much as on the primary.

Only X-ray diffraction methods seem capable, even in principle, of unravelling the tertiary and secondary structures. But the great efforts which have been devoted to the study of proteins by X-rays, while achieving successes in clarifying the secondary (though not yet the tertiary) structures of fibrous proteins, have hitherto paid small dividends among the metabolically more important globular, or crystalline, proteins. Progress here has been slow because globular proteins are much more complicated then the organic molecules which are the normal objects of X-ray analysis (not counting hydrogens, myoglobin contains 1,200 atoms, whereas the most complicated molecule the structure of which has been completely determined by X-rays, vitamin B_{12}, contains 93). Until five years ago, no one knew how, in practice, the complete structure of a crystalline protein might be found by X-rays, and it was realized that the methods then in vogue among protein crystallographers could at best give the most sketchy indications about the structure of the molecule. This situation was transformed by the discovery, made by Perutz and his colleagues[2], that heavy atoms could be attached to protein molecules in specific sites and that the resulting complexes gave diffraction patterns sufficiently different from normal to enable a classical method of structure analysis, the so-called 'method of isomorphous replacement', to be used to determine the relative phases of the reflexions. This method can most easily be applied in two dimensions, giving a projection of the contents of the unit cell along one of its axes. Perutz attached a *p*-chloromercuri-benzoate molecule to each of two free sulphydryl groups in hæmoglobin and used the resulting changes in certain of the reflexions to prepare a projection along the *y*-axis of the unit cell[3]. Disappointingly, the projection was largely uninterpretable. This was because the thickness of the molecule along the axis of projection was 63 A. (corresponding to some 40 atomic diameters), so that the various features of the molecule were superposed in inextricable confusion, and even at the increased resolution of 2.7 A. it has proved impossible to disentangle them[4]. It was clear that further progress could only be made if the analysis were extended to three dimensions. As we shall see, this involves the collection of many more observations and the production of three or four different isomorphous replacements of the same unit cell, a requirement which presents great technical difficulties in most proteins.

The present article describes the application, at low resolution, of the isomorphous replacement method in three dimensions to type *A* crystals of sperm whale myoglobin[5]. The result is a three-dimensional Fourier, or electron-density, map of the unit cell, which for the first time reveals the general nature of the tertiary structure of a protein molecule.

Isomorphous replacement in myoglobin

No type of myoglobin has yet been found to contain free sulphydryl groups, so that the method of attaching heavy atoms used by Perutz for hæmoglobin could not be employed. Eventually, we were able to attach several heavy atoms to the myoglobin molecule at different specific sites by crystallizing it with a variety of heavy ions chosen because they might be expected, on general chemical grounds, to possess affinity for protein side-chains. X-ray, rather than chemical, methods were used to determine whether combination had taken place, and, if so, whether the ligand was situated predominantly at a single site on the surface of the molecule. Among others, the following ligands were found to combine in a way suitable for the present purpose: (i) potassium mercuri-iodide and auri-iodide; (ii) silver nitrate, potassium auri-chloride; (iii) *p*-chloro-mercuri-benzene sulphonate; (iv) mercury diammine ($Hg(NH_3)^{2+}$, prepared by dissolving mercuric oxide in hot strong ammonium sulphate), *p*-chloro-aniline; (v) *p*-iodo-phenylhydroxylamine. Each group of ligands combined specifically at a particular site, five distinct sites being found in all. The substituted phenylhydroxylamine is a specific reagent for the iron atom of the hæm group[6], and may be assumed to combine with that group; in none of the other ligands have we any certain knowledge of the mechanism of attachment or of the chemical nature of the site involved.

Methods of X-ray analysis

Type *A* crystals of myoglobin are monoclinic (space group $P2_1$) and contain two protein molecules per unit cell. Only the *hol* reflexions are 'real', that is, can be regarded as having relative phase angles limited to o or π, or positive or negative signs, rather than general phases; when introduced into a Fourier synthesis, these reflexions give a projection of the contents of the cell along its *y*-axis. In two dimensions the analysis followed lines[7] similar to that of hæmoglobin. First, the heavy atom was located by carrying out a so-called difference-Patterson synthesis; if all the heavy atoms are located at the same site on every molecule in the crystal, this synthesis will contain only one peak, from the position of which the *x*- and *z*-co-ordinates of the heavy atom can be deduced, and the signs of the *hol* reflexions determined. These signs were cross-checked by repeating the analysis for each separate isomorphous replacement in turn; we are sure of almost all of them to a resolution of 4 A., and of most to 1.9 A. Using the signs, together with the measured amplitudes, we may, finally, compute an electron-density projection of the contents of the unit cell along *y;* but, as in hæmoglobin and for the same reasons, the projection is in most respects uninterpretable (even though here the axis of projection is only 31 A.). On the other hand, knowledge of the signs of the *hol* reflexions to high resolution enabled us to determine the *x*- and *z*-co-ordinates of all the heavy atoms with some precision. This was the starting point for the three-dimensional analysis now to be described.

In three dimensions the procedure is much more lengthy because all the general reflexions *hkl* must be included in the synthesis, and more complicated because these reflexions may have any relative phase angles, not only 0 or π. Furthermore, we need to know all three co-ordinates of the heavy atoms; the two-dimensional analysis gives *x* and *z,* but to find *y* is more difficult, and details of the methods used will be published elsewhere, including among others two proposed by Perutz[8] and one proposed by Bragg[9]. Finally, a formal ambiguity enters into the deduction of general phase angles if only one isomorphous replacement is available; this can be resolved by using several replacements[10], such as are available in the present case. Once the phases of the general reflexions have been determined, one can carry out a three-dimensional Fourier synthesis which will be a representation of the electron density at every point in the unit cell.

Before such a programme is embarked upon, however, the resolution to be aimed at must be decided. The number of reflexions needed, and hence the amount of labour, is proportional to the cube of the resolution. To resolve individual atoms it would be necessary to include at least all terms of the series with spacings greater than 1.5 A.—some 20,000 in all; and it is to be remembered that the intensities of all the reflexions would have to be measured for *each* isomorphous derivative. Besides this, introduction of a heavy group may cause slight distortion of the crystal lattice; as the resolution is increased, this distortion has an increasingly serious effect on the accuracy of phase determination. In the present stage of the analysis the most urgent objective was an electron-density map detailed enough to show the general layout of the molecule—in other words, its tertiary structure. If the α-helix, or something like it, forms the basis of the structure, we need only work to a resolution sufficient to show up a helical chain as a rod of high electron density. For this purpose we require only reflexions with spacings greater than about 6 A.; in all there are some 400 of these, of which about 100 are *hol*'s already investigated in the two-dimensional study. The Fourier synthesis described here is computed from these 400 reflexions only, and is in consequence blurred; besides this, it is distorted by an unknown amount of experimental error, believed to be small but at the moment difficult to estimate. Thus while the general features of the synthesis are undoubtedly correct, there may be some spurious detail which will require correction at a later stage.

The three-dimensional Fourier synthesis

The synthesis was computed in 70 min. on the EDSAC Mark I electronic computer at Cambridge (as a check, parts of the computation were repeated on DEUCE at the National Physical Laboratory). It is in the form of sixteen sections perpendicular to *y* and spaced nearly 2 A. apart; these must be piled on top of one another to represent the electron density throughout the cell, containing two myoglobin molecules together with associated mother liquor (which amounts to nearly half the whole). Unfortunately, the synthesis cannot be so represented within the two-dimensional pages of a journal; furthermore, if the sections are displayed side by side, they give no useful idea of the structure they represent. The examples reproduced in Fig. 1 illustrate some of the more striking features.

A first glance at the synthesis shows that it contains a number of prominent rods of high electron density; these usually run fairly straight for distances of 20, 30 or 40 A., though there are some curved ones as well. Their cross-section is generally nearly circular, their diameter about 5 A., and they tend to lie at distances from their neighbours of 8–10 A. (axis to axis). In some instances two segments of rod are joined by fairly sharp corners. Fig. 1*a* shows several rods—three of them (*A*, *B* and *C*)

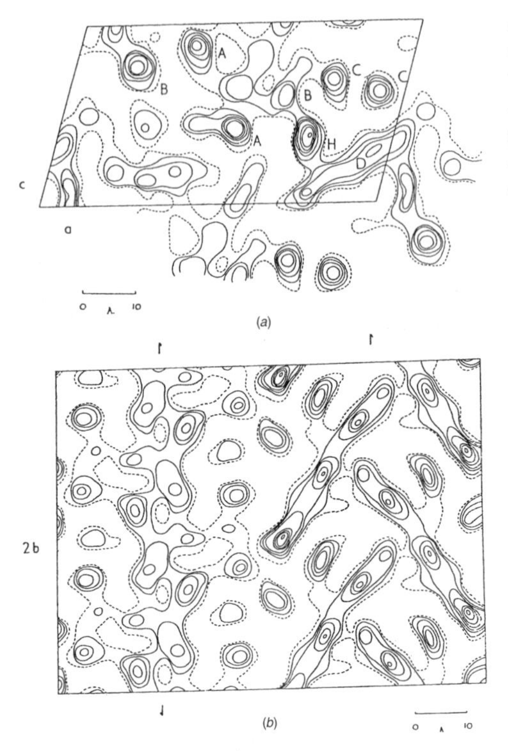

Fig. 1. (*a*) Section of three-dimensional Fourier synthesis of type *A* myoglobin at $y = -1/8\ b$. *A–D*, polypeptide chains; *H*, hæm group. (*b*) Section parallel to [201] at $x = 0$, showing polypeptide chain *A* (on the right).

cross the plane of the section almost at right angles, while one (*D*) lies nearly in that plane. *D* is part of a nearly straight segment of chain about 40 A. long, of which some 20 A. is visible in this section. It seems virtually certain that these rods of high density are the polypeptide chains themselves—indeed, it is hard to think of any other features of the structure which they could possibly be. Their circular cross-section is what would be expected if the configuration were helical, and the electron density along their axes is of the right order for a helical arrangement such as the α-helix. The various rods in the structure are intertwined in a very complex manner, the nature of which we shall describe later.

Another prominent feature is a single disk-shaped region of high electron density which reaches a peak value greater than at any other point in the cell. A section through this disk is shown at *H* in Fig. 1*a*. We identify this feature as the hæm group itself, for the following reasons: (i) the hæm group is a flat disk of about the same size; (ii) its centre is occupied by an iron atom and therefore has a higher electron density than any other point in the whole molecule; (iii) a difference-Fourier projection of the *p*-iodo-phenylhydroxylamine derivative shows that, at least in *y*-projection, the position of the iodine atom is near that of our group; this is what we should expect, since this reagent specifically combines with the hæm group; (iv) the orientation of the disk corresponds, as closely as the limited resolution of the map allows one to determine it, with the orientation of the hæm group deduced from measurements of electron spin resonance[5,11].

We cannot understand the structure of the molecules in the crystal unless we can decide where one ends and its neighbours begin. In a protein crystal the interstices are occupied by mother liquor, in this case strong ammonium sulphate, the electron density of which is nearly equal to the average for the whole cell. Hence it is to be expected that in the intermolecular regions the electron density will be near average (the density of coiled polypeptide chains is much above average, and that of side-chains well below). It should also be fairly uniform; these regions should not be crossed by major features such as polypeptide chains. Using these criteria, it is possible to outline the whole molecule with minor uncertainties. It was gratifying to find that the result agreed very well, in projection, with a salt-water difference-Fourier projection made as part of the two-dimensional programme (for the principles involved, see ref. 12). Moreover, the dimensions of the molecule agreed closely with those deduced from packing considerations in various types of unit cell.

The myoglobin molecule

We are now in a position to study the tertiary structure of a single myoglobin molecule separated from its neighbours. Fig. 2 illustrates various views of a three-dimensional model constructed to show the regions of high electron density in the isolated molecule. Several points must be noticed. First, the model shows only the general distribution of dense regions. The core of a helical polypeptide chain would be such a region; but if the chain were pulled out, into a β-configuration, for example, its mean density would drop to near the average for the cell and the chain would fade out at this resolution. Similarly, side-chains should, in general, scarcely show up, so that the polypeptide rods in the model must be imagined as clothed in an invisible integument of side-chains, so thick that neighbouring chains in reality touch. Third, features other than polypeptide chains may be responsible for some of the regions of high density; patches of adsorbed salt, for example. Fourth, the surface chosen to

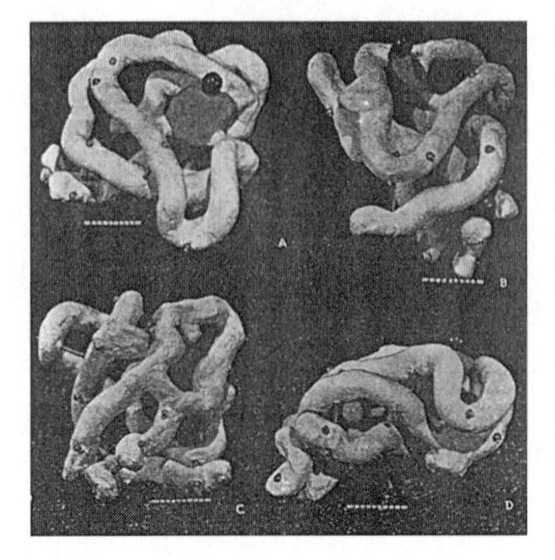

Fig. 2. Photographs of a model of the myoglobin molecule. Polypeptide chains are white; the grey disk is the hæm group. The three spheres show positions at which heavy atoms were attached to the molecule (black: Hg of *p*-chloro-mercuri-benzene-sulphonate; dark grey: Hg of mercury diammine; light grey: Au of auri-chloride). The marks on the scale are 1 A. apart.

demarcate a molecule cannot be traced everywhere with certainty, so it is possible that the molecule shown contains parts of its neighbours, and correspondingly lacks part of its own substance.

Making due allowance for these difficulties, we may note the main features. It is known[13] that myoglobin has only one terminal amino-group: it is simplest to suppose that it consists of a single polypeptide chain. This chain is folded to form a flat disk of dimensions about 43 A. × 35 A. × 23 A. Within the disk chains pursue a complicated course, turning through large angles and generally behaving so irregularly that it is difficult to describe the arrangement in simple terms; but we note the strong tendency for neighbouring chains to lie 8–10 A. apart in spite of the irregularity. One might loosely say that the molecule consists of two layers of chains, the predominant directions of which are nearly at right angles in the two layers. If we attempt to trace a single continuous chain throughout the model, we soon run into difficulties and ambiguities, because we must follow it around corners, and it is precisely at corners that the chain must lose the tightly packed configuration which alone makes it visible at this resolution (an α-helix, for example, cannot turn corners without its helical configuration being disrupted). Also, there are several apparent bridges between neighbouring chains, perhaps due to the apposition of bulky side-chains. The model is certainly compatible with a single continuous chain, but there are at least two alternative ways of tracing it through the molecule, and it will not be possible to ascertain which (if either) is correct until the resolution has been improved. Of the secondary structure we can see virtually nothing directly at this stage. Owing to the corners, the chain cannot be in helical configuration throughout; in fact, the total length of chain in the model is 300 A., whereas an α-helix of 152 residues would be only 228 A. long. The 300 A. might correspond, for example, to 70 per cent α-helix and 30 per cent fully extended chain, but of course intermediate configurations are probably present, too. The hæm group is held in the structure by links to at least four neighbouring chains; nevertheless, one side of it is readily accessible from the environment to oxygen and to larger reagents such as *p*-iodo-phenylhydroxylamine (in the difference-Fourier projection of this complex, referred to above, the position of the iodine atom indicates that the ligand is attached to the outside of the group). Clearly, however, the model cannot at present be correlated in detail with what we know of the chemistry of myoglobin; this must await further refinement.

Perhaps the most remarkable features of the molecule are its complexity and its lack of symmetry. The arrangement seems to be almost totally lacking in the kind of regularities which one instinctively anticipates, and it is more complicated than has been predicated by any theory of protein structure. Though the detailed principles of construction do not yet emerge, we may hope that they will do so at a later stage of the analysis. We are at present engaged in extending the resolution to 3 A., which should show us something of the secondary structure; we anticipate that still further extensions will later be possible—eventually, perhaps, to the point of revealing even the primary structure.

Full details of this work will be published elsewhere. We wish to record our debt to Miss Mary Pinkerton for assistance of all kinds; to the Mathematical Laboratory, University of Cambridge, for computing facilities on the EDSAC; to Dr. J. S. Rollett and the National Physical Laboratory for similar facilities on the DEUCE; to Mrs. Joan Blows and Miss Ann Mansfield for assistance in computing; for fellowships to the U.S. Public Health Service (H. W.), the Merck Fellowship Board (R. G. P.), the U.S. National Science Foundation (R. G. P. and H. M. D.), and the Rockefeller Foundation (H. M. D.); and to Sir Lawrence Bragg for his interest and encouragement. Finally, we wish to express our profound gratitude to the Rockefeller Foundation, which has actively supported this research from its earliest beginnings.

J.C.K., G.B., H.M.D., R.G.P., H.W.: *Medical Research Council Unit for Molecular Biology, Cavendish Laboratory, Cambridge;* D.C.P.: *Davy Faraday Laboratory, The Royal Institution, London*

[Published 8 March 1958]

References

1. Sanger, F., and Tuppy, H., *Biochem. J.*, **49**, 481 (1951). Sanger, F., and Thompson, E. O. P., *ibid.*, **53**, 353, 366 (1953).
2. Green, D. W., Ingram, V. M., and Perutz, M. F., *Proc. Roy. Soc.*, A, **225**, 287 (1954).
3. Bragg, W. L., and Perutz, M. F., *Proc. Roy. Soc.*, A, **225**, 315 (1954).

4. Dintzis, H. M., Cullis, A. F., and Perutz, M. F. (in the press).
5. Kendrew, J. C., and Parrish, R. G., *Proc. Roy. Soc.* A, **238**, 305 (1956).
6. Jung, F., *Naturwiss.*, **28**, 264 (1940). Keilin, D., and Hartree, E. F., *Nature*, **151**, 390 (1943).
7. Bluhm, M. M., Bodo, G., Dintzis, H. M., and Kendrew, J. C. (in the press).
8. Perutz, M. F., *Acta Cryst.*, **9**, 867 (1956).
9. Bragg, W. L. (in the press).
10. Bokhoven, C., Schoone, J. C., and Bijvoet, J. M., *Acta Cryst.*, **4**, 275 (1951).
11. Ingram, D. J. E., and Kendrew, J. C., *Nature*, **178**, 905 (1956).
12. Bragg, W. L., and Perutz, M. F., *Acta Cryst.*, **5**, 277 (1952).
13. Schmid, K., *Helv. Chim. Acta*, **32**, 105 (1949). Ingram, V. M. (unpublished work).

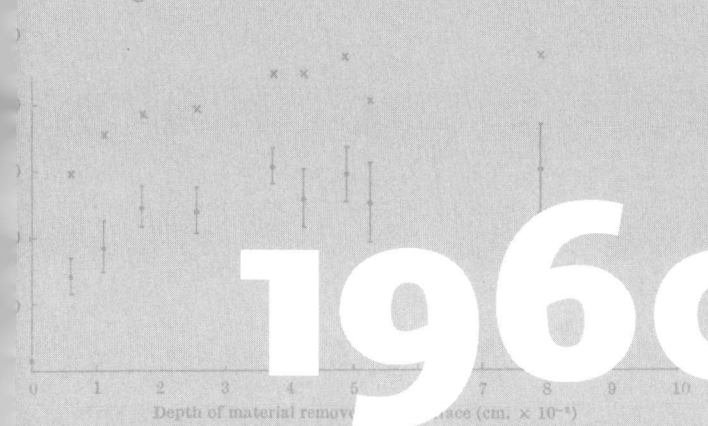

Breaking strength as a function of depth of material removed from the surface. ... ghest strength value in one group of rods ; ●, mean strength and 95 per cent confidence limits for one group of rods

...lar etching solution removes material from a ...urface it is possible to study the strength of ... specimens as a function of the depth of ...al removed from the surface. Such a study ...ive some information about the size and nature ... surface imperfections.

...mercially available soda-glass rods, of 6–8 ...iameter, have been etched and broken in four ...bending over a constant bending moment span ... The rod diameters and loads at fracture were ...red and the breaking stresses calculated using ...mple bending formula. Groups of rods (con... 16–32 rods) were given different periods of ...g, and the depth of material removed from the ... of the rods was calculated for each group.

... variation of the mean breaking strength of ...groups of rods, with depth of material removed ...he surface, is shown in Fig. 1. Also shown on ... are the 95 per cent confidence limits on the ... strength and the highest strength value ...ed in each group of rods. Fig. 2 is a histogram ...ring the distribution of breaking stresses for ...p of rods which have been etched for 40 min. ...hat for unetched rods. The maximum strengths obtained with these bulk g... specimens are in the region 450,0... 500,000 lb./sq. in. and clo... approach the value obtained... Thomas[2] for fine glass fibres.

The glass rod used in these ...periments had the following appr...imate composition by weight (...centages): SiO_2, 69; Na_2O, ...O, 4; Al_2O_3, 3; MgO, 3. ... etching solution conta... abo... 15 per cent hydroflu... aci... 5 per cent sulphuric ... ...eight and the remai... ...ter.

Experiments are being contin... to determine the effect on t... results of varying the concen...tion, temperature and nature ... the etchant; and of changing ... thermal history, size and composition of the g...

B. A. Procto...

Rolls-Royce, Ltd.,
Aerophysics Laboratory,
Littleover,
Derby.

June 22.

[1] Greene, C. H., *J. Amer. Cer. Soc.*, 39, 66 (1956).
[2] Thomas, W. F., *Nature*, 181, 1006 (1958); *Phys. and Chem. Gl...* 1, 4 (1960).

Stimulated Optical Radiation in Ruby

Schawlow and Townes[1] have proposed a techni... for the generation of very monochromatic radia... in the infra-red optical region of the spectrum u... an alkali vapour as the active medium. Javan[2] ... Sanders[3] have discussed proposals involving elect... excited gaseous systems. In this laboratory ... optical pumping technique has been successf... applied to a fluorescent solid resulting in the att...ment of negative temperatures and stimulated opt... emission at a wave-length of 6943 Å.; the ac... material used was ruby (chromium... corundum).

A simplified energy-level diag... for triply ionized chromium in ... crystal is shown in Fig. 1. W... this material is irradiated with ene... at a wave-length of about 5500... chromium ions are excited to ... $^4F_2$ state and then quickly lose so... of their excitation energy thro... non-radiative transitions to the ... state[4]. This state then slowly dec... by spontaneously emitting a sh... doublet the components of wh... at 300° K. are at 6943 Å. and 692... (Fig. 2*a*). Under very intense exc...tion the population of this me...stable state (^2E) can become grea... than that of the ground-state; thi... the condition for negative tempe...tures and consequently amplificat... via stimulated emission.

To demonstrate the above effec... ruby crystal of 1-cm. dimensions coa... ...

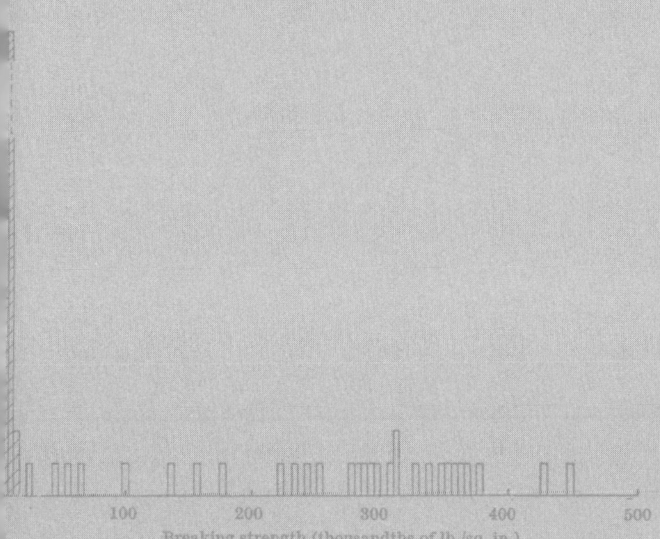

...2. Frequency of results as a function of breaking strength for etched (un...

EIGHT

The first laser

Charles H. Townes

When the first working laser was reported in 1960, it was described as "a solution looking for a problem." But before long the laser's distinctive qualities—its ability to generate an intense, very narrow beam of light of a single wavelength—were being harnessed for science, technology and medicine. Today, lasers are everywhere: from research laboratories at the cutting edge of quantum physics to medical clinics, supermarket checkouts and the telephone network.

Theodore Maiman made the first laser operate on 16 May 1960 at the Hughes Research Laboratory in California, by shining a high-power flash lamp on a ruby rod with silver-coated surfaces. He promptly submitted a short report of the work to the journal *Physical Review Letters*, but the editors turned it down. Some have thought this was because the *Physical Review* had announced that it was receiving too many papers on masers—the longer-wavelength predecessors of the laser—and had announced that any further papers would be turned down. But Simon Pasternack, who was an editor of *Physical Review Letters* at the time, has said that he turned down this historic paper because Maiman had just published, in June 1960, an article on the excitation of ruby with light, with an examination of the relaxation times between quantum states,[1] and that the new work seemed to be simply more of the same. Pasternack's reaction perhaps reflects the limited understanding at the time of the nature of lasers and their significance. Eager to get his work quickly into publication, Maiman then turned to *Nature*, usually even more selective than *Physical Review Letters*, where the paper was better received and published on 6 August.[2]

With official publication of Maiman's first laser under way, the Hughes

Research Laboratory made the first public announcement to the news media on 7 July 1960. This created quite a stir, with front-page newspaper discussions of possible death rays, but also some skepticism among scientists, who were not yet able to see the careful and logically complete *Nature* paper. Another source of doubt came from the fact that Maiman did not report having seen a bright beam of light, which was the expected characteristic of a laser. I myself asked several of the Hughes group whether they had seen a bright beam, which surprisingly they had not. Maiman's experiment was not set up to allow a simple beam to come out of it, but he analyzed the spectrum of light emitted and found a marked narrowing of the range of frequencies that it contained. This was just what had been predicted by the theoretical paper on optical masers (or lasers) by Art Schawlow and myself,[3] and had been seen in the masers that produced the longer-wavelength microwave radiation. This evidence, presented in figure 2 of Maiman's *Nature* paper (see p. 114), was definite proof of laser action. Shortly afterward, both in Maiman's laboratory at Hughes and in Schawlow's at Bell Laboratories in New Jersey, bright red spots from ruby laser beams hitting the laboratory wall were seen and admired.

Maiman's laser had several aspects not considered in our theoretical paper,[3] nor discussed by others before the ruby demonstration. First, Maiman used a pulsed light source, lasting only a few milliseconds, to excite (or "pump") the ruby. The laser thus produced only a short flash of light rather than a continuous wave, but because substantial energy was released during a short time, it provided much more power than had been envisaged in most of the earlier discussions. Before long, a technique known as "Q switching" was introduced at the Hughes Laboratory, shortening the pulse of laser light still further and increasing the instantaneous power to millions of watts and beyond. Lasers now have powers as high as a million billion (10^{15}) watts! The high intensity of pulsed laser light allowed a wide range of new types of experiment, and launched the now-burgeoning field of nonlinear optics. Nonlinear interactions between light and matter allow the frequency of light to be doubled or tripled, so for example an intense red laser can be used to produce green light.

I had a busy job in Washington at the time when various groups were trying to make the earliest lasers. But I was also supervising graduate students at Columbia University who were trying to make continuously pumped infrared lasers. Shortly after the ruby laser came out I advised them to stop this work and instead capitalize on the power of the new ruby laser to do an experiment on two-photon excitation of atoms. This was one of the early experiments in nonlinear optics, and two-photon excitation is now widely used to study atoms and molecules.

Lasers work by adding energy to atoms or molecules, so that there are more in a high-energy ("excited") state than in some lower-energy state; this is known as a "population inversion." When this occurs, light waves passing through the material stimulate more radiation from the excited states than they lose by absorption due to atoms or molecules in the lower state (fig. 8.1). This "stimulated emission" is the basis of masers (whose name stands for "microwave amplification by stimulated emission of radiation") and lasers (the same, but for light instead of microwaves).

Before Maiman's paper, ruby had been widely used for masers, which produce waves at microwave frequencies, and had also been considered for lasers producing infrared or visible light waves. But the second surprising

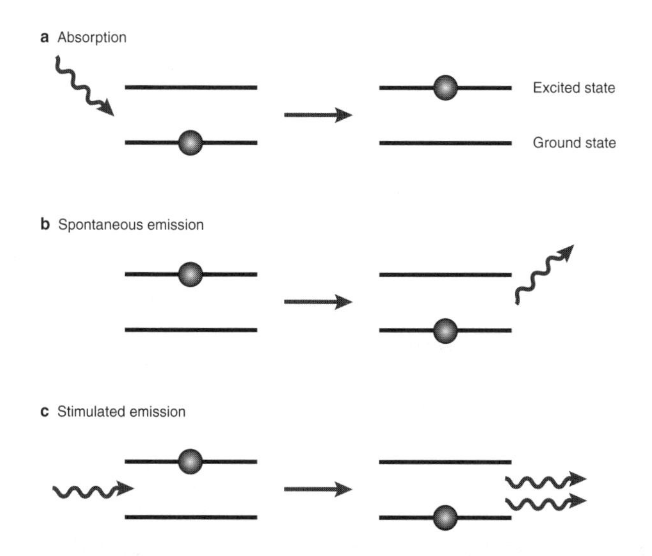

Fig. 8.1. Stimulated emission of radiation, the basis of laser operation. These schematic diagrams contrast stimulated emission from an excited atom, c, with the more usual cases of absorption, a, and spontaneous emission, b. When an atom in the lowest-energy ("ground") state (black dot in a, left) absorbs a quantum of light, or photon (wavy arrow), it is raised to a higher-energy ("excited") state (dot in a, right). The excited atom (b, left) may then radiate energy spontaneously, emitting a photon and reverting to the ground state (b, right). In stimulated emission, c, an incoming photon stimulates an excited atom to emit another photon of the same wavelength, as the atom reverts to the ground state; the result is amplification of the radiation (two photons instead of one). The new photon travels in exactly the same direction as the initial photon. In a working laser, reflective surfaces keep photons trapped for many round-trips through the lasing material, so that the light intensity builds up rapidly.

feature of Maiman's laser, in addition to the pulsed source, was that he was able to empty the lowest-energy ("ground") state of ruby enough so that stimulated emission could occur from an excited to the ground state. This was unexpected. In fact, Schawlow, who had worked on ruby, had publicly commented that transitions involving the ground state of ruby would not be suitable for lasers because it would be difficult to empty adequately. He recommended a different transition in ruby, which was indeed made to work, but only after Maiman's success. Maiman, who had been carefully studying the relaxation times of excited states of ruby,[1] came to the conclusion that the ground state might be sufficiently emptied by a flash lamp to provide laser action—and it worked.

The ruby laser was used in many early spectacular experiments. One amusing one, in 1969, sent a light beam to the Moon, where it was reflected back from a retro-reflector placed on the Moon's surface by astronauts in the U.S. Apollo program. The round-trip travel time of the pulse provided a measurement of the distance to the Moon. Later, ruby laser beams sent out and received by telescopes measured distances to the Moon with a precision of about three centimeters—a great use of the ruby laser's short pulses.

When the first laser appeared, scientists and engineers were not really prepared for it. Many people said to me—partly as a joke but also as a challenge—that the laser was "a solution looking for a problem." But by bringing together optics and electronics, lasers opened up vast new fields of science and technology. And many different laser types and applications came along quite soon. At IBM's research laboratories in Yorktown Heights, New York, Peter Sorokin and Mirek Stevenson demonstrated two lasers that used techniques similar to Maiman's but with calcium fluoride, instead of ruby, as the lasing substance. Following that—and still in 1960—was the very important helium-neon laser of Ali Javan, William Bennett, and Donald Herriott at Bell Laboratories. This produced continuous radiation at low power but with a very pure frequency and the narrowest possible beam. Then came semiconductor lasers, first made to operate in 1962 by Robert Hall and his associates at the General Electric laboratories in Schenectady, New York. Semiconductor lasers now involve many different materials and forms, can be quite small and inexpensive, and are by far the most common type of laser. They are used, for example, in supermarket bar-code readers, in optical-fiber communications, and in laser pointers.

By now, lasers come in countless varieties. They include the "edible" laser, made as a joke by Schawlow out of flavored gelatin (but not in fact eaten because of the dye that was used to color it), and its companion the

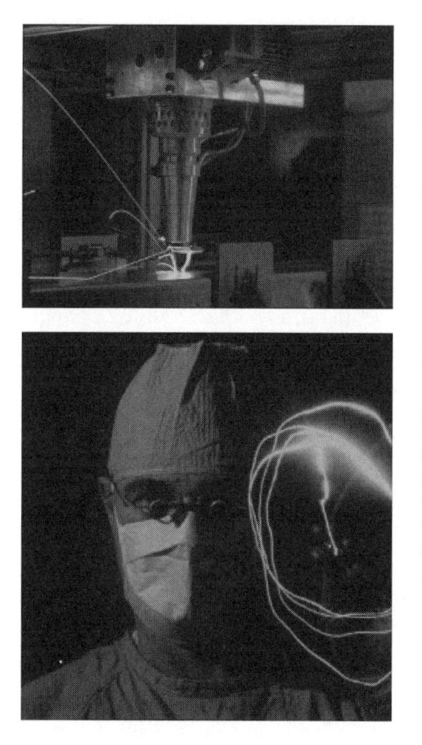

Fig. 8.2. Lasers in industry, science, and health care: just a few examples of their applications today. Clockwise from top left, a carbon dioxide laser for welding metal; a laser beam used to produce an artificial guide "star," which in astronomy allows corrections for atmospheric distortion to be made; a "laser surgeon," holding a coiled optical fiber that conducts the laser light. (Figures courtesy of Maximilian Stock Ltd/Science Photo Library; Lawrence Livermore National Laboratory; Alexander Tsiaras/Science Photo Library.)

"drinkable" laser, made of an alcoholic mixture at Eastman Kodak's laboratories in Rochester, New York. Natural lasers have now been found in astronomical objects; for example, infrared light is amplified by carbon dioxide in the atmospheres of Mars and Venus, excited by solar radiation, and intense radiation from stars stimulates laser action in hydrogen atoms in circumstellar gas clouds. This raises the question: why weren't lasers invented long ago, perhaps by 1930 when all the necessary physics was already understood, at least by some people? What other important phenomena are we blindly missing today?

Maiman's paper is so short, and has so many powerful ramifications, that I believe it might be considered the most important per word of any of the wonderful papers in *Nature* over the past century. Lasers today (fig. 8.2) produce much higher power densities than were previously possible, more precise measurements of distances, gentle ways of picking up and moving small objects such as individual microorganisms, the lowest temperatures ever achieved, new kinds of electronics and optics, and many billions of dollars worth of new industries. The U.S. National Academy of Engineering has chosen the combination of lasers and fiber optics—which has revolutionized communications—as one of the twenty most important engineering developments of the twentieth century. Personally, I am particularly pleased with lasers as invaluable medical tools (for example, in laser eye surgery), and as scientific instruments—I use them now to make observations in astronomy. And there are already at least ten Nobel Prize winners whose work was made possible by lasers.

There have been great and good developments since Ted Maiman, probably a bit desperately, mailed off a short paper on what was then a somewhat obscure subject, hoping to get it published quickly in *Nature*. Fortunately, *Nature*'s editors accepted it, and the rest is history.

References

1. Maiman, T. H. Optical and microwave-optical experiments in ruby. *Phys. Rev. Lett.* **4,** 564–566 (1960).
2. Maiman, T. H. Stimulated optical radiation in ruby. *Nature* **187,** 493–494 (1960).
3. Schawlow, A. L. & Townes, C. H. Infrared and optical masers. *Phys. Rev.* **112,** 1940–1949 (1958).

Further reading

Nobel e-Museum. *The Nobel Prize in Physics 1964* (http://www.nobel.se/physics/laureates/1964/).

Townes, C. H. *How the Laser Happened: Adventures of a Scientist* (Oxford University Press, 1999).

1960

Stimulated optical radiation in ruby

T. H. Maiman

Schawlow and Townes[1] have proposed a technique for the generation of very monochromatic radiation in the infra-red optical region of the spectrum using an alkali vapour as the active medium. Javan[2] and Sanders[3] have discussed proposals involving electron-excited gaseous systems. In this laboratory an optical pumping technique has been successfully applied to a fluorescent solid resulting in the attainment of negative temperatures and stimulated optical emission at a wave-length of 6943 Å.; the active material used was ruby (chromium in corundum).

A simplified energy-level diagram for triply ionized chromium in this crystal is shown in Fig. 1. When this material is irradiated with energy at a wave-length of about 5500 Å., chromium ions are excited to the $^4F_2$ state and then quickly lose some of their excitation energy through non-radiative transitions to the ^2E state[4]. This state then slowly decays by spontaneously emitting a sharp doublet the components of which at 300° K. are at 6943 Å. and 6929 Å. (Fig. 2*a*). Under very intense excitation the population of this metastable state (^2E) can become greater than that of the ground-state; this is the condition for negative temperatures and consequently amplification via stimulated emission.

To demonstrate the above effect a ruby crystal of 1-cm. dimensions coated on two parallel faces with silver was irradiated by a high-power flash lamp; the emission spectrum obtained under these conditions is shown in Fig. 2*b*. These results can be explained on the basis that negative tempera-

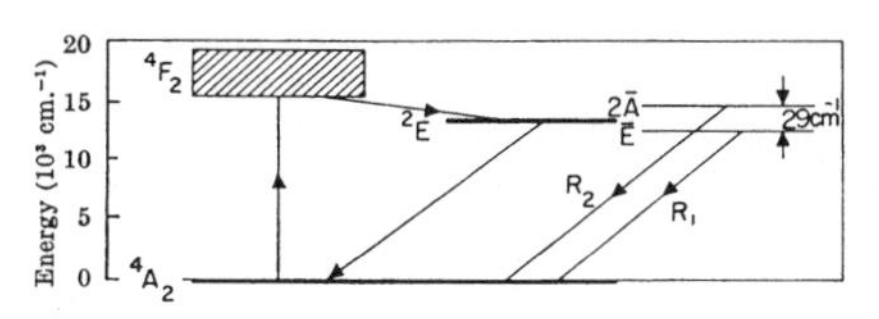

Fig. 1. Energy-level diagram of Cr^{3+} in corundum, showing pertinent processes.

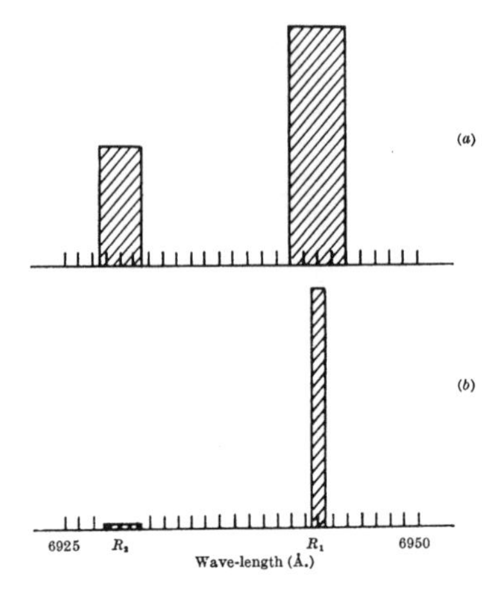

Fig. 2. Emission spectrum of ruby: *a*, low-power excitation; *b*, high-power excitation.

tures were produced and regenerative amplification ensued. I expect, in principle, a considerably greater ($\sim 10^8$) reduction in line width when mode selection techniques are used[1].

I gratefully acknowledge helpful discussions with G. Birnbaum, R. W. Hellwarth, L. C. Levitt, and R. A. Satten and am indebted to I. J. D'Haenens and C. K. Asawa for technical assistance in obtaining the measurements.

Hughes Research Laboratories, A Division of Hughes Aircraft Co., Malibu, California

[Published 6 August 1960]

References

1. Schawlow, A. L., and Townes, C. H., *Phys. Rev.*, 112, 1940 (1958).
2. Javan, A., *Phys. Rev. Letters*, 3, 87 (1959).
3. Sanders, J. H., *Phys. Rev. Letters*, 3, 86 (1959).
4. Maiman, T. H., *Phys. Rev. Letters*, 4, 564 (1960).

1963

3C 273 : A STAR-LIKE OBJECT WITH LARGE RED-SHIFT

By Dr. M. SCHMIDT

Vilson and Palomar Observatories, Carnegie Institution of Washington, California Institute of Technology, P

nly objects seen on a 200-in. plate near the ns of the components of the radio source ported by Hazard, ar ains ceding article are a star ab m th and a faint wisp or jet. j s a h nd extends away from st n It is not visible within rom th d tly at 20″ from the star. The positio he y furnished by Dr. T. A. Matth R.A. 3·35s ± 0·04s, Decl. +2° 19′ 42·0″ ± 0·5″ (1950), of component B of the radio source. The end s 1″ east of component A. The close correla- n the radio structure and the star with the jet ve and intriguing.

of the star were taken with the prime-focus h at the 200-in. telescope with dispersions of 0 Å per mm. They show a number of broad atures on a rather blue continuum. The most features, which have widths around 50 Å, ler of strength, at 5632, 3239, 5792, 5032 Å. other weaker emission bands are listed in the first Table 1. For three faint bands with widths of the total range of wave-length is indicated.

explanation found for the spectrum involves able red-shift. A red-shift $\Delta\lambda/\lambda_0$ of 0·158 tification of four emission bands as Balmer dicated in Table 1. Their relative strengths are nt with this explanation. Other identifications e above red-shift involve the Mg II lines around s far only found in emission in the solar chromo- d a forbidden line of [O III] at 5007 Å. On nother [O III] line is expected at 4959 Å with one-third of that of the line at 5007 Å. Its y in the spectrum would be marginal. A weak and suspected at 5705 Å, or 4927 Å reduced for does not fit the wave-length. No explanation is the three very wide emission bands.

appears that six emission bands with widths Å can be explained with a red-shift of 0·158. nces between the observed and the expected hs amount to 6 Å at the most and can be entirely l in terms of the uncertainty of the measured hs. The present explanation is supported by ns of the infra-red spectrum communicated by

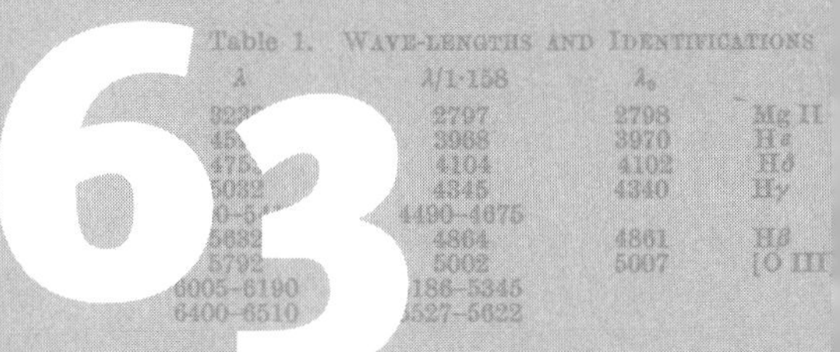

Table 1. WAVE-LENGTHS AND IDENTIFICATIONS

λ	$\lambda/1·158$	λ_0	
323	2797	2798	Mg II
45	3968	3970	Hε
475	4104	4102	Hδ
5032	4345	4340	Hγ
0–5	4490–4675		
5632	4864	4861	Hβ
5792	5002	5007	[O III]
6005–6190	186–5345		
6400–6510	527–5622		

Oke in a following article, and by the spectrum o star-like object associated with the radio sour discussed by Greenstein and Matthews in anot munication.

The unprecedented identification of the spectr apparently stellar object in terms of a large suggests either of the two following explanations

(1) The stellar object is a star with a large grav red-shift. Its radius would then be of the order o Preliminary considerations show that it would be e difficult, if not impossible, to account for the oc of permitted lines and a forbidden line with the s shift, and with widths of only 1 or 2 per cent of t length.

(2) The stellar object is the nuclear region of with a cosmological red-shift of 0·158, correspond apparent velocity of 47,400 km/sec. The distan be around 500 megaparsecs, and the diamete nuclear region would have to be less than 1 ki This nuclear region would be about 100 times optically than the luminous galaxies which ha identified with radio sources thus far. If the op and component A of the radio source are associa the galaxy, they would be at a distance of 50 kil implying a time-scale in excess of 10^5 years. T energy radiated in the optical range at constant lu would be of the order of 10^{59} ergs.

Only the detection of an irrefutable proper m parallax would definitively establish 3C 273 as a within our Galaxy. At the present time, howe explanation in terms of an extragalactic origin see direct and least objectionable.

I thank Dr. T. A. Matthews, who directed my a to the radio source, and Drs. Greenstein and valuable discussions.

UTE ENERGY DISTRIBUTION IN THE OPTICAL SPECTRUM OF 3

By Dr. J. B. OKE

Wilson and Palomar Observatories, Carnegie Institution of Washington, California Institute of Technology, Pa

dio source 3C 273 has recently been identified a thirteenth magnitude star-like object. The given by M. Schmidt in the preceding com- . Since 3C 273 is relatively bright, photo- ectrophotometric observations were made with telescope at Mount Wilson to determine the stribution of energy in the optical region of the such observations are useful for determining

selected points is approximately 2 per cent. Th emission features found by Schmidt were readily d other very faint features not apparent on S spectra may be present.

The source 3C 273 is considerably bluer than t known star-like objects 3C 48, 3C 196, and 3C 28 have been studied in detail[2]. The absolute ene tribution of the apparent continuum can be ac

NINE

The quasar enigma

Malcolm Longair

In the early 1960s, astronomers were puzzled by "quasars"—sources of intense radio emission that seemed to be stars, but had unintelligible optical spectra. In March 1963, Maarten Schmidt solved the problem, reporting that the quasar 3C 273 was not a star, but the nucleus of a distant galaxy. The prodigious energy release of 3C 273—a thousand times that of our own Galaxy—and its rapid variability led to the realization that supermassive black holes lurk in the centers of some galaxies.

The quasar story goes back to May 1933, when radio astronomy was born with Karl Jansky's discovery[1] of radio emission from the Milky Way—the plane of our Galaxy. Nearly twenty years later, Hannes Alfvén and Nicolai Herlofson explained this emission[2] as "synchrotron" radiation, emitted by very-high-energy electrons gyrating in the interstellar magnetic field of our Galaxy. By the mid-1950s, the shape of the spectrum of the radiation and its highly polarized nature confirmed the synchrotron hypothesis, and the same mechanism was invoked to explain the radio and optical emissions of other energetic phenomena outside our Galaxy.

Surveys of the radio sky had revealed a population of extragalactic radio sources, and a few of them were associated with nearby galaxies that were abnormal in some way—for example, the famous "jet" in the nearby giant elliptical galaxy M87. Then, in 1954, one of the brightest radio sources in the northern sky, Cygnus A, was identified with a distant galaxy. The radio luminosity of Cygnus A—its intrinsic power at radio wavelengths—was enormous, more than a million times greater than that of our Galaxy. From Einstein's relation between mass and energy, $E = mc^2$, the energy in high-energy particles and magnetic fields that had to be present in Cygnus A corresponded to at least a million times the mass of the Sun. An astrophysical mechanism had to be found by which a significant fraction of the mass of a galaxy could be converted into high-energy particles and magnetic fields.

This was not the only puzzle, for Roger Jennison and M. K. Das Gupta, at the Jodrell Bank Observatory, U.K., had shown that the emission originated not from the galaxy itself, but from two huge lobes located on either side of it (fig. 9.1). Thus, not only must the galaxy accelerate enormous fluxes of electrons to speeds approaching that of light, but this "relativistic" material had also to be ejected enormous distances into intergalactic space.

These discoveries stimulated significant investment in the construction of radio telescopes, leading to the discovery of many more radio sources outside our Galaxy. By determining accurate positions for the radio sources, their optical counterparts could be identified. Many of these turned out to be among the most luminous galaxies known, and hence, because more luminous objects can be seen from farther away, also the most distant. By 1960, the radio galaxy 3C 295 held the distance record, with a "redshift" of 0.46 (see box 9.1).

By 1962, Thomas Matthews and Allan Sandage had identified three of the brightest radio sources in the northern sky—3C 48, 196, and 286—with stellar objects of an unknown type,[3] with quite unrecognizable optical

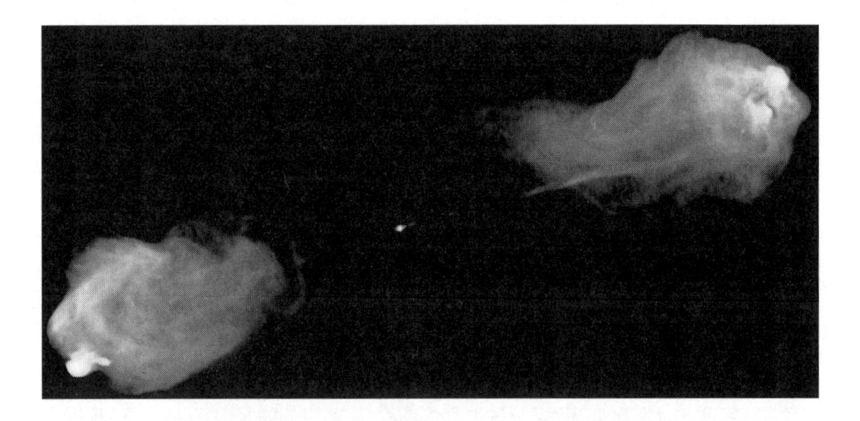

Fig. 9.1. A radio image of the powerful double radio source Cygnus A, obtained by the U.S. National Radio Astronomy Observatory's Very Large Array. A faint jet can be seen extending from the galaxy's nucleus, the central bright spot, to one of the outer lobes. This jet, and its counterpart, which is not visible, consists of very-high-energy particles moving at speeds close to that of light, which power the bright "hot-spots" that can be seen in the lobes. The dim, wispy features are artifacts of the imaging process. (Image courtesy of NRAO, the VLA, and Chris Carilli.)

Box 9.1 Redshifts and their interpretation

On the largest scales we can investigate, our Universe is the same in all directions with quite remarkable precision—technically, we say that the Universe is **isotropic** on the largest scales. If the distribution of galaxies in such an isotropic Universe expands uniformly, just like stretching a rubber sheet, then the farther away a galaxy is from any other galaxy, the greater the speed with which they recede from each other. Recession speed is exactly proportional to distance—a relation known as **Hubble's law**, after Edwin Hubble, who in 1929 discovered that galaxies obey this relation. Thus, the distance to a galaxy can be estimated by measuring its recession speed.

The speed can be measured by the change in frequency of features in the optical spectrum of a galaxy as compared with their frequencies in a stationary object. Just like the pitch of a receding siren, the frequency of light from a distant galaxy is lowered, or **redshifted**, by the galaxy's motion.

The simplest way of understanding the cosmological meaning of redshift is that it is a measure of the past separation between galaxies that partake in the universal expansion, relative to their separation today. If the present separation of two galaxies is s, then at a redshift z, their separation would have been only $s/(1+z)$. Thus, galaxies at redshift 1 emitted their light when the galaxies were all closer together by a factor of two. The most distant quasars have redshifts greater than six, and they emitted their light when the galaxies were all closer together by a factor of seven.

spectra; they termed these objects "quasi-stellar radio sources," soon to be abbreviated to "quasars." They were naturally assumed to be stars because their images were pointlike on photographic plates, and because the optical light of 3C 48 was variable over a period of months—both characteristics unknown in galaxies.

Also in 1962, the position and structure of another bright radio source, 3C 273, was determined from observations with the Parkes 210-foot radio telescope in Australia. The Moon passed in front of the radio source on three occasions in that year, and these "occultations" allowed Cyril Hazard and colleagues[4] to locate 3C 273 with sufficient precision to identify it with a "star" with a remarkable optical jet, which also emitted radio waves (fig. 9.2). This was another example of a quasar, and it is still the brightest known example in the sky.

The nature of the quasars remained a mystery, but the discovery of this

Fig. 9.2. An optical image of the quasar 3C 273 observed by the Hubble Space Telescope. (The six bright rays emerging from the quasar are optical artifacts caused by the extreme brightness of the quasar.) A jet extends from the quasar toward the lower right of the image, and the object at the upper left is a nearby star. The faint, fuzzy objects toward the bottom are galaxies similar to our own, located at the same distance as the quasar. With a luminosity one thousand times that of these ordinary galaxies, 3C 273 is still the brightest quasar in the sky. (Image courtesy of John Bahcall, Zolt Levay, NASA, and the Space Telescope Science Institute.)

very bright example enabled Maarten Schmidt to obtain a high-quality optical spectrum of 3C 273 with the Palomar two-hundred-inch telescope, in California. The spectrum again was unrecognizable, except that there was a prominent series of emission lines that displayed a familiar pattern. Schmidt realized that these were emission lines from hydrogen, the so-called Balmer series, but shifted to longer wavelengths. The inferred redshift of 0.158 was almost as large as those of the most distant galaxies known at that time, but its luminosity was hundreds of times greater—this was no ordinary star!

Schmidt published his discovery[5] in the 16 March 1963 issue of *Nature,* on the page after the paper by Hazard *et al.*[4] There, he mentions two possible explanations for the large redshift, but chooses as the "most direct and least objectionable" the interpretation of 3C 273 as the bright center, or "nucleus," of a distant galaxy. The alternative explanation—that 3C 273 was a nearby star with a large redshift caused by a strong gravitational field—was ruled out by Schmidt and Jesse Greenstein the following year.

All such discoveries require a bit of luck. In this case, it was the fact that the Balmer series of hydrogen was clearly present in the spectrum of 3C 273—it turns out that this is quite a rare occurrence in quasar spectra. Once the large redshifts of the quasars were appreciated, the inexplicable lines in the spectra of other quasars could be understood as strong emission lines of the common elements. Within a year, the redshifts of several other quasars were determined: 3C 48 at 0.368, 3C 47 at 0.425, and 3C 147 at 0.545.

In 1965, the quasar 3C 9 was found to have a redshift of 2.016; the current record stands at 6.42, for a quasar discovered in 2002 by the Sloan Digital Sky Survey.

The great distance of 3C 273 implied that its optical luminosity was about a thousand times that of our own Galaxy. Searches back through the Harvard photographic plate archives showed that this prodigious luminosity varied by 50 percent or more on a timescale of months to years. Nothing like this had been observed in extragalactic astronomy before.

One of the first discussions of these remarkable discoveries was held in Dallas in 1963 at the First Texas Symposium on Relativistic Astrophysics. For the first time, the optical and radio astronomers got together with the theoretical astrophysicists and, in particular, with experts in general relativity. The combination of the quasars' very small dimensions (less than a light year in size) and enormous luminosities—which implied similarly enormous masses—indicated that quasars must involve strong gravitational fields. Therefore, general relativity must play a central role in understanding their properties. At the closing dinner, Thomas Gold remarked, "Everyone is pleased: the relativists who feel they are being appreciated, who are suddenly experts in a field which they hardly knew existed; the astrophysicists for having enlarged their domain, their empire, by the annexation of another subject—general relativity."

For some time, there was a vigorous debate about whether the quasar redshifts really were of cosmological origin—that is, whether they reflected the quasars' location in the expanding Universe, and hence their distance from us. Their extreme luminosities, and particularly the short timescales of their variability, had no natural explanation in the traditional picture of galaxies as objects composed only of stars and interstellar material. To produce the observed properties, it would have to be assumed that millions of supernovae exploded simultaneously and that these explosions were synchronized. These properties seemed so extreme that some astronomers, including Fred Hoyle, Geoffrey Burbidge, and Halton Arp, suggested that quasars were in fact relatively nearby objects and that their large redshifts had some explanation other than distance.[6] But no such explanation was forthcoming, and the hypothesis never attracted wide support.

In 1965, "radio-quiet" quasi-stellar objects, with the same optical properties as quasars but with no appreciable radio emission, were discovered by Sandage, and these turned out to be much more common than the radio-loud quasars. It slowly dawned on astronomers that quasars are simply the most extreme examples of what are now called "active galactic nuclei." The first of these had been discovered in the early 1940s by Carl Seyfert, who had studied the spectra of spiral galaxies that had starlike nuclei. He found that the nuclei of these galaxies had very intense, broad emission lines, quite unlike those of the clouds of ionized hydrogen found in normal galaxies. In addition, the radiation underlying these emission lines (the "continuum emission") had a smooth spectrum, unlike that of starlight. Seyfert's pioneering work was largely neglected until the 1960s.

The similarity between quasars and the nuclei of Seyfert galaxies was reinforced by the discovery in 1968 that the continuum emission of the latter is variable, like that of quasars. It became apparent that there is a continuous sequence of high-energy astrophysical activity in the nuclei of galaxies, from weak examples, such as the center of our own Galaxy, to the most extreme examples of the quasars. Initially, confirmation that quasars are galactic nuclei was hindered by the fact that, on a photographic plate, the starlight of a quasar's host galaxy is totally swamped by the intense radiation from the nucleus. But observations with the superb imaging capabilities of the Hubble Space Telescope have enabled the host galaxies to be clearly seen and studied.

But what powered the high-energy activity in quasars, radio galaxies, and Seyfert galaxies? In 1964 Yakov Zel'dovich and Igor Novikov in Moscow and Edwin Salpeter at Cornell University were the first to point out that the accretion of matter onto black holes is a very powerful energy source.[7–9] As a black hole sucks in gas from the surrounding galaxy, the infalling material heats up and emits vast quantities of radiation. Zel'dovich and Novikov[9] showed that the masses of the nuclei of quasars had to be very large in order for gravity to hold the source together against the pressure of the emitted radiation. For 3C 273, the source of energy had to be a billion times more massive than the Sun. Donald Lynden-Bell subsequently showed how thin "accretion discs" around black holes could in principle account for the most extreme properties of active galactic nuclei and quasars.[10]

In parallel with these insights, remarkable progress was being made in understanding the physics of matter falling into black holes. Up to 42 percent of the mass energy of the infalling matter can be released in this process—an enormous conversion efficiency, far greater even than the best that can be achieved with nuclear energy, which is just less than 1 percent. Moreover, most of the energy is released very close to the black hole's "event horizon"—the radius at which matter and radiation disappear down the black hole. The radius of the event horizon is the smallest size that an object of a given mass can have: if it attempted to become any smaller, it would collapse into a black hole. Therefore, matter falling into a black hole liberates energy on the smallest space and timescales possible for an object of a given mass. This can account for the rapid variability of the huge luminosities of quasars.

These ideas paved the way for today's picture of quasars, which involves supermassive black holes as an essential ingredient; indeed, these objects have become part of the standard furniture of modern astrophysics. The way in which this came about may seem to be rather circuitous, but the reasons are easy to understand. In the 1950s, no one realized that there might be supermassive black holes in the nuclei of galaxies—astronomy was all about stars and gas clouds. The contribution of radio astronomy was to draw attention to the fact that a key component was missing from this picture: large amounts of relativistic material, ejected from the nuclei of active galaxies. Radio sources such as Cygnus A, which are now regarded as quasars in which the optical emission is obscured by dust surrounding the nucleus, were the first direct evidence that galaxies could generate huge fluxes of high-energy particles.

The discovery of quasars came about because the intense radio emission acted as a beacon signaling that something exotic was going on. Without this clue, searching for quasars in the 1950s would have been like looking for a needle in a haystack—on any photographic plate, there would be about a hundred thousand stars for every quasar. That needle was found by radio location, and Maarten Schmidt's elucidation of its nature opened completely new perspectives for astronomy and high-energy astrophysics.

References

1. Jansky, K. G. Electrical disturbances apparently of extraterrestrial origin. *Proc. Inst. Radio Engineers* **21,** 1387–1398 (1933).
2. Alfvén, H. & Herlofson, N. Cosmic radiation and radio stars. *Phys. Rev.* **78,** 616 (1950).
3. Matthews, T. A. & Sandage, A. R. Optical identification of 3C 48, 3C 96, and 3C 286 with stellar objects. *Astrophys. J.* **138,** 30–56 (1963).
4. Hazard, C., Mackey, M. B. & Shimmins, A. J. Investigation of the radio source 3C 273 by the method of lunar occultations. *Nature* **197,** 1037–1039 (1963).
5. Schmidt, M. 3C 273: a star-like object with large red-shift. *Nature* **197,** 1040 (1963).
6. Field, G. B., Arp, H. C. & Bahcall, J. (eds) *The Redshift Controversy* (W. A. Benjamin, Reading, MA, 1973).
7. Zel'dovich, Ya. B. The fate of a star and the evolution of gravitational energy upon accretion. *Sov. Phys. Dokl.* **9,** 195–197 (1964). (In Russian original, **155,** 67–69, 1964.)
8. Salpeter, E. E. Accretion of interstellar matter by massive objects. *Astrophys. J.* **140,** 796–800 (1964).
9. Zel'dovich, Ya. B. & Novikov, I. D. Mass of quasi-stellar objects. *Sov. Phys. Dokl.* **9,** 834–837 (1965). (In Russian original, **158,** 811–814, 1964.)
10. Lynden-Bell, D. Galactic nuclei as collapsed old quasars. *Nature* **223,** 690–694 (1969).

Further reading

Longair, M. S. "Astrophysics and Cosmology" in *Twentieth Century Physics* Vol. 3 (eds Brown, L. M., Pais, A. & Pippard, A. B.) Ch. 23 (Institute of Physics Publishing, Bristol, 1995).

Longair, M. S. *Our Evolving Universe* (Cambridge University Press, 1997).

Robinson, I., Schild, A. & Schucking, E. L. (eds) *Quasi-stellar Sources and Gravitational Collapse* (University of Chicago Press, 1965).

Thorne, K. S. *Black Holes and Time Warps: Einstein's Outrageous Legacy* (Norton, New York, 1995).

1963

3*C* 273: a star-like object with large red-shift

M. Schmidt

The only objects seen on a 200-in. plate near the positions of the components of the radio source 3*C* 273 reported by Hazard, Mackey and Shimmins in the preceding article are a star of about thirteenth magnitude and a faint wisp or jet. The jet has a width of 1″–2″ and extends away from the star in position angle 43°. It is not visible within 11″ from the star and ends abruptly at 20″ from the star. The position of the star, kindly furnished by Dr. T. A. Matthews, is R.A. 12h 26m 33.35s ± 0.04s, Decl. +2° 19′ 42.0″ ± 0.5″ (1950), or 1″ east of component *B* of the radio source. The end of the jet is 1″ east of component *A*. The close correlation between the radio structure and the star with the jet is suggestive and intriguing.

Spectra of the star were taken with the prime-focus spectrograph at the 200-in. telescope with dispersions of 400 and 190 Å per mm. They show a number of broad emission features on a rather blue continuum. The most prominent features, which have widths around 50 Å, are, in order of strength, at 5632, 3239, 5792, 5032 Å. These and other weaker emission bands are listed in the first column of Table 1. For three faint bands with widths of 100–200 Å the total range of wave-length is indicated.

The only explanation found for the spectrum involves a considerable red-shift. A red-shift $\Delta\lambda/\lambda_0$ of 0.158 allows identification of four emission bands as Balmer lines, as indicated in Table 1. Their relative strengths are in agreement with this explanation. Other identifications based on the above red-shift involve the Mg II lines around 2798 Å, thus far only found in emission in the solar chromosphere, and a forbidden line of

Table 1 Wave-lengths and Identifications

λ	$\lambda/1.158$	λ_0	
3239	2797	2798	Mg II
4595	3968	3970	Hε
4753	4104	4102	Hδ
5032	4345	4340	Hγ
5200–5415	4490–4675		
5632	4864	4861	Hβ
5792	5002	5007	[O III]
6005–6190	5186–5345		
6400–6510	5527–5622		

[O III] at 5007 Å. On this basis another [O III] line is expected at 4959 Å with a strength one-third of that of the line at 5007 Å. Its detectability in the spectrum would be marginal. A weak emission band suspected at 5705 Å, or 4927 Å reduced for red-shift, does not fit the wave-length. No explanation is offered for the three very wide emission bands.

It thus appears that six emission bands with widths around 50 Å can be explained with a red-shift of 0.158. The differences between the observed and the expected wave-lengths amount to 6 Å at the most and can be entirely understood in terms of the uncertainty of the measured wave-lengths. The present explanation is supported by observations of the infra-red spectrum communicated by Oke in a following article, and by the spectrum of another star-like object associated with the radio source 3*C* 48 discussed by Greenstein and Matthews in another communication.

The unprecedented identification of the spectrum of an apparently stellar object in terms of a large red-shift suggests either of the two following explanations.

(1) The stellar object is a star with a large gravitational red-shift. Its radius would then be of the order of 10 km. Preliminary considerations show that it would be extremely difficult, if not impossible, to account for the occurrence of permitted lines and a forbidden line with the same red-shift, and with widths of only 1 or 2 per cent of the wave-length.

(2) The stellar object is the nuclear region of a galaxy with a cosmological red-shift of 0.158, corresponding to an apparent velocity of 47,400 km/sec. The distance would be around 500 megaparsecs, and the diameter of the nuclear region would have to be less than 1 kiloparsec. This nuclear region would be about 100 times brighter optically than the luminous galaxies which have been identified with radio sources thus far. If the optical jet and component *A* of the radio source are associated with

the galaxy, they would be at a distance of 50 kiloparsecs, implying a time-scale in excess of 10^5 years. The total energy radiated in the optical range at constant luminosity would be of the order of 10^{59} ergs.

Only the detection of an irrefutable proper motion or parallax would definitively establish 3*C* 273 as an object within our Galaxy. At the present time, however, the explanation in terms of an extragalactic origin seems most direct and least objectionable.

I thank Dr. T. A. Matthews, who directed my attention to the radio source, and Drs. Greenstein and Oke for valuable discussions.

Mount Wilson and Palomar Observatories, Carnegie Institution of Washington, California Institute of Technology, Pasadena

[Published 16 March 1963]

1963

MAGNETIC ANOMALIES OVER OCEANIC RIDGES

By F. J. VINE and Dr. D. H. MATTHEWS
Department of Geodesy and Geophysics, University of Cambridge

L profiles showing bat[illegible]y and [illegible]oci-
tal magnetic field anoma[illegible]se[illegible]s-
rth Atlantic and North-[illegible] n [illegible]s
n Fig. 1. They illustrate [illegible]ess[illegible]l f[illegible]s
anomalies over the oce[illegible] ridge[illegible]g-
malies over the exposed or buried foot[illegible] of
(2) shorter-period anomalies ove[illegible]ged
he ridge; (3) a pronounced cen[illegible]nomaly
with the median valley. This pattern has now
ved in the North Atlantic[1,2], the Antarctic[3],
dian Oceans[4,5]. In this article we describe an
account for it.

ral increase in wave-length of the anomalies
the crest of the ridge is almost certainly
with the increase in depth to the magnetic
terial[1]. Local anomalies of short-period may
orrelated with bathymetry, and explained in
easonable susceptibility contrasts and crustal
ons; but the long-period anomalies of category
so readily explained. The central anomaly
oduced if it is assumed that a block of material
gly magnetized in the present direction of the
d underlies the median valley and produces a
sceptibility contrast with the adjacent crust.
lear, however, why this considerable suscepti-
rast should exist beneath the median valley
ewhere under the ridge. Recent work in this
t has suggested a new mechanism.

mber 1962, H.M.S. *Owen* made a detailed
survey over a central part of the Carlsberg
art of the International Indian Ocean Expedi-
area (50 × 40 nautical miles; centred on
1° 45′ E.) is predominantly mountainous, depths
m 900 to 2,200 fathoms, and the topographic
generally elongated parallel to the trend of the
his elongation is more marked on the total
ield anomaly map where a trough of negative
flanked by steep gradients, separates two areas
anomalies. The trough of negative anomalies
s to a general depression in the bottom topo-
ich represents the median valley of the Ridge.

[illegible] positive [illegible]alies correspond to moun
[illegible]o o[illegible]y.
[illegible]w ma[illegible] latitude (inclination − 6°)
a [illegible] m[illegible]ed in the present directio
[illegible]ld is to[illegible]e the strength of the field
p[illegible]ing a nega[illegible] anomaly over the body an
positive ano[illegible]e north. Here, over the
the Ridge, t[illegible]tom topography indicates the
basic extrusives such as volcanoes and fissure e
and there is little sediment fill. The bathymetry, t
defines the upper surface of magnetic material
considerable intensity of magnetization, poten
high as any known igneous rock type[6], and
higher, because it is extrusive, than the mai
layer beneath. That the topographic features *ar*
of producing anomalies is immediately appa
comparing the bathymetric and the anomaly
several have well-defined anomalies associated wi

Two comparatively isolated volcano-like featu
singled out and considered in detail. One has a
ated negative anomaly as one would expect fo
magnetization, the other, completely the reverse
pattern, that is, a pronounced positive anomaly su
reversed magnetization. Data on the topography
feature and its associated anomaly were fed int
puter and an intensity and direction of magneti
each obtained. Fig. 2 shows the directions of the
vectors plotted on a stereographic projection.
computed the magnetic vector by a 'best fit' pro
computer recalculated the anomaly over the body
ing this vector, thus giving an indication of the
of fit. The fit was good for the case of revers
netization but poor for that of approximately
magnetization. The discrepancy is scarcely surpris
we have ignored the effects of adjacent topogra
the interference of other anomalies in the vici
addition, the example of normal magnetization i
corner of the area where the control of contouri
precise. The other example is central where th
is good. In both cases the intensity of magn
deduced was about 0·005 e.m.u.; this is equival

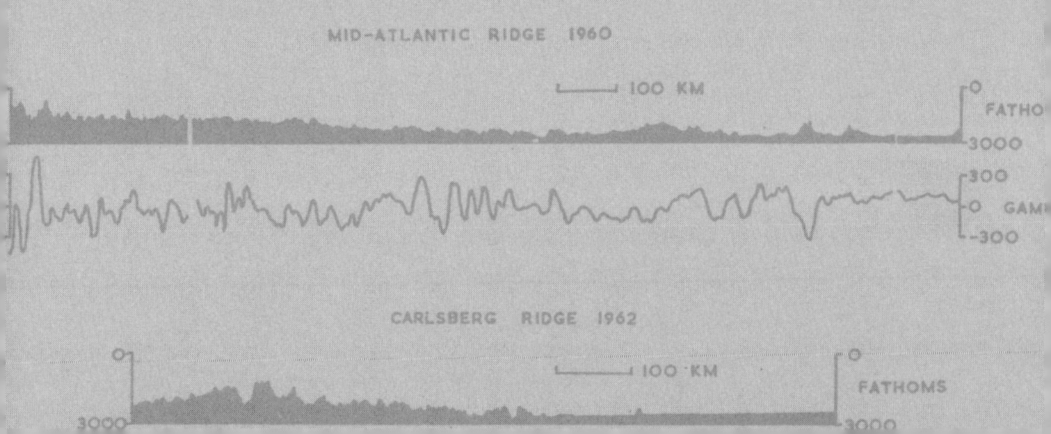

TEN

Seafloor magnetism and drifting continents

Dan McKenzie

In 1960, almost no one believed in the idea of continental drift—that the continents move across the face of the Earth. Within a decade, a complete shift of opinion had occurred: continental drift, or plate tectonics, as it was by then called, had become almost universally accepted. Central to this revolution in the Earth sciences was a speculative proposal by two young geophysicists that explained how magnetic stripes on the sea floor are produced.

The accepted way of writing scientific papers gives only veiled glimpses of the heated arguments behind the measured prose, and even these are visible only to those who know the subject well. Fred Vine and Drum Matthews' paper[1] of 7 September 1963 on the magnetic stripes formed at underwater features known as oceanic ridges is an excellent example of such restraint. Its title gives no indication that it is concerned with continental drift, the slow relative movement of landmasses over geological time. Nor would any casual reader be likely to recognize the paper's importance. Nonetheless, it provided one of the two clues that resolved the continental drift controversy within three years, to (almost) everyone's satisfaction.

Serious scientific discussion of continental drift started with Alfred Wegener's book *The Origin of Continents and Oceans,* first published in 1915, which was widely read in its English translation.[2] Many who did so found Wegener's style of argument uncongenial. Some of the observations that he used to support his ideas were wrong. But others are now known to be correct—for example, the geometric fit of the continents surrounding the Atlantic Ocean, and the presence of identical fossil species on continents now separated by oceans. He also proposed that the force driving the motions resulted from the tendency of continents to move towards the Earth's Equator, driven by the Earth's rotation. Such a force certainly exists, but is too small by many orders of magnitude to account for continental drift. Unfortunately, the confusion caused by his uncritical advocacy produced bitter controversy for forty years, and delayed, rather than advanced, the acceptance of continental drift.[3]

In retrospect, the key observations that demonstrated that continental drift had occurred were made in the mid-1950s by a group working in paleomagnetism—the study of the Earth's ancient magnetic field. This group included Keith Runcorn and Ted Irving, and, like Vine and Matthews, was based in the Department of Geodesy and Geophysics at the University of Cambridge. Paleomagnetism involves the measurement of the orientation of the magnetization in rocks of known age. At Cambridge it was carried out in a specially constructed nonmagnetic hut that still exists.

The Earth's present magnetic field is similar to that of a bar magnet, with the north and south magnetic poles approximately coinciding with the two ends of the axis on which the Earth spins (the north and south rotational poles). If one assumes that this has always been the case, measurement of the tilt of the magnetic field from the horizontal gives the geographical latitude at which the rock formed. A rock formed at one of the poles should preserve a vertically oriented magnetic field, whereas one formed at the Equator should be magnetized horizontally.

The rotational axis of the Earth must remain fixed in space, but the geographic positions of the north and south poles can change if the whole Earth rotates as a rigid body, with no internal deformation. This process, called polar wandering, was first discussed in the nineteenth century. If polar wandering has occurred with no continental drift, rocks of the same age from all continents should give the same pole positions. But if the pole positions differ, continental drift must have taken place.

At first Runcorn interpreted the paleomagnetic observations from Europe and North America as evidence for polar wandering. But during 1956 he made more measurements on North American rocks, and became convinced that the pole positions for the two continents had not always been the same. He then argued[4] that continental drift must have occurred during the past 200 million years. The difference between the pole positions was not large, and the argument was not at first very convincing. But Irving[5] had also made a series of measurements on rocks which were about 100 million years old from India and Australia. He found that the differences between the pole positions from these two continents and those from Europe and North America were much larger than the measurement errors, and clearly required the continents to have drifted.

We now know that Runcorn and Irving were correct. Yet these paleomagnetic results of the 1950s had almost no impact on the work, by Vine and Matthews and others, that led to the general acceptance of such motions ten years later. This is particularly surprising because much of the later work was carried out in the same small department in Cambridge, which in the 1960s still contained people who had been there ten years earlier. The paleomagnetic work was also ignored by those working on seafloor spreading and plate tectonics in the United States.[6,7] For me this failure of communication remains very puzzling, and is one of the most interesting aspects of the development of plate tectonics.

In 1963 it was not generally accepted that continental drift had occurred. Partly for this reason, Vine and Matthews were careful about what they said and how they said it. Their immediate purpose was to explain the magnetic observations that had been made over underwater ridges in several different oceans. As a ship travels across a ridge, it measures magnetic fields that are alternately stronger and weaker than normal—a pattern that Vine and Matthews interpreted as being produced by blocks of ocean crust in which the rocks are magnetized either along the direction of the Earth's present-day magnetic field or opposite to it. As Vine and Matthews pointed out in their paper,[1] such an arrangement of alternately magnetized blocks of crust is the inevitable outcome of two processes: seafloor spreading and reversals of the Earth's magnetic field.

Seafloor spreading was first proposed by Harry Hess in 1960 and published two years later.[8] He suggested that new ocean crust is formed at ridges by the upwelling of molten rock from below. The newly formed crust then moves away from the ridge like a conveyor belt. Vine and Matthews realized that, as the molten rock cools, iron-bearing minerals in the rock must become magnetized along the direction of the Earth's magnetic field. If the field reverses its polarity from time to time (geographic north becoming magnetic south for certain intervals), the result would be alternating stripes of normal and reversed magnetization on the sea floor (fig. 10.1). This process could therefore produce the pattern of positive and negative magnetic anomalies measured by survey ships in the Atlantic and Indian oceans (see p. 139, figure 1).[1]

I have often wondered why Vine and Matthews' discovery was not made

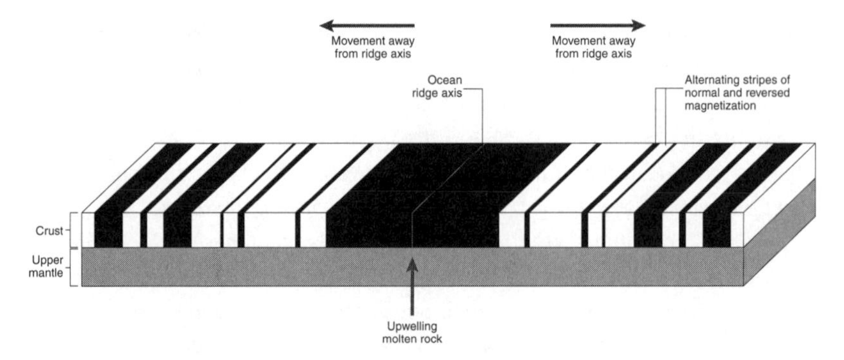

Fig. 10.1. How magnetic stripes are produced at ocean ridges by seafloor spreading. Upwelling molten rock creates new ocean crust on both sides of a ridge. As the rock cools, and moves away from the ridge axis, it becomes magnetized in a direction that depends on the state of the Earth's magnetic field, normal or reversed. The result is a striped pattern of positive and negative magnetic anomalies, with mirror-image symmetry across the ridge axis, as shown here schematically and in the more complicated case depicted in fig. 10.2.

earlier,* using the well-known magnetic survey that had been carried out in the 1950s by Arthur Raff and Ronald Mason[9] in the northeast Pacific Ocean. This survey revealed a remarkable zebra-striped pattern of magnetic field (fig. 10.2), with mirror symmetry across an oceanic ridge. With hindsight this pattern provides more obvious support for Vine and Matthews' proposal than do the data presented in their paper. But the main purpose of Raff and Mason's survey, which was funded by the U.S. Navy, was to make a map of the topography of the sea floor, rather than of the magnetic field. The topography was classified as "Secret," so only those who had a security clearance (which did not include Vine and Matthews) realized that there was a ridge in the survey area. Another problem was that the ridge is not a simple one, but consists of three offset segments (fig. 10.2). No one, including Vine and Matthews, at first realized that this survey beautifully confirmed their suggestion, even though they were well aware of its existence and refer to it in their paper. Two years were to elapse before Vine recognized the symmetry and identified the ridge segments shown here in fig. 10.2.[10]

* In 1963 Lawrence Morley also proposed that magnetic stripes in the oceans were formed by seafloor spreading and reversals of the Earth's magnetic field. When he did so neither he nor Vine and Matthews were aware that they had made the same proposal at the same time. His paper was rejected by *Nature*, and by the *Journal of Geophysical Research*, and was eventually published in 1967.[6]

When Vine and Matthews wrote their paper, their two main assumptions were controversial. Few geophysicists, other than those who were interested in how the Earth's magnetic field is generated, accepted that the polarity of the field could reverse. Some rocks were known to be reversely magnetized, but the situation was confused by the existence of (rare) lavas whose magnetization direction reverses as they cool. Another problem was that Hess's proposal,[8] that new sea floor is formed only in a narrow zone on the axes of ocean ridges, had little observational support. Such behavior is necessary for the formation of well-defined magnetic stripes. But the overwhelming observational support for the Vine-Matthews hypothesis conclusively showed that both assumptions were correct. Indeed, magnetic anomaly profiles now provide the best record of the history of geomagnetic reversals over the past 80 million years, and also the standard way of deducing the geological history of ocean basins.

Vine and Matthews' hypothesis attracted little interest for two years after their paper was published. At first even they themselves were not convinced that their idea was correct. But in 1966, many groups recognized that mag-

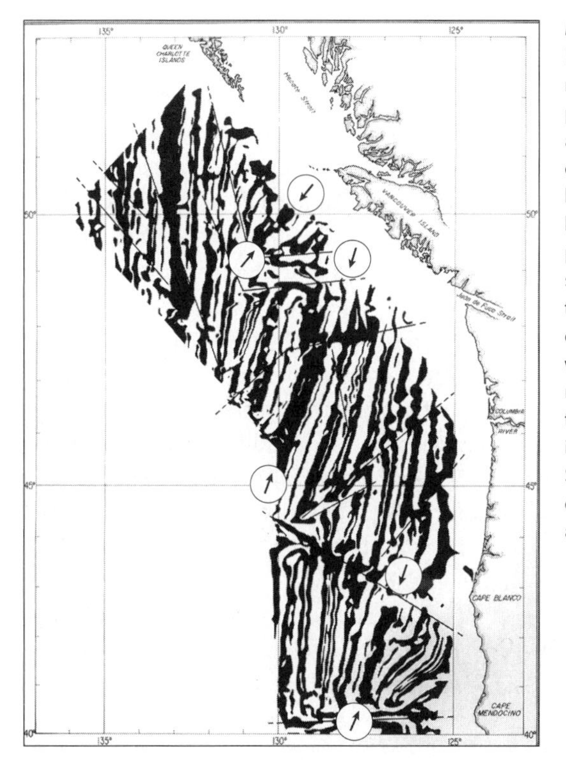

Fig. 10.2. The survey of Raff and Mason,[9] showing magnetic anomalies (black, positive; white, negative) in an area southwest of Vancouver Island. Only with hindsight did it become clear how well the zebra-striped pattern seen here provides support for Vine and Matthews' proposal.[1] In part, the difficulties in interpretation were because the ocean ridge in this area consists of three offset ridge segments, indicated here by arrows. Straight lines on the map indicate faults offsetting the anomaly pattern.

netic profiles from all of the world's oceans clearly confirmed the Vine-Matthews hypothesis, and this led to widespread acceptance of the ideas of seafloor spreading and continental drift.

Our present theoretical framework, known as plate tectonics, was proposed during the following year.[11,12] It differs from the earlier ideas only in the explicit recognition that the surface of the Earth is divided into rigid plates in relative motion. The plate boundaries are outlined by earthquakes, generated by slip between the plates. The direction of such slip can be determined from worldwide seismic recordings, which, together with the Vine-Matthews hypothesis, provided the two clues that led to the present theory. Seafloor spreading occurs where plates are separating, generating new crust that becomes magnetized in the way Vine and Matthews proposed. Island arcs (such as the Japanese islands) and deep-sea trenches (such as the Marianas trench) mark the places where plates move towards each other and one is destroyed. Mountain ranges such as the Himalayas are formed where colliding plates both carry continental crust. The plate velocities, which are as large as 100 millimeters per year, can now be measured directly using orbiting satellites, and these measurements have precisely confirmed the rates and directions obtained earlier from seafloor magnetic anomaly patterns.

The general acceptance of continental drift and plate tectonics that followed from Vine and Matthews' work has dominated later research in many areas; indeed, it is largely responsible for the emergence of the new discipline of Earth science, from the fusion of fields such as geology, geophysics, paleontology, petrology, and geochemistry. Despite the overwhelming acceptance of plate tectonics, a number of important problems remain poorly understood. Although there is general agreement that convection in the Earth's mantle maintains the plate movements, we still have only a sketchy idea of how this process works. Another problem concerns continental deformation. Large areas of the continents have no earthquakes, and their rigid motions can be accurately described by plate tectonics. But the deforming belts between them are unlike the narrow oceanic plate boundaries, and are sometimes as wide as 2,000 kilometers. We are not yet confident that we can describe the complicated motions that occur within these belts, or understand the forces that drive them.

Vine and Matthews' paper was important because it showed how a number of ideas and observations could be combined to provide conclusive quantitative evidence for seafloor spreading and continental drift. Once their hypothesis was shown to be correct, the theory of plate tectonics was an obvious generalization.

References

1. Vine, F. J. & Matthews, D. H. Magnetic anomalies over oceanic ridges. *Nature* **199,** 947–949 (1963).
2. Wegener, A. *The Origin of Continents and Oceans* (Methuen, London, 1924).
3. Menard, H. W. in *Proc. New Hampshire Bicentennial Conf. on the History of Geology* (ed. Schneer, C.) 19–30 (University Press of New England, Hanover, 1979).
4. Runcorn, S. K. Palaeomagnetic comparisons between Europe and North America. *Geol. Ass. Canada Proc.* **8,** 77–85 (1956).
5. Irving, E. Palaeomagnetic and palaeoclimatological aspects of polar wandering. *Geofis. Pura Appl.* **33,** 23–41 (1956).
6. Cox, A. *Plate Tectonics and Geomagnetic Reversals* (Freeman, San Francisco, 1973).
7. Menard, H. W. *The Ocean of Truth* (Princeton University Press, 1986).
8. Hess, H. H. in *Petrologic Studies: A Volume to Honor A. F. Buddington* (eds Engel, A. E. J., James, H. L. & Leonard, B. F.) 599–620 (Geol. Soc. Am., Boulder, 1962).
9. Raff, A. D. & Mason, R. G. A magnetic survey off the west coast of North America 40 °N latitude to 50 °N latitude. *Bull. Geol. Soc. Am.* **72,** 1267–1270 (1961).
10. Vine, F. J. Spreading of the ocean floor: new evidence. *Science* **154,** 1405–1415 (1966).
11. McKenzie, D. P. & Parker, R. L. The North Pacific: an example of tectonics on a sphere. *Nature* **216,** 1276–1280 (1967).
12. Morgan, W. J. Rises, trenches, great faults, and crustal blocks. *J. Geophys. Res.* **73,** 1959–1982 (1968).

Further reading

Le Grand, H. E. *Drifting Continents and Shifting Theories* (Cambridge University Press, 1988).

Oreskes, N. (ed.) *Plate Tectonics: An Insider's History of the Modern Theory of the Earth* (Westview, Boulder, 2002).

Oreskes, N. *The Rejection of Continental Drift* (Cambridge University Press, 1999).

1963

Magnetic anomalies over oceanic ridges

F. J. Vine and D. H. Matthews

Typical profiles showing bathymetry and the associated total magnetic field anomaly observed on crossing the North Atlantic and North-West Indian Oceans are shown in Fig. 1. They illustrate the essential features of magnetic anomalies over the oceanic ridges: (1) long-period anomalies over the exposed or buried foothills of the ridge; (2) shorter-period anomalies over the rugged flanks of the ridge; (3) a pronounced central anomaly associated with the median valley. This pattern has now been observed in the North Atlantic[1,2], the Antarctic[3], and the Indian Oceans[4,5]. In this article we describe an attempt to account for it.

The general increase in wave-length of the anomalies away from the crest of the ridge is almost certainly associated with the increase in depth to the magnetic crustal material[1]. Local anomalies of short-period may often be correlated with bathymetry, and explained in terms of reasonable susceptibility contrasts and crustal configurations; but the long-period anomalies of category (1) are not so readily explained. The central anomaly can be reproduced if it is assumed that a block of material very strongly magnetized in the present direction of the Earth's field underlies the median valley and produces a positive susceptibility contrast with the adjacent crust. It is not clear, however, why this considerable susceptibility contrast should exist beneath the median valley but not elsewhere under the ridge. Recent work in this Department has suggested a new mechanism.

In November 1962, H.M.S. *Owen* made a detailed magnetic survey over a central part of the Carlsberg Ridge as part of the International Indian Ocean Expedition. The area (50 × 40 nautical miles; centred on

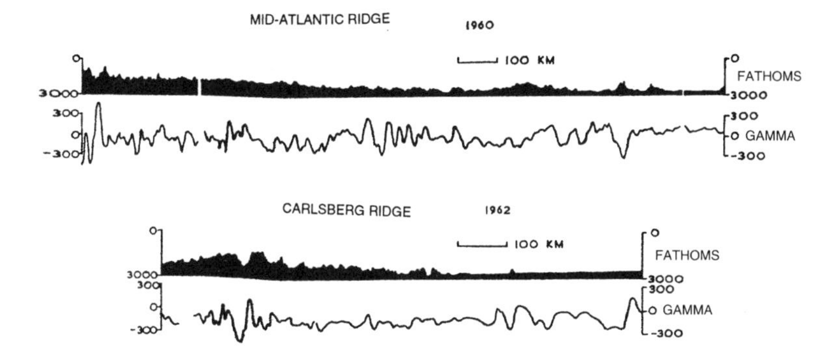

Fig. 1. Profiles showing bathymetry and the associated total magnetic field anomaly observed on crossing the North Atlantic and the north-west Indian Oceans. Upper profile from 45° 17′ N. 28° 27′ W. to 45° 19′ N. 11° 29′ W. Lower profile from 30° 5′ N. 61° 57′ E. to 10° 10′ N. 66° 27′ E.

5° 25′ N., 61° 45′ E.) is predominantly mountainous, depths ranging from 900 to 2,200 fathoms, and the topographic features are generally elongated parallel to the trend of the Ridge. This elongation is more marked on the total magnetic field anomaly map where a trough of negative anomalies, flanked by steep gradients, separates two areas of positive anomalies. The trough of negative anomalies corresponds to a general depression in the bottom topography which represents the median valley of the Ridge. The positive anomalies correspond to mountains on either side of the valley.

In this low magnetic latitude (inclination −6°) the effect of a body magnetized in the present direction of the Earth's field is to reduce the strength of the field above it, producing a negative anomaly over the body and a slight positive anomaly to the north. Here, over the centre of the Ridge, the bottom topography indicates the relief of basic extrusives such as volcanoes and fissure eruptives, and there is little sediment fill. The bathymetry, therefore, defines the upper surface of magnetic material having a considerable intensity of magnetization, potentially as high as any known igneous rock type[6], and probably higher, because it is extrusive, than the main crustal layer beneath. That the topographic features *are* capable of producing anomalies is immediately apparent on comparing the bathymetric and the anomaly charts; several have well-defined anomalies associated with them.

Two comparatively isolated volcano-like features were singled out and considered in detail. One has an associated negative anomaly as one would expect for normal magnetization, the other, completely the

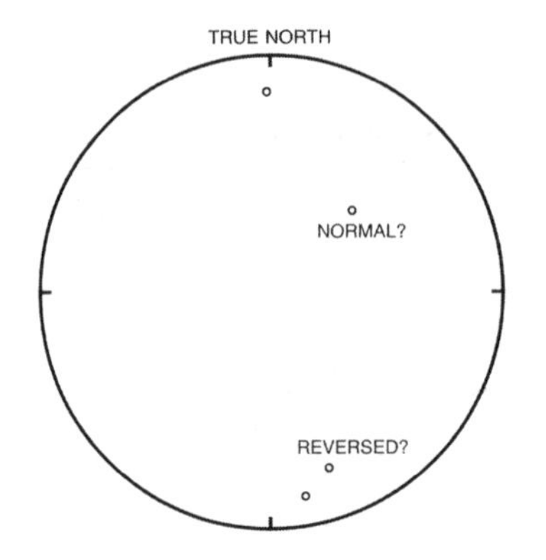

Fig. 2. Directions of the magnetic vectors obtained by the computer programme plotted on a stereographic projection, together with the present field vector and its reverse. Bearings and inclinations: present field vector 356°; −6° (up); computed vectors 038°; −40° (up); 166° 30′; +13° (down).

reverse anomaly pattern, that is, a pronounced positive anomaly suggesting reversed magnetization. Data on the topography of each feature and its associated anomaly were fed into a computer and an intensity and direction of magnetization for each obtained. Fig. 2 shows the directions of the resulting vectors plotted on a stereographic projection. Having computed the magnetic vector by a 'best fit' process, the computer recalculated the anomaly over the body, assuming this vector, thus giving an indication of the accuracy of fit. The fit was good for the case of reversed magnetization but poor for that of approximately normal magnetization. The discrepancy is scarcely surprising since we have ignored the effects of adjacent topography, and the interference of other anomalies in the vicinity. In addition, the example of normal magnetization is near a corner of the area where the control of contouring is less precise. The other example is central where the control is good. In both cases the intensity of magnetization deduced was about 0.005 e.m.u.; this is equivalent to an effective susceptibility of ± 0.0133: (effective susceptibility = total intensity of magnetization (remanent + induced)/present total magnetic field intensity: mean value for basalts of the order of 0.01).

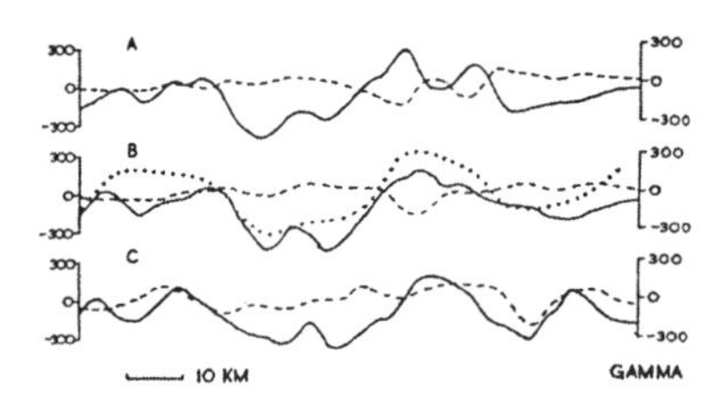

Fig. 3. Observed and computed profiles across the crest of the Carlsberg Ridge. Solid lines, observed anomaly; broken lines, computed profile assuming uniform normal magnetization and an effective susceptibility of 0.0133; dotted line, assuming reversals—see text. The computed profiles were obtained assuming infinite lateral extent of the bathymetric profiles.

In addition, three profiles, perpendicular to the trend of the Ridge, have been considered. Computed profiles along these, assuming infinite lateral extent of the bathymetric profile, and uniform normal magnetization, bear little resemblance to the observed profiles (Fig. 3). These results suggested that

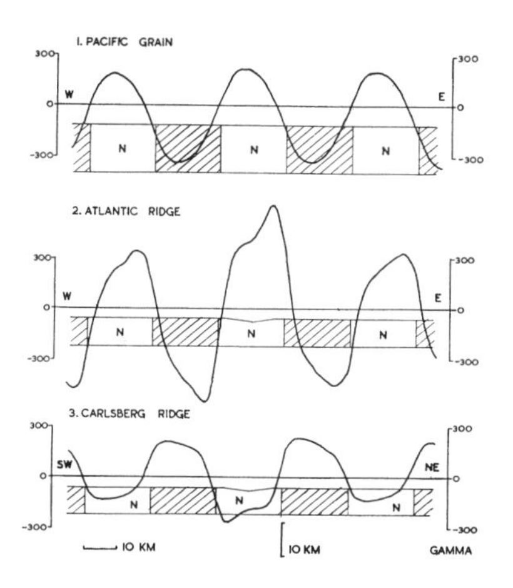

Fig. 4. Magnetic profiles computed for various crustal models. Crustal blocks marked *N*, normally magnetized; diagonally shaded blocks, reversely magnetized. Effective susceptibility of blocks, 0.0027, except for the block under the median valley in profiles 2 and 3, 0.0053. (1) Pacific Grain. Total field strength, $T = 0.5$ œrsted; inclination, $I = 60°$; magnetic bearing of profile, $\theta = 073°$. (2) Mid-Atlantic Ridge, $T = 0.48$ œrsted; $I = 65°$; $\theta = 120°$. (3) Carlsberg Ridge, $T = 0.376$ œrsted; $I = -6°$; $\theta = 044°$.

whole blocks of the survey area might be reversely magnetized. The dotted curve in Fig. 3 *B* was computed for a model in which the main crustal layer and overlying volcanic terrain were divided into blocks about 20 km wide, alternately normally and reversely magnetized. The blocks were given the effective susceptibility values shown in the caption to Fig. 4 (3).

Work on this survey led us to suggest that some 50 per cent of the oceanic crust might be reversely magnetized and this in turn has suggested a new model to account for the pattern of magnetic anomalies over the ridges.

The theory is consistent with, in fact virtually a corollary of, current ideas on ocean floor spreading[7] and periodic reversals in the Earth's magnetic field[8]. If the main crustal layer (seismic layer 3) of the oceanic crust is formed over a convective up-current in the mantle at the centre of an oceanic ridge, it will be magnetized in the current direction of the Earth's field. Assuming impermanence of the ocean floor, the whole of the oceanic crust is comparatively young, probably not older than 150 million years, and the thermo-remanent component of its magnetization is therefore either essentially normal, or reversed with respect to the present field of the Earth. Thus, if spreading of the ocean floor occurs, blocks of alternately normal and reversely magnetized material would drift away from the centre of the ridge and parallel to the crest of it.

This configuration of magnetic material could explain the lineation or 'grain' of magnetic anomalies observed over the Eastern Pacific to the west of North America[6] (probably equivalent to the long-period anomalies of category (1)). Here north–south highs and lows of varying width, usually of the order of 20 km, are bounded by steep gradients. The amplitude and form of these anomalies have been reproduced by Mason[9,10], but the most plausible of the models used involved very severe restrictions on the distribution of lava flows in crustal layer 2. They are readily explained in terms of reversals assuming the model shown in Fig. 4 (1). It can be shown that this type of anomaly pattern will be produced for virtually all orientations and magnetic latitudes, the amplitude decreasing as the trend of the ridge approaches north–south or the profile approaches the magnetic equator. The pronounced central anomaly over the ridges is also readily explained in terms of reversals. The central block, being most recent, is the only one which has a uniformly directed magnetic vector. This is comparable to the area of normally magnetized late Quaternary basics in Central Iceland[11,12] on the line of the Mid-Atlantic Ridge. Adjacent and all other blocks have doubtless been subjected to subsequent vulcanism in the form of volcanoes, fissure eruptions, and lava flows, often oppositely magnetized and hence reducing the effective susceptibility of the block, whether initially normal or reversed. The effect of assuming a reduced effective susceptibility for the adjacent blocks is illustrated for the North Atlantic and Carlsberg Ridges in Fig. 4 (2, 3).

In Fig. 4, no attempt has been made to reproduce observed profiles in detail, the computations simply show that the essential form of the anomalies is readily achieved. The whole of the magnetic material of the oceanic crust is probably of basic igneous composition; however, variations in its intensity of magnetization and in the topography and direction of magnetization of surface extrusives could account for the complexity of the observed profiles. The results from the preliminary Mohole drilling[13,14] are considered to substantiate this conception. The drill penetrated 40 ft. into a basalt lava flow at the bottom of the hole, and this proved to be reversely magnetized[13]. Since the only reasonable explanation of the magnetic anomalies mapped near the site of the drilling is that the area is underlain by a block of normally magnetized crustal material[14], it appears that the drill penetrated a layer of reversely magnetized lava overlying a normally magnetized block.

In Fig. 4 it will also be noticed that the effective susceptibilities assumed are two to five times less than that derived for the isolated features in the survey area described. Although no great significance can be attached to this derived intensity it is suggested that the fine-grained extrusives (basalts) of surface features are more highly magnetized than the intrusive material of the main crustal layer which, in the absence of evidence to the contrary, we assume to be of analogous chemical composition (that is, gabbros). This would appear to be consistent with recent investigations of the magnetic properties of basic rocks[6].

The vertical extent of the magnetic crust is defined by the depth to the curie-point isotherm. In the models this has been assumed to be at 20 km below sea-level over the deep ocean but at a depth of 11 km beneath the centre of the ridges where the heat flow and presumably the thermal gradient are higher. These assumptions are questionable but not critical because the amplitude of the simulated anomaly depends on both the thickness of the block and its effective susceptibility, and, although the thickness is in doubt by a factor of two, the susceptibility is in doubt by a factor of ten. Present magnetic declination has been assumed throughout the calculations: it would probably have been better to have ignored this, as in palæomagnetism, assuming that true north approximates to the mean of secular variations; but this is unimportant and in no way affects the essential features of the computations.

In order to explain the steep gradients and large amplitudes of magnetic anomalies observed over oceanic ridges all authors have been compelled to assume vertical boundaries and high-susceptibility contrasts between adjacent crustal blocks. It is appreciated that magnetic contrasts within the oceanic crust can be explained without postulating reversals of the Earth's magnetic field; for example, the crust might contain blocks of very strongly magnetized material adjacent to blocks of material weakly magnetized in the same direction. However, the model suggested in this article seems to be more plausible because high susceptibility contrasts between adjacent blocks can be explained without recourse to major inhomogeneities of rock type within the main crustal layer or to unusually strongly magnetized rocks.

We thank Dr. R. G. Mason and K. Kunaratnam of the Imperial College of Science and Technology, London, for details of the three-dimensional programme used in this work. The programme was originally devised by K. Kunaratnam for a Ferranti *Mercury* Computer. It has been rewritten for use on *Edsac* 2. We also thank the Director of the Cambridge University Mathematical Laboratory for permission to use *Edsac* 2, and Sir Edward Bullard for his advice and encouragement throughout.

This work was partly supported by a grant from the U.S. Office of Naval Research (Contract No. *N*62558–3542).

Department of Geodesy and Geophysics, University of Cambridge

[Published 7 September 1963]

References

1. Heezen, B. C., Ewing, M., and Miller, E. T., *Deep Sea Res.*, 1, 25 (1953).
2. Keen, M. J., *Nature*, **197**, 888 (1963).
3. Adams, R. D., and Christoffel, D. A., *J. Geophys. Res.*, **67**, 805 (1962).

4. Heirtzler, J. R., *Tech. Rep., No.* 2, *Lamont Geol. Obs., New York* (1961).
5. Matthews, D. H., *et al., Admiralty Marine Sci. Pub. No.* 4 (in the press).
6. Bullard, E. C., and Mason, R. G., *The Sea,* **3**, edit. by Hill, M. N. (in the press).
7. Dietz, R. S., *Nature,* **190**, 854 (1961).
8. Cox, A., Doell, R. R., and Dalrymple, G. B., *Nature,* **198**, 1049 (1963).
9. Mason, R. G., *Geophys. J.*, 1, 320 (1958).
10. Mason, R. G., and Raff, A. D., *Bull. Geol. Soc. Amer.*, **72**, 1259 (1961).
11. Hospers, J., *Geol. Mag.*, **91**, 352 (1954).
12. Thorarinsson, S., Einarsson, T., and Kjartansson, G., *Intern. Geog. Cong. (Norden)*, Excursion E.I.1 (1960).
13. Cox, A., and Doell, R. R., *J. Geophys. Res.*, **67**, 3997 (1962).
14. Raff, A. D., *J. Geophys. Res.*, **68**, 955 (1963).

f aspects of the interaction, and in particu- arked contrast, for example, with ribo- hows that the entry of the substrate pro- eciable local changes in disposition of side here is no evidence of specific interactions y groups in the enzyme and the C-terminal of the substrate, which fits easily into the owever, the terminal carboxylate group into contact with an arginine side chain, es 2 Å in the pro It i ively moreover, that the c en m he eptide bond may bec a d he It was previously ob d nd that certain tyrosine dues seem be enzymatic activity, and inde d it und rosine moves a distance of no l han 14 Å

when the substrate is introduced so as to brin phenolic hydroxyl group close to the substrate bond. The motion occurs in a non-helical the chain, and evidently involves a cons disturbance of the backbone conformation as rotation about side chain carbon–carbon A carboxylate group also seems to form par a te. e deta so available are consiste emical da ner a more precise deline actio een the active centre and rat l ea y suggest a catalytic me to e here are grounds for op the kable failure of known structures so far to reveal the mechanism function unequivocably.

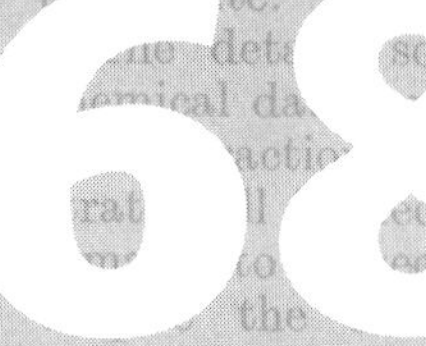

ervation of a Rapidly Pulsating Radio Sou

H

LKINGTON
T
LINS

Astronomy Observatory,
oratory,
Cambridge

Unusual signals from pulsating radio sources have been reco the Mullard Radio Astronomy Observatory. The radiation s come from local objects within the gala<y, and may be ass with oscillations of white dwarf or neutron stars.

7, a large radio telescope operating at a fre- 1·5 MHz was brought into use at the Mullard nomy Observatory. This instrument was investigate the angular structure of compact s by observing the scintillation caused by r structure of the interplanetary medium[1]. urvey includes the whole sky in the declination $< \delta < 44°$ and this area is scanned once a rge fraction of the sky is thus under regular Soon after the instrument was brought into was noticed that signals which appeared at weak sporadic interference were repeatedly a fixed declination and right ascension; this d that the source could not be terrestrial in

e investigations were started in November eed records showed that the signals, when sisted of a series of pulses each lasting ~0·3 s repetition period of about 1·337 s which was to be maintained with extreme accuracy. ervations have shown that the true period is better than 1 part in 10^7 although there is a ariation which can be ascribed to the orbital e Earth. The impulsive nature of the recorded used by the periodic passage of a signal of frequency through the 1 MHz pass band of

kable nature of these signals at first suggested erms of man-made transmissions which might

of three others having remarkably similar pr which suggests that this type of source may be re common at a low flux density. A tentative expl of these unusual sources in terms of the stable osc of white dwarf or neutron stars is proposed.

Position and Flux Density

The aerial consists of a rectangular array co 2,048 full-wave dipoles arranged in sixteen rows elements. Each row is 470 m long in an E.–W. d and the N.–S. extent of the array is 45 m. Phase-s is employed to direct the reception pattern in dec and four receivers are used so that four different tions may be observed simultaneously. Phase-sw receivers are employed and the two halves of th are combined as an E.–W. interferometer. Each dipole elements is backed by a tilted reflecting sc that maximum sensitivity is obtained at a declin approximately +30°, the overall sensitivity being by more than one-half when the beam is scan declinations above +90° and below −5°. The bea of the array to half intensity is about ±½° in right sion and ±3° in declination; the phasing arrange designed to produce beams at roughly 3° inter declination. The receivers have a bandwidth of centred at a frequency of 81·5 MHz and routine rec are made with a time constant of 0·1 s; the r.m.s fluctuations correspond to a flux density of 0·5 W m⁻² Hz⁻¹. For detailed studies of the pulsating

ELEVEN

Stellar timekeepers

Joseph H. Taylor

Pulsars are rapidly spinning, collapsed stars—with the mass of the Sun but only ten kilometers across—that emit lighthouse-like beams of radio noise. Their discovery, reported in 1968, came as a complete surprise, and astonishingly their radio signals behave like the ticks of a super-accurate clock. The unique characteristics of pulsars have since been used to explore the behavior of gravity, the nature of nuclear matter, the late evolutionary stages of massive stars, and the character of the interstellar medium.

The mid-1960s arguably marked the dawn of the modern astrophysical era. New windows in the electromagnetic spectrum had been opened for routine astronomical observations. Radio astronomy, in particular, was in a sort of adolescent stage, its practitioners seemingly rushing from one unexpected discovery to the next. Electronic technology was advancing rapidly: vacuum tubes had largely given way to transistors, and new methods for recording and analyzing large quantities of data were in hand. X-ray observations made from rockets and satellites, their development aided by the Cold War space program in the United States, were yielding exciting results to complement those at radio wavelengths.

Much of the scientific activity was related to the study of newly recognized types of "compact object," including quasars (see chapter 9) and what would later be called black holes. The possibility of compact objects with the mass of stars was of particular interest: the idea that massive stars might collapse under the force of gravity to form extremely dense objects composed mainly of neutrons ("neutron stars") had been proposed in the 1930s,[1] shortly after the discovery of the neutron. Moreover, a general understanding of the nuclear equation of state—that is, the compressibility of nuclear matter—showed that such objects should be stable as long as their masses are less than a critical limit, probably around twice the mass of the Sun. (Above this limit, a neutron star should collapse further, to form a black hole.) But no such object had ever been observed, and the prospects did not look good. The expected size of a neutron star is about twenty kilometers across—so tiny that no plausible temperature would make one bright enough to be detected at the typical distances of nearby stars.

I defended my Ph.D. thesis at Harvard University in January 1968, and was staying on as a postdoctoral fellow. In the next few weeks I busied myself by looking for a new research project in radio astronomy. Imagine my delight when the 24 February issue of *Nature* arrived in the observatory library, bearing the enticing words "Possible neutron star" on its cover, alerting the reader to a paper[2] entitled "Observation of a rapidly pulsating radio source" by radio astronomers at the University of Cambridge. Quick scrutiny showed that the piece had been submitted only fifteen days earlier—an almost unheard-of short publication time—so it seemed the work had to be important. And indeed it was. Antony Hewish, his student S. Jocelyn Bell, and their colleagues presented a clear and succinct description of the new class of radio source they had found, how the observations were made, what it must mean, and what should come next. The implications were so fascinating that I decided on the spot to attempt follow-up observations of my own.

More than a thousand discrete radio sources had been catalogued by 1968, and nearly all had radio brightnesses that did not change over timescales of years and longer. By contrast, the strength of the source detected by Hewish *et al.* at a radio frequency of 81.5 megahertz (MHz) varied widely over intervals of only a few milliseconds. Stranger still, the variations came in the form of periodic pulses that repeated every 1.337 seconds, keeping time to an accuracy of one part in ten million, or better, over many weeks! A man-made origin of the signals had seemed the most likely explanation at first, but was ruled out because the source maintained its apparent celestial position over several months, showing that it must lie far outside the Solar System. A throwaway line near the end of an introductory paragraph mentions that three more objects with similar properties had already been found, and suggests with some prescience that "this type of source may be relatively common."

Other characteristics of the signals were also outlined in some detail. The pulses consisted of short bursts containing a narrow range of frequencies, sweeping downward in frequency at a rate of 5 MHz per second. The pulse duration at a single frequency was about sixteen milliseconds, from which the authors concluded that the source could be no bigger than a small planet. (The pulse duration cannot in general be shorter than the time it takes light to travel from one edge of the source to the other.) The downward-sweeping frequency behavior was attributed to the effect of dispersion (higher-frequency waves traveling more quickly) as the radio waves traveled through the interstellar gas between the source and the Earth; this yielded an estimate for the distance to the source of about 65 parsecs, or 210 light years, comparable with the distances to some nearby stars.

Hewish *et al.* pointed out that with the positional accuracy available to them, there was little hope of matching the source to any optically visible object. In any event, the upper size limit, at least two orders of magnitude smaller than a star, offered little prospect of detectable optical emission. Feeling compelled to speculate on the physical nature of the pulsating sources, despite a conviction that only the most tentative suggestions were yet warranted, the authors state that the most significant feature to be explained is the extreme regularity of the pulses. They suggest that radial pulsations of compact stars, such as white dwarfs or neutron stars, could be the cause, and they conclude by mentioning that if one of the suggested origins is confirmed, further study may "throw valuable light on the behaviour of compact stars and also on the properties of matter at high density."

Antony Hewish was awarded a share of the 1974 Nobel Prize in Physics for his role in the discovery of pulsars. (The word "pulsar" does not appear in the discovery paper, but it came into general use within a few months, probably invented in more than one place.) Hewish's Nobel lecture[3] includes some reflections on the weeks immediately surrounding the discovery. One was that "We had to face the possibility that the signals were . . . generated on a planet circling some distant star, and that they were artificial." This idea was very much in the air at the time of the original announcement, and had certainly fueled much of the attention given the discovery by the popular press. In the written version of a 1976 after-dinner speech,[4] Jocelyn Bell (Burnell) describes an incident that occurred in December 1967, during the interval of several weeks when the Cambridge group knew they were onto something important, but before they had told anyone else about it. She went to see Hewish in his office, and walked into "a high-level conference on how to present these results. We did not really believe that we had picked up signals from another civilization, but . . . if one thinks one may have detected life elsewhere in the universe how does one announce the results responsibly? Who does one tell first?"

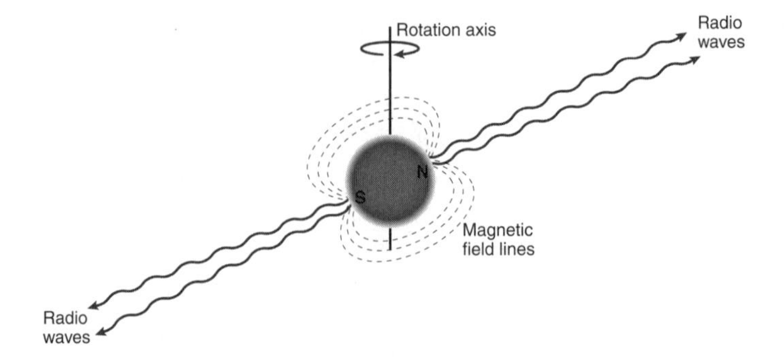

Fig. 11.1. Radio emission from a pulsar. Rapid rotation of a strongly magnetized neutron star leads to very high voltages near the star's magnetic poles. Electrical discharges then create radio noise, which is beamed outward in the polar directions and is detectable over interstellar distances. The sweeping of the beam across the observer leads to detection of the emission as discrete pulses.

Needless to say, I was not the only reader of the *Nature* paper who quickly formed ideas about follow-up studies. Furious activity began at most of the world's large radio telescopes to put together suitable equipment for observing pulsars. Theorists began to work on models that might explain the clocklike precision of the pulses as well as the generation of the radio noise. For many months a torrent of papers rained down on the journals. Thomas Gold was the first to propose[5] that pulsars are strongly magnetized neutron stars whose rapid spins, together with tightly beamed radio emission, cause the observed periodic signals (fig. 11.1). As a necessary consequence, Gold predicted that the pulsation periods should gradually lengthen over time, as the star loses rotational energy. The prediction was confirmed within a few months, after the discovery of pulsars in the Crab nebula (fig. 11.2) and the Vela-X supernova remnant. As neutron stars are believed to be created in supernova explosions, these young pulsars and their clear associations with supernovae lent much weight to Gold's neutron-star hypothesis.

Many of us radio observers quickly set about finding additional pulsars. An ad hoc group that I helped to form at Harvard found the first one after the original Cambridge four; we used digital recording techniques and a computer-based method that I devised to search for periodic, dispersed signals buried in the otherwise random receiver noise. Interestingly, for several years such computer methods competed more or less equally with simple visual inspection of paper chart recordings, as had been done by the Cambridge group. But computers and digital recording hardware were rapidly becoming more powerful, and computerized searches soon became the norm. More than thirteen hundred pulsars have now been detected in and near our Galaxy, nearly all of them by using refinements of the search method that I designed in 1968.

Although the early levels of intense activity have abated, pulsar research has remained active and highly productive for a third of a century. As foreseen by the original authors, valuable light has, indeed, been thrown on the behavior of compact stars and the properties of matter at high density. But other unforeseen dividends have appeared as well, with perhaps even greater significance. My 1974 discovery with Russell Hulse of a pulsar orbiting another neutron star[6] opened the way for a new test of gravitation theory,[7,8] establishing the existence of gravitational radiation (or "gravity waves"), in accord with Einstein's general theory of relativity. This has turned out to be but the first example of a "recycled" pulsar—a neutron star in mutual orbit with another star from which mass is transferred to the pulsar, "spinning up" the pulsar and prolonging its life as an active radio source. Like the first one, most of the recycled pulsars have periods of milliseconds or tens of milliseconds. Millisecond pulsars appear to deserve the title "nature's most precise clocks," as they keep time (after the predictable steady slowdown is measured and accounted for) with stabilities comparable to the very best man-made atomic clocks.

Perhaps surprisingly, the physics of the region immediately surrounding a pulsar is not yet understood in detail. In particular, we do not have a

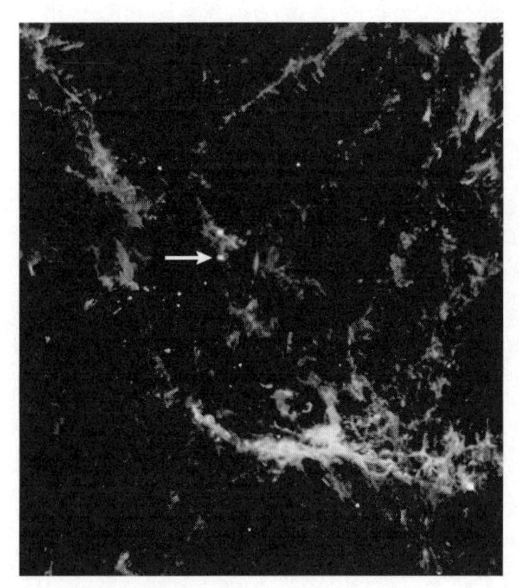

Fig. 11.2. Hubble Space Telescope image of the heart of the Crab nebula, glowing clouds of gas left behind by a supernova explosion that took place in A.D. 1054. When a massive star explodes in a supernova, its core can collapse to form a neutron star, a compact object with the density of nuclear, rather than atomic, matter. A typical neutron star has 1.4 times the mass of the Sun, but a radius of only ten kilometers. Unusually, the Crab pulsar (arrowed) emits light, as well as radio waves—a consequence of its extremely strong magnetic fields and very high rotation rate (thirty times a second). (Image courtesy of NASA and the Hubble Heritage Team [STScI/AURA].)

reliable and self-consistent model for generation of the detected radio noise. One reason is that the radio emissions comprise an insignificant fraction, typically less than 0.01 percent, of the rotational energy losses; most of the rest probably goes into undetectable electromagnetic radiation and into accelerating charged particles that escape and become cosmic rays, streaming through the Galaxy. My good friend Venkataraman Radhakrishnan has described trying to understand the pulsar radio emission mechanism as akin to standing in the parking lot outside a windowless factory, trying to understand its workings by listening to the squeaks of machinery inside.

One of the recycled pulsars turns out to be accompanied by at least two planet-sized companions, and this realization[9] marked the first definitive detection of planets outside our Solar System (see p. 317). Several others have dwarf companions that are being evaporated by high-energy radiation from the pulsar, thus clearly illustrating an important stage in the evolution of closely spaced double-star systems. The masses of a number of pulsars in orbiting systems have been measured with high accuracy, providing experimental data with which to evaluate calculations of the equation of state for bulk nuclear matter, as well as elucidating the processes of stellar collapse, supernova explosions, and neutron-star formation. Comparisons of "pulsar time" with terrestrial atomic time, carried out over many years, have provided a stringent upper limit to any background of gravity waves, such as might be rattling around in the Universe as a distant relic of the Big Bang. Pulsar signals have been used with great effect to trace the Galactic magnetic field and the density and distribution of interstellar matter.

What further astrophysical bounties might these remarkable objects have in store for us? One likely prospect is the role they may play in the direct observation of gravity waves, so far only inferred to exist from the analysis of pulsar orbits. Orbiting pairs of neutron stars must gradually spiral closer together, their orbits losing energy in the form of gravity waves. When the two stars ultimately coalesce, a huge burst of gravity waves is generated, which should be detectable at distances of up to hundreds of millions of light years. Gravity-wave detectors are now being built and put into operation in the United States, Europe, and Japan, with the aim of opening yet another type of window on the Universe—this time involving an entirely new form of radiant energy.

References

1. Baade, W. & Zwicky, F. On super-novae. *Proc. Natl Acad. Sci. USA* **20,** 254–259 (1934).
2. Hewish, A., Bell, S. J., Pilkington, J. D. H., Scott, P. F. & Collins, R. A. Observation of a rapidly pulsating radio source. *Nature* **217,** 709–713 (1968).

3. Hewish, A. Pulsars and high density physics. *Science* **188,** 1079–1083 (1975).
4. Bell Burnell, S. J. Petits fours. *Ann. NY Acad. Sci.* **302,** 685–689 (1977).
5. Gold, T. Rotating neutron stars as the origin of pulsating radio sources. *Nature* **218,** 731–732 (1968).
6. Hulse, R. A. & Taylor, J. H. Discovery of a pulsar in a binary system. *Astrophys. J.* **195,** L51–L53 (1975).
7. Taylor, J. H., Fowler, L. A. & McCulloch, P. M. Measurements of general relativistic effects in the binary pulsar PSR1913+16. *Nature* **277,** 437–440 (1979).
8. Taylor, J. H. Binary pulsars and relativistic gravity. *Rev. Mod. Phys.* **66,** 711–719 (1994).
9. Wolszczan, A. & Frail, D. A. A planetary system around the millisecond pulsar PSR1257+12. *Nature* **355,** 145–147 (1992).

Further reading

Nobel e-Museum. *The Nobel Prize in Physics 1974* (http://www.nobel.se/physics/laureates/1974/). (Also physics prize in 1993.)

Weisberg, J. M., Taylor, J. H. & Fowler, L. A. Gravitational waves from an orbiting pulsar. *Scientific American* 74–82 (April 1981).

1968

Observation of a rapidly pulsating radio source

A. Hewish, S. J. Bell, J. D. H. Pilkington, P. F. Scott,
and R. A. Collins

Unusual signals from pulsating radio sources have been recorded at the Mullard Radio Astronomy Observatory. The radiation seems to come from local objects within the galaxy, and may be associated with oscillations of white dwarf or neutron stars.

In July 1967, a large radio telescope operating at a frequency of 81.5 MHz was brought into use at the Mullard Radio Astronomy Observatory. This instrument was designed to investigate the angular structure of compact radio sources by observing the scintillation caused by the irregular structure of the interplanetary medium[1]. The initial survey includes the whole sky in the declination range $-08° < \delta < 44°$ and this area is scanned once a week. A large fraction of the sky is thus under regular surveillance. Soon after the instrument was brought into operation it was noticed that signals which appeared at first to be weak sporadic interference were repeatedly observed at a fixed declination and right ascension; this result showed that the source could not be terrestrial in origin.

Systematic investigations were started in November and high speed records showed that the signals, when present, consisted of a series of pulses each lasting ~0.3 s and with a repetition period of about 1.337 s which was soon found to be maintained with extreme accuracy. Further observations have shown that the true period is constant to better than 1 part in 10^7 although there is a systematic variation which can be ascribed to the orbital motion of the Earth. The impulsive nature of the recorded signals is caused by the periodic passage of a signal of descending frequency through the 1 MHz pass band of the receiver.

The remarkable nature of these signals at first suggested an origin in terms of man-made transmissions which might arise from deep space probes, planetary radar or the reflexion of terrestrial signals from the Moon. None of these interpretations can, however, be accepted because the absence of any parallax shows that the source lies far outside the solar system. A preliminary search for further pulsating sources has already revealed the presence of three others having remarkably similar properties which suggests that this type of source may be relatively common at a low flux density. A tentative explanation of these unusual sources in terms of the stable oscillations of white dwarf or neutron stars is proposed.

Position and flux density

The aerial consists of a rectangular array containing 2,048 full-wave dipoles arranged in sixteen rows of 128 elements. Each row is 470 m long in an E.–W. direction and the N.–S. extent of the array is 45 m. Phase-scanning is employed to direct the reception pattern in declination and four receivers are used so that four different declinations may be observed simultaneously. Phase-switching receivers are employed and the two halves of the aerial are combined as an E.–W. interferometer. Each row of dipole elements is backed by a tilted reflecting screen so that maximum sensitivity is obtained at a declination of approximately +30°, the overall sensitivity being reduced by more than one-half when the beam is scanned to declinations above +90° and below −5°. The beamwidth of the array to half intensity is about ±1/2° in right ascension and ± 3° in declination; the phasing arrangement is designed to produce beams at roughly 3° intervals in declination. The receivers have a bandwidth of 1 MHz centred at a frequency of 81.5 MHz and routine recordings are made with a time constant of 0.1 s; the r.m.s. noise fluctuations correspond to a flux density of 0.5×10^{-26} W m^{-2} Hz^{-1}. For detailed studies of the pulsating source a time constant of 0.05 s was usually employed and the signals were displayed on a multi-channel 'Rapidgraph' pen recorder with a time constant of 0.03 s. Accurate timing of the pulses was achieved by recording second pips derived from the *MSF* Rugby time transmissions.

A record obtained when the pulsating source was unusually strong is shown in Fig. 1*a*. This clearly displays the regular periodicity and also the characteristic irregular variation of pulse amplitude. On this occasion the largest pulses approached a peak flux density (averaged over the

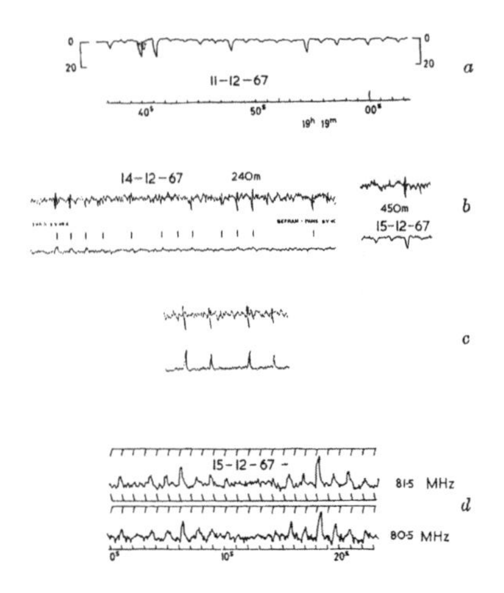

Fig. 1. *a*, A record of the pulsating radio source in strong signal conditions (receiver time constant 0.1 s). Full scale deflexion corresponds to 20 × 10^{-26} W m^{-2} Hz^{-1}. *b*, Upper trace: records obtained with additional paths (240 m and 450 m) in one side of the interferometer. Lower trace: normal interferometer records. (The pulses are small for l = 240 m because they occurred near a null in the interference pattern; this modifies the phase but not the amplitude of the oscillatory response on the upper trace.) *c*, Simulated pulses obtained using a signal generator. d, Simultaneous reception of pulses using identical receivers tuned to different frequencies. Pulses at the lower frequency are delayed by about 0.2 s.

1 MHz pass band) of 20 × 10^{-26} W m^{-2} Hz^{-1}, although the mean flux density integrated over one minute only amounted to approximately 1.0 × 10^{-26} W m^{-2} Hz^{-1}. On a more typical occasion the integrated flux density would be several times smaller than this value. It is therefore not surprising that the source has not been detected in the past, for the integrated flux density falls well below the limit of previous surveys at metre wavelengths.

The position of the source in right ascension is readily obtained from an accurate measurement of the "crossover" points of the interference pattern on those occasions when the pulses were strong throughout an interval embracing such a point. The collimation error of the instrument was determined from a similar measurement on the neighbouring source 3*C* 409 which transits about 52 min later. On the routine recordings which first revealed the source the reading accuracy was only ± 10 s and the earliest record suitable for position measurement was obtained on August 13, 1967. This and all subsequent measurements agree within the error limits. The position in declination is not so well determined and relies on the relative amplitudes of the signals obtained when the reception pattern is centred on declinations of 20°, 23° and 26°. Combining the measurements yields a position

$$\alpha_{1950} = 19\text{h } 19\text{m } 38\text{s} \pm 3\text{s}$$

$$\delta_{1950} = 22^\circ\ 00' \pm 30'$$

As discussed here, the measurement of the Doppler shift in the observed frequency of the pulses due to the Earth's orbital motion provides an alternative estimate of the declination. Observations throughout one year should yield an accuracy of ±1′. The value currently attained from observations during December–January is $\delta = 21° 58' \pm 30'$, a figure consistent with the previous measurement.

Time variations

It was mentioned earlier that the signals vary considerably in strength from day to day and, typically, they are only present for about 1 min, which may occur quite randomly within the 4 min interval permitted by the reception pattern. In addition, as shown in Fig. 1*a*, the pulse amplitude may vary considerably on a time-scale of seconds. The pulse to pulse variations may possibly be explained in terms of interplanetary scintillation[1], but this cannot account for the minute to minute variation of mean pulse amplitude. Continuous observations over periods of 30 min have been made by tracking the source with an E.–W. phased array in a 470 m × 20 m reflector normally used for a lunar occultation programme. The peak pulse amplitude averaged over ten successive pulses for a period of 30 min is shown in Fig. 2*a*. This plot suggests the possibility of periodicities of a few minutes duration, but a correlation analysis

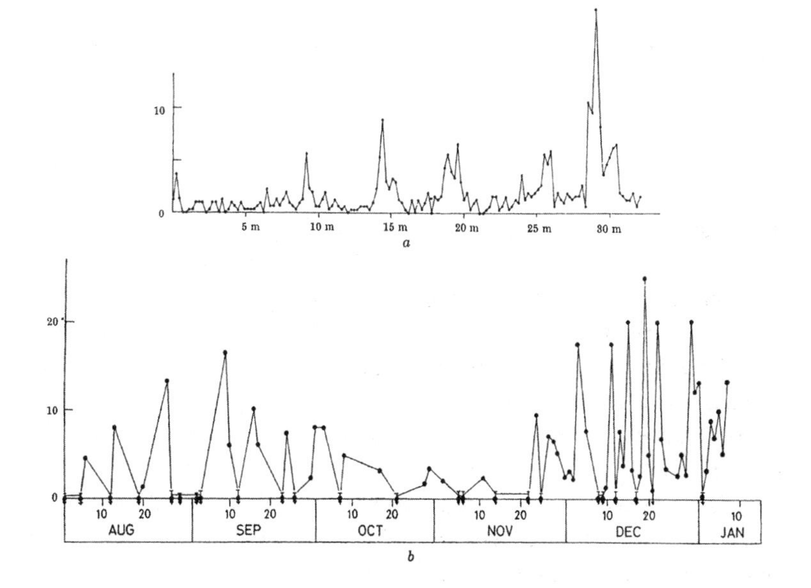

Fig. 2. *a*, The time variation of the smoothed (over ten pulses) pulse amplitude. *b*, Daily variation of peak pulse amplitude. (Ordinates are in units of W m^{-2} Hz^{-1} × 10^{-26}.)

yields no significant result. If the signals were linearly polarized, Faraday rotation in the ionosphere might cause the random variations, but the form of the curve does not seem compatible with this mechanism. The day to day variations since the source was first detected are shown in Fig. 2*b*. In this analysis the daily value plotted is the peak flux density of the greatest pulse. Again the variation from day to day is irregular and no systematic changes are clearly evident, although there is a suggestion that the source was significantly weaker during October to November. It therefore appears that, despite the regular occurrence of the pulses, the magnitude of the power emitted exhibits variations over long and short periods.

Instantaneous bandwidth and frequency drift

Two different experiments have shown that the pulses are caused by a narrow-band signal of descending frequency sweeping through the 1 MHz band of the receiver. In the first, two identical receivers were used, tuned to frequencies of 80.5 MHz and 81.5 MHz. Fig. 1*d*, which illustrates a record made with this system, shows that the lower frequency pulses are delayed by about 0.2 s. This corresponds to a frequency drift of ~ -5 MHz s^{-1}. In the second method a time delay was introduced into the signals reaching the receiver from one-half of the aerial by incorporating an extra cable of known length l. This cable introduces a phase shift proportional to frequency so that, for a signal the coherence length of which exceeds l, the output of the receiver will oscillate with period

$$t_0 = \frac{c}{l}\left(\frac{d\nu}{dt}\right)^{-1}$$

where $d\nu/dt$ is the rate of change of signal frequency. Records obtained with $l = 240$ m and 450 m are shown in Fig. 1*b* together with a simultaneous record of the pulses derived from a separate phase-switching receiver operating with equal cables in the usual fashion. Also shown, in Fig. 1*c*, is a simulated record obtained with exactly the same arrangement but using a signal generator, instead of the source, to provide the swept frequency. For observation with $l > 450$ m the periodic oscillations were slowed down to a low frequency by an additional phase shifting device in order to prevent severe attenuation of the output signal by the time constant of the receiver. The rate of change of signal frequency has been deduced from the additional phase shift required and is $d\nu/dt = -4.9 \pm 0.5$ MHz s^{-1}. The direction of the frequency drift can be obtained from

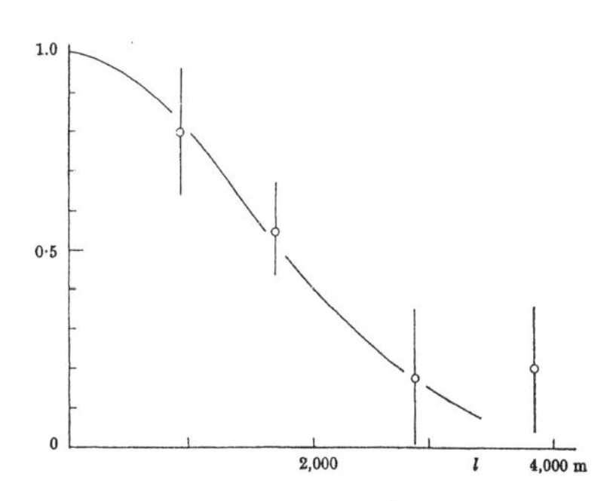

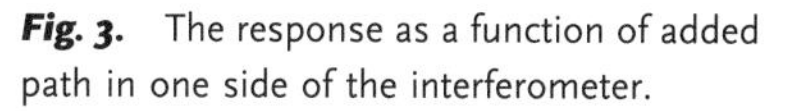
Fig. 3. The response as a function of added path in one side of the interferometer.

the phase of the oscillation on the record and is found to be from high to low frequency in agreement with the first result.

The instantaneous bandwidth of the signal may also be obtained from records of the type shown in Fig. 1*b* because the oscillatory response as a function of delay is a measure of the autocorrelation function, and hence of the Fourier transform, of the power spectrum of the radiation. The results of the measurements are displayed in Fig. 3 from which the instantaneous bandwidth of the signal to exp (-1), assuming a Gaussian energy spectrum, is estimated to be 80 ± 20 kHz.

Pulse recurrence frequency and Doppler shift

By displaying the pulses and time pips from *MSF* Rugby on the same record the leading edge of a pulse of reasonable size may be timed to an accuracy of about 0.1 s. Observations over a period of 6 h taken with the tracking system mentioned earlier gave the period between pulses as $P_{obs} = 1.33733 \pm 0.00001$ s. This represents a mean value centred on December 18, 1967, at 14 h 18 m UT. A study of the systematic shift in the frequency of the pulses was obtained from daily measurements of the time interval T between a standard time and the pulse immediately following it as shown in Fig. 4. The standard time was chosen to be 14 h 01 m 00 s UT on December 11 (corresponding to the centre of the reception pattern) and subsequent standard times were at intervals of 23 h 56 m 04 s (approximately one sidereal day). A plot of the variation of T from day to day is shown in Fig. 4. A constant pulse recurrence frequency would show a linear increase or decrease in T if care was taken to add or subtract one period where necessary. The observations, however, show a marked curvature in the sense of a steadily increasing frequency. If we assume a Doppler shift due to the Earth alone, then the number of pulses received per day is given by

$$N = N_0\left(1 + \frac{v}{c}\cos\varphi\sin\frac{2\pi n}{366.25}\right)$$

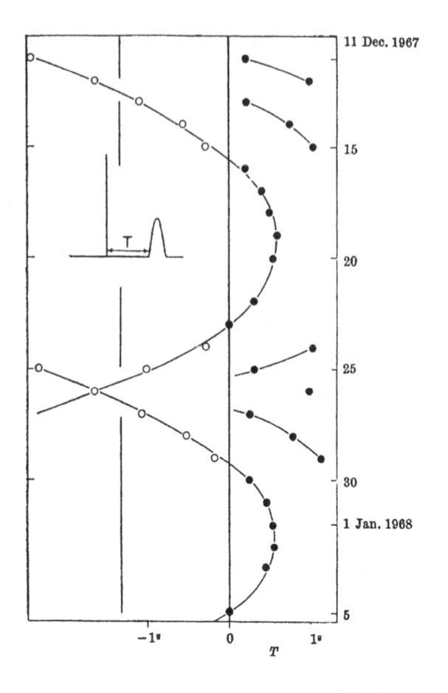

Fig. 4. The day to day variation of pulse arrival time.

where N_0 is the number of pulses emitted per day at the source, v the orbital velocity of the Earth, φ the ecliptic latitude of the source and n an arbitrary day number obtained by putting $n = 0$ on January 17, 1968, when the Earth has zero velocity along the line of sight to the source. This relation is approximate since it assumes a circular orbit for the Earth and the origin $n = 0$ is not exact, but it serves to show that the increase of N observed can be explained by the Earth's motion alone within the accuracy currently attainable. For this purpose it is convenient to estimate the values of n for which $\delta T/\delta n = 0$, corresponding to an exactly integral value of N. These occur at $n_1 = 15.8 \pm 0.1$ and $n_2 = 28.7 \pm 0.1$, and since N is increased by exactly one pulse between these dates we have

$$1 = \frac{N_0 v}{c} \cos \varphi \left[\sin \frac{2\pi n_2}{366.25} - \sin \frac{2\pi n_1}{366.25} \right]$$

This yields $\varphi = 43° \, 36' \pm 30'$ which corresponds to a declination of 21° 58′ ± 30′, a value consistent with the declination obtained directly. The true periodicity of the source, making allowance for the Doppler shift and using the integral condition to refine the calculation, is then

$$P_0 = 1.3372795 \pm 0.0000020 \text{ s}$$

By continuing observations of the time of occurrence of the pulses for a year it should be possible to establish the constancy of N_0 to about 1 part in 3×10^8. If N_0 is indeed constant, then the declination of the source may be estimated to an accuracy of ±1′; this result will not be affected by ionospheric refraction.

It is also interesting to note the possibility of detecting a variable Doppler shift caused by the motion of the source itself. Such an effect might arise if the source formed one component of a binary system, or if the signals were associated with a planet in orbit about some parent star. For the present, the systematic increase of N is regular to about 1 part in 2×10^7 so that there is no evidence for an additional orbital motion comparable with that of the Earth.

The nature of the radio source

The lack of any parallax greater than about 2′ places the source at a distance exceeding 10^3 A.U. The energy emitted by the source during a single pulse, integrated over 1 MHz at 81.5 MHz, therefore reaches a value which must exceed 10^{17} erg if the source radiates isotropically. It is also possible to derive an upper limit to the physical dimension of the source. The small instantaneous bandwidth of the signal (80 kHz) and the rate of sweep (-4.9 MHz s^{-1}) show that the duration of the emission at any given frequency does not exceed 0.016 s. The source size therefore cannot exceed 4.8×10^3 km.

An upper limit to the distance of the source may be derived from the observed rate of frequency sweep since impulsive radiation, whatever its origin, will be dispersed during its passage through the ionized hydrogen in interstellar space. For a uniform plasma the frequency drift caused by dispersion is given by

$$\frac{d\nu}{dt} = -\frac{c}{L}\frac{\nu^3}{\nu_p^2}$$

where L is the path and ν_p the plasma frequency. Assuming a mean density of 0.2 electron cm^{-3} the observed frequency drift (-4.9 MHz s^{-1}) corresponds to $L \sim 65$ parsec. Some frequency dispersion may, of course, arise in the source itself; in this case the dispersion in the interstellar medium must be smaller so that the value of L is an upper limit. While the interstellar electron density in the vicinity of the Sun is not well known, this result is important in showing that the pulsating radio sources so far detected must be local objects on a galactic distance scale.

The positional accuracy so far obtained does not permit any serious attempt at optical identification. The search area, which lies close to the galactic plane, includes two twelfth magnitude stars and a large number of weaker objects. In the absence of further data, only the most tentative suggestion to account for these remarkable sources can be made.

The most significant feature to be accounted for is the extreme regularity of the pulses. This suggests an origin in terms of the pulsation of an entire star, rather than some more localized disturbance in a stellar atmosphere. In this connexion it is interesting to note that it has already been suggested[2,3] that the radial pulsation of neutron stars may play an important part in the history of supernovae and supernova remnants.

A discussion of the normal modes of radial pulsation of compact stars has recently been given by Meltzer and Thorne[4], who calculated the periods for stars with central densities in the range 10^5 to 10^{19} g cm^{-3}. Fig. 4 of their paper indicates two possibilities which might account for the observed periods of the order 1 s. At a density of 10^7 g cm^{-3}, corresponding to a white dwarf star, the fundamental mode reaches a minimum period of about 8 s; at a slightly higher density the period increases again as the system tends towards gravitational collapse to a neutron star. While the fundamental period is not small enough to account for the observations the higher order modes have periods of the correct order of magnitude. If this model is adopted it is difficult to understand why the fundamental period is not dominant; such a period would have readily been detected in the present observations and its absence cannot be ascribed to observational effects. The alternative possibility occurs at a density of 10^{13} g cm^{-3}, corresponding to a neutron star; at this density the fundamental has a period of about 1 s, while for densities in excess of 10^{13} g cm^{-3} the period rapidly decreases to about 10^{-3} s.

If the radiation is to be associated with the radial pulsation of a white dwarf or neutron star there seem to be several mechanisms which could account for the radio emission. It has been suggested that radial pulsation would generate hydromagnetic shock fronts at the stellar surface which might be accompanied by bursts of X-rays and energetic electrons[2,3]. The radiation might then be likened to radio bursts from a solar flare occurring over the entire star during each cycle of the oscillation. Such a model would be in fair agreement with the upper limit of $\sim 5 \times 10^3$ km for the dimension of the source, which compares with the mean value of 9×10^3 km quoted for white dwarf stars by Greenstein[5]. The energy requirement for this model may be roughly estimated by noting that the total energy emitted in a 1 MHz band by a type III solar burst would produce a radio flux of the right order if the source were at a distance of $\sim 10^3$ A.U. If it is assumed that the radio energy may be related to the total flare energy ($\sim 10^{32}$ erg)[6] in the same manner as for a solar flare and supposing that each pulse corresponds to one flare, the required energy would be $\sim 10^{39}$ erg yr^{-1}; at a distance of 65 pc the corresponding value would be $\sim 10^{47}$ erg yr^{-1}. It has been estimated that a neutron star may contain $\sim 10^{51}$ erg in vibrational modes so the energy requirement does not appear unreasonable, although other damping mechanisms are likely to be important when considering the lifetime of the source[4].

The swept frequency characteristic of the radiation is reminiscent of type II and type III solar bursts, but it seems unlikely that it is caused in the same way. For a white dwarf or neutron star the scale height of any atmosphere is small and a travelling disturbance would be expected to produce a much faster frequency drift than is actually observed. As has been mentioned, a more likely possibility is that the impulsive radiation suffers dispersion during its passage through the interstellar medium.

More observational evidence is clearly needed in order to gain a better understanding of this strange new class of radio source. If the suggested origin of the radiation is confirmed further study may be expected to throw valuable light on the behaviour of compact stars and also on the properties of matter at high density.

We thank Professor Sir Martin Ryle, Dr J. E. Baldwin, Dr P. A. G. Scheuer and Dr J. R. Shakeshaft for helpful discussions and the Science Research Council who financed this work. One of us (S. J. B.) thanks the Ministry of Education of Northern Ireland and another (R. A. C.) the SRC for a maintenance award; J. D. H. P. thanks ICI for a research fellowship.

Mullard Radio Astronomy Observatory, Cavendish Laboratory, University of Cambridge
Received February 9, 1968 [Published 24 February]

References

1. Hewish, A., Scott, P. F., and Wills, D., *Nature,* **203**, 1214 (1964).
2. Cameron, A. G. W., *Nature,* **205**, 787 (1965).
3. Finzi, A., *Phys. Rev. Lett.*, **15**, 599 (1965).
4. Meltzer, D. W., and Thorne, K. S., *Ap. J.*, **145**, 514 (1966).
5. Greenstein, J. L., in *Handbuch der Physik, L.*, **161** (1958).
6. Fichtel, C. E., and McDonald, F. B., in *Annual Review of Astronomy and Astrophysics,* **5**, 351 (1967).

RNA-dependent DNA Polymerase

1970

wo ependent grou of in igators have found evic
ne in ons mour viruses which syn
DN om a NA template. This discovery, if upheld, w
cant cations not only for carcinogenesis by RNA
but also for the general understanding of genetic transc
apparently the classical process of information transfer fro
to RNA can be inverted.

pendent DNA Polymerase
ns of RNA Tumour Viruses

s to have a critical role in the multiplication and
ng ability of RNA tumour viruses[1]. Infection
ormation by these viruses can be prevented by
of DNA synthesis added during the first 8–12 h
ure of cells to the virus[1–4]. The necessary DNA
eems to involve the production of DNA which is
specific for the infecting virus[5,6], although
on studies intended to demonstrate virus-
A have been inconclusive[1]. Also, the formation
by the RNA tumour viruses is sensitive to
n D and therefore seems to involve DNA-
RNA synthesis[1–4,7]. One model which explains
postulates the transfer of the information of
ng RNA to a DNA copy which then serves as
or the synthesis of viral RNA[1,2,7]. This model
unique enzyme, an RNA-dependent DNA
e.
yme which synthesizes DNA from an RNA
has been found in any type of cell. Unless
nzyme exists in uninfected cells, the RNA
uses must either induce its synthesis soon after
r carry the enzyme into the cell as part of the
ecedents exist for the occurrence of nucleotide
es in the virions of animal viruses. Vaccinia[8,9]—
rus, Reo[10,11]—a double-stranded RNA virus,
lar stomatitis virus (VSV)[12]—a single-stranded
s, have all been shown to contain RNA poly-
This study demonstrates that an RNA-depen-
polymerase is present in the virions of two
our viruses: Rauscher mouse leukaemia virus
and Rous sarcoma virus. Temin[13] has also
this activity in Rous sarcoma virus.

tion of Radioactivity from ^{3}H-TTP by

aration of purified R-MLV was incubated in
of DNA polymerase assay. The preparation
ed radioactivity from ^{3}H-TTP into an acid-
product (Table 1). The reaction required
ough Mn^{2+} could partially substitute and each

course may indicate the occurrence of a slow a
of the polymerase in the reaction mixture. The
is approximately proportional to the amount o
virus.

For other viruses which have nucleotide poly
in their virions, there is little or no activity demo
unless the virions are activated by heat, pr
enzymes or detergents[8–12]. None of these tr
increased the activity of the R-MLV DNA poly
In fact, incubation at 50° C for 10 min totally ina
the R-MLV enzyme as did inclusion of trypsin (5
in the reaction mixture. Addition of as little
per cent 'Triton N-101' (a non-ionic deterge
markedly depressed activity.

Table 1. PROPERTIES OF THE RAUSCHER MOUSE LEUKAEMIA V POLYMERASE

Reaction system	pmoles ^{3}H-T incorporated in
Complete	3·31
Without magnesium acetate	0·04
Without magnesium acetate + 6 mM $MnCl_2$	1·59
Without dithiothreitol	0·38
Without NaCl	2·18
Without dATP	< 0·10
Without dCTP	0·12
Without dGTP	< 0·10

A preparation of R-MLV was provided by the Viral Resources
the National Cancer Institute. The virus had been purified from
of infected Swiss mice by differential centrifugation. The prepar
titre of $10^{4·85}$ spleen enlarging doses (50 per cent end point) per
use the preparation was centrifuged at 105,000*g* for 30 min and
was suspended in 0·137 M NaCl–0·003 M KCl–0·01 M phosp
(*p*H 7·4)–0·6 mM EDTA (PBS–EDTA) at 1/20 of the initial vo
concentrated virus suspension contained 3·1 mg/ml. of protein.
mixture contained, in 0·1 ml., 5 μmoles Tris-HCl (*p*H 8·3) at 37° C
magnesium acetate, 6 μmoles NaCl, 2 μmoles dithiothreitol, 0·08
of dATP, dCTP and dGTP, 0·001 μmole [^{3}H-*methyl*]-TTP (708
pmole) (New England Nuclear) and 15 μg viral protein. T
mixture was incubated for 45 min at 37° C. The acid-insoluble ra
in the sample was then determined by addition of sodium pyro
carrier yeast RNA and trichloroacetic acid followed by filtration
membrane filter and counting in a scintillation spectrometer, all as
described[15]. The radioactivity of an unincubated sample was
from each value (less than 7 per cent of the incorporation in th
reaction mixture).

Characterization of the Product

The nature of the reaction product was investi

TWELVE

Viruses reverse the genetic flow

Robin A. Weiss

In 1970 it emerged that certain viruses that have their genes in the form of RNA can copy the RNA "backward" into DNA in infected cells. This discovery has had an immense impact. The enzyme concerned, reverse transcriptase, makes possible the manufacture of specific proteins for use as medicines. It is also the best target for attacking HIV, the cause of AIDS.

With the rise of molecular biology in the 1950s and 1960s, it became widely believed that the flow of biological information goes in one direction only—that RNA is made according to information stored in the genetic material DNA, and that protein is then created from the instructions in the RNA. On 27 June 1970, two short papers published in *Nature* by David Baltimore[1] and by Howard Temin and Satoshi Mizutani[2] showed that the flow can be reversed. As with many key discoveries, the experiments were quite simple. The authors showed that purified particles of two cancer-causing (tumor) viruses—Rous sarcoma virus of chickens and Rauscher leukemia virus of mice—contain a reverse transcription enzyme activity that makes DNA using RNA as the template.

Living cells faithfully replicate their genetic material, the DNA, so that an exact copy is reproduced in each daughter cell (see p. 234). Replication is catalyzed by an enzyme, DNA polymerase. To read out the genetic information, another enzyme, called RNA polymerase or transcriptase, makes RNA that contains the same genetic sequence as one of the DNA strands. This "messenger" RNA is then translated into proteins, the molecules that form the structure of living cells and carry out their functions.

But the genetic material of certain viruses—very small parasites inside living cells—is in the form of RNA. This was first shown in 1955 for a virus of plants, tobacco mosaic virus, and it is true of many human viruses too. These viruses encode an RNA polymerase that can copy new RNA from an RNA template. Although this phenomenon is today restricted to viruses, life's very beginnings may have started with RNA rather than DNA—the so-called RNA world.

The discovery of reverse transcription showed that a different group of RNA viruses, RNA tumor viruses, can make a DNA copy of the viral RNA and incorporate that copy into the host's genome. The authors[1,2] called the relevant enzyme RNA-dependent DNA polymerase (the popular name for the enzyme, reverse transcriptase, was coined later). To complete replication, the DNA that has become incorporated into the host's genome is transcribed into messenger RNA using the enzymes of the infected cell. That RNA is then translated into viral proteins, again using the host-cell machinery, so creating a new generation of the virus (fig. 12.1). The RNA tumor viruses that use reverse transcriptase to make DNA were later renamed retroviruses, to denote the backward flow of genetic information.

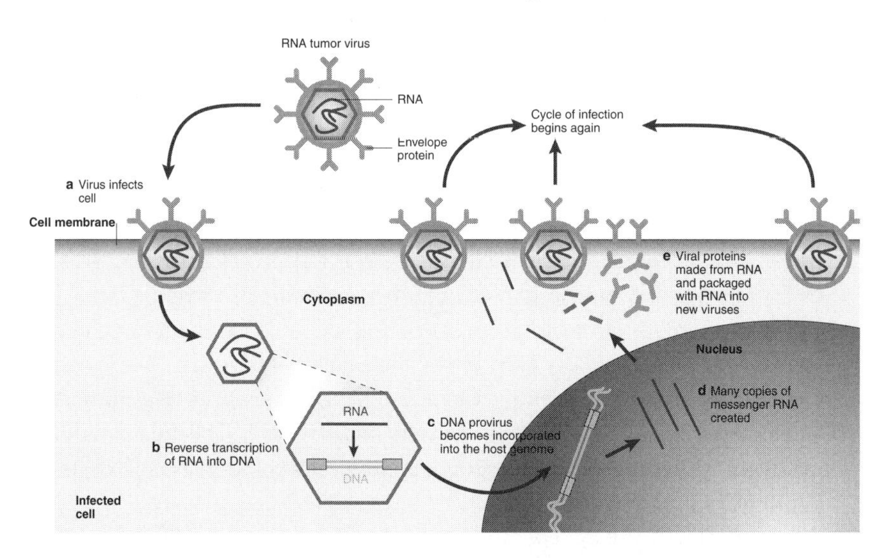

Fig. 12.1. Life cycle of an RNA tumor virus. a, The virus infects a cell by crossing the cell membrane, in the process losing its outer envelope. b, The viral RNA is reverse transcribed, using the virus's reverse transcriptase enzyme, into DNA. c, This DNA becomes incorporated into the host genome as a DNA "provirus." d, Many copies of messenger RNA are then made from the DNA. e, Using host-cell enzymes, the RNA is translated into the various protein components of the virus that, together with the RNA, are packaged into a new generation of viruses that bud off from the infected cell's membrane and begin the cycle of infection again. RNA tumor viruses are part of a broader group, the retroviruses (including HIV), that replicate in this way.

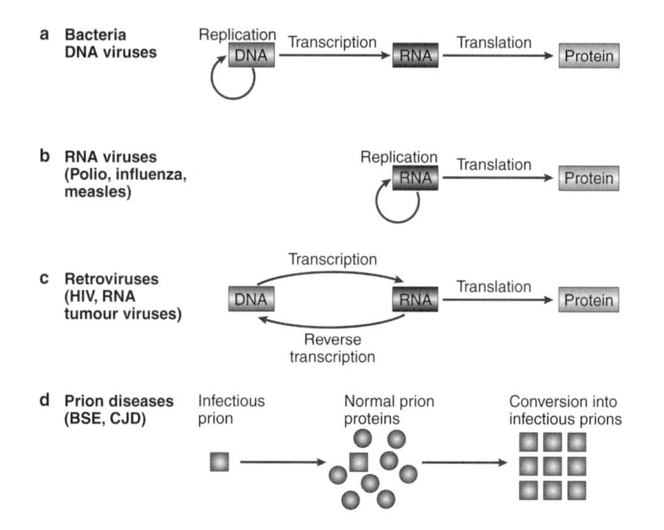

Fig. 12.2. How different infections propagate. a, DNA is replicated and transcribed into RNA, which is then translated into protein. Bacteria reproduce in this way, as do DNA viruses using the machinery of the host cell (herpes viruses for example). b, RNA is replicated by a virus-specific RNA replicase, and is translated into protein. This procedure is used by RNA viruses such as those causing polio, influenza, and measles. c, The viral RNA is reverse transcribed into DNA and incorporated into the host genome before being transcribed back into RNA for translation into protein (see fig. 12.1). This strategy is the one described in the 1970 *Nature* papers,[1,2] and is used by RNA tumor viruses and other retroviruses such as HIV. d, A more recent discovery is that of prion proteins. When an infectious prion is introduced into a population of normal proteins it can, in as yet unknown ways, cause their conversion into the infectious form. Examples of prion diseases are bovine spongiform encephalopathy (BSE) in cattle (or "mad cow disease") and Creutzfeldt-Jakob disease (CJD) in humans.

Fig. 12.2 shows how various infectious agents use these genetic mechanisms to propagate themselves. It includes the latest discovery in this area, that of certain proteins—called prions—that are now known to be infectious and cause disease.

So how did reverse transcriptase come to light? In the early 1960s, Temin proposed[3] that Rous sarcoma virus replicated by forming DNA, but his hypothesis was generally regarded with suspicion and even ridicule. Noting that the virus persists in infected cells even in the absence of viral replication, Temin also proposed that the DNA "provirus" becomes inserted into the host cell's DNA in the chromosomes. Temin came to this extraordinarily prescient conclusion from his observation that DNA inhibitors block an early step in infection by Rous sarcoma virus, an observation also made by John

Bader in 1965. Later, David Boettiger and Temin showed that viral DNA is the target for inhibition. This important piece of evidence was submitted for publication before the discovery of reverse transcriptase, but was not published until later,[4] as *Nature* was reluctant to accept the finding until reverse transcriptase made it plausible.

Support for Temin's provirus hypothesis came in 1969, when the Czech scientist Jan Svoboda showed that Rous sarcoma virus latent in rat tumor cells could be activated by fusing them with chicken cells. But it was not until 1972 that formal proof of the existence of the DNA provirus came, when Svoboda's former students, Miroslav Hill and Jana Hillova, extracted DNA from cells containing Rous sarcoma virus and recovered infectious virus from it.[5]

The discovery of reverse transcription immediately turned the heresy of the DNA provirus hypothesis into the new orthodoxy. Temin announced his discovery in May 1970 on the first day of the International Congress of Cancer held in Houston, Texas, in a session chaired by Svoboda. One anecdote holds that Sol Spiegelman of Columbia University immediately sought to confirm the findings and one month later was able to tell an audience at the Royal Society in London that Temin was indeed correct.

How did it come about that two laboratories independently demonstrated reverse transcription activity at exactly the same time? The key to the identification of the enzyme was to search for it in the virus particles themselves rather than in infected cells, an idea that occurred to both Baltimore and Temin early in 1970. Baltimore had recently shown that the RNA polymerase of vesicular stomatitis virus was contained in the virus particles, which, he soon pointed out in another classic paper,[6] must also be the case for other types of RNA viruses in which the viral genes cannot be used as messenger RNA. These findings made it logical to search for an RNA-dependent DNA polymerase in purified particles of RNA tumor viruses.

Temin and Baltimore made the same discovery at the same time and with essentially the same experimental approach, yet they did not regard it as a competitive race because they were unaware of each other's progress until their work was largely completed. Baltimore then telephoned Temin to tell him the exciting findings, only to learn that Temin already had the same results! The two authors came to this discovery from different starting points in their scientific outlooks. Temin had long sought evidence for his postulated DNA intermediate but had only lately come to search for the enzyme that made it. Baltimore, on the other hand, started with a biochemical approach based on his work with other RNA viruses and initially used RNA rather than DNA precursors, rather expecting to find an RNA polymerase. Looking for the enzyme in the virus particles rather than in the infected cell is what turned speculation into rapid experimental success for both of them.

The Nobel Prize in Physiology or Medicine was awarded to Baltimore and Temin in 1975, an unusually short incubation period. The third person who shared the prize was Renato Dulbecco of the Salk Institute, San Diego, in recognition of his pioneering work on poliovirus and DNA tumor viruses (both Baltimore and Temin, incidentally, had earlier worked in his lab). Perhaps the unlucky person here was the Japanese scientist Satoshi Mizutani. He was the postdoctoral scientist in Temin's laboratory who conducted the key enzymological experiments. Their manuscript was originally submitted to *Nature* with Mizutani as the first author. He told me: "*Nature* sent a letter to Dr. Temin saying that the discovery is important to Temin's provirus hypothesis and therefore the editor has the authority to change the order of authors. By the time he received the letter it was too late to do anything because the paper was already in print." Temin himself was always generous about Mizutani's crucial contribution.

As tiny parasites that propagate inside living cells, viruses have thrown light on many fundamental aspects of molecular biology. For example, viruses led not only to the discovery of reverse transcriptase, but also to that of oncogenes (genes that when activated lead to cancer) and of messenger RNA splicing (a process by which noncoding material is excised before protein production). In his Nobel lecture, Baltimore noted that "we derive much of our pleasure as biologists from the continuing realization of how economical, elegant and intelligent are the accidents of evolution that have been maintained by selection. A virologist is among the luckiest of biologists because he can see into his chosen pet down to the details of all of its molecules."

Even before the discovery of reverse transcriptase, Jim Payne and I had observed the genetic transmission in chickens of a virus related to Rous sarcoma virus; in Amsterdam, Peter Bentvelzen similarly showed the inheritance of mammary tumor virus in mice. Now we know that most vertebrate species harbor retroviral genomes that have become incorporated into their own genomes: a significant proportion of human DNA (about 8 percent) is derived by reverse transcriptase from retroviruses and from RNA called retrotransposons. It also became evident later that retroviruses are members of a broader group of viruses and genetic elements that replicate in this way. Human hepatitis B virus undergoes RNA-to-DNA reverse transcription during replication, as does cauliflower mosaic virus.

In 1983 came a further development. This was the discovery of the cause of AIDS, when Françoise Barré-Sinoussi and colleagues in Paris employed a reverse transcriptase assay on samples from patients to detect a virus, now called HIV, that destroyed human white blood cells. Antiviral drugs that inhibit reverse transcription are crucial for treating people infected with HIV. Thus the discovery of reverse transcriptase has been of great practical value. Without it we would not have anti-HIV drugs. Nor would we have sophisticated methods to monitor gene activity and to make proteins to use as drugs (interferon, for instance, which is used in the treatment of cancer and viral hepatitis), or the specially designed retroviruses that are used for gene delivery in gene therapy.

And what of the protagonists in this story? After receiving the Nobel Prize at the early age of thirty-seven, Baltimore continued to make major research contributions to biology, as well as to public policy on medical research and the search for an AIDS vaccine. He founded the Whitehead Institute at the Massachusetts Institute of Technology, and at the time of writing is president of the California Institute of Technology, in Pasadena. Despite his fame through the Nobel Prize, formally awarded on his forty-first birthday, Temin did not change his lifestyle at the University of Wisconsin, Madison, where he had worked since 1960. Sadly he died in February 1994 from a nonsmoker's lung cancer. A symposium intended to celebrate his sixtieth birthday instead became his epitaph.[7]

References

1. Baltimore, D. RNA-dependent DNA polymerase in virions of RNA tumour viruses. *Nature* **226,** 1209–1211 (1970).
2. Temin, H. M. & Mizutani, S. RNA-dependent DNA polymerase in virions of Rous sarcoma virus. *Nature* **226,** 1211–1213 (1970).
3. Temin, H. M. The effects of actinomycin D on growth of Rous sarcoma virus *in vitro*. *Virology* **20,** 577–582 (1963).
4. Boettiger, D. & Temin, H. M. Light inactivation of focus formation by chicken embryo fibroblasts infected with avian sarcoma virus in the presence of 5-bromodeoxyuridine. *Nature* **228,** 622–624 (1970).
5. Hill, M. & Hillova, J. Virus recovery in chicken cells tested with Rous sarcoma cell DNA. *Nature New Biol.* **237,** 35–39 (1972).
6. Baltimore, D. Expression of animal virus genomes. *Bacteriol. Rev.* **35,** 235–241 (1971).
7. Cooper, G. M., Temin, R. G. & Sugden, W. (eds.) *The DNA Provirus: Howard Temin's Scientific Legacy* (American Society for Microbiology, Washington, DC, 1995).

Further reading

Nobel e-Museum. *The Nobel Prize in Physiology or Medicine 1975* (http://www.nobel.se/medicine/laureates/1975/).

1970

RNA-dependent DNA polymerase in virions of RNA tumour viruses

David Baltimore

DNA seems to have a critical role in the multiplication and transforming ability of RNA tumour viruses[1]. Infection and transformation by these viruses can be prevented by inhibitors of DNA synthesis added during the first 8–12 h after exposure of cells to the virus[1–4]. The necessary DNA synthesis seems to involve the production of DNA which is genetically specific for the infecting virus[5,6], although hybridization studies intended to demonstrate virus-specific DNA have been inconclusive[1]. Also, the formation of virions by the RNA tumour viruses is sensitive to actinomycin D and therefore seems to involve DNA-dependent RNA synthesis[1–4,7]. One model which explains these data postulates the transfer of the information of the infecting RNA to a DNA copy which then serves as template for the synthesis of viral RNA[1,2,7]. This model requires a unique enzyme, an RNA-dependent DNA polymerase.

No enzyme which synthesizes DNA from an RNA template has been found in any type of cell. Unless such an enzyme exists in uninfected cells, the RNA tumour viruses must either induce its synthesis soon after infection or carry the enzyme into the cell as part of the virion. Precedents exist for the occurrence of nucleotide polymerases in the virions of animal viruses. Vaccinia[8,9]—a DNA virus, Reo[10,11]—a double-stranded RNA virus, and vesicular stomatitis virus (VSV)[12]—a single-stranded RNA virus, have all been shown to contain RNA polymerases. This study demonstrates that an RNA-dependent DNA polymerase is present in the virions of two RNA tumour viruses: Rauscher mouse leukaemia virus (R-MLV) and Rous sarcoma virus. Temin[13] has also identified this activity in Rous sarcoma virus.

Incorporation of radioactivity from ^{3}H-TTP by R-MLV

A preparation of purified R-MLV was incubated in conditions of DNA polymerase assay. The preparation incorporated radioactivity from ^{3}H-TTP into an acid-insoluble product (Table 1). The reaction required Mg^{2+}, although Mn^{2+} could partially substitute and each of the four deoxyribonucleoside triphosphates was necessary for activity. The reaction was stimulated strongly by dithiothreitol and weakly by NaCl (Table 1). The kinetics of incorporation of radioactivity from ^{3}H-TTP by R-MLV are shown in Fig. 1, curve 1. The reaction rate accelerates for about 1 h and then declines. This time-course may indicate the occurrence of a slow activation of the polymerase in the reaction mixture. The activity is approximately proportional to the amount of added virus.

For other viruses which have nucleotide polymerases in their virions, there is little or no activity demonstrable unless the virions are activated by heat, proteolytic enzymes or detergents[8–12]. None of these treatments increased the activity of the R-MLV DNA polymerase. In fact, incubation at 50° C for 10 min totally inactivated the R-MLV enzyme as did inclusion of trypsin (50 μg/ml.) in the reaction mixture. Addition of as little as 0.01

Table 1 Properties of the Rauscher mouse leukaemia virus DNA polymerase

Reaction system	pmoles ^{3}H-TMP incorporated in 45 min
Complete	3.31
Without magnesium acetate	0.04
Without magnesium acetate + 6 mM $MnCl_2$	1.59
Without dithiothreitol	0.38
Without NaCl	2.18
Without dATP	<0.10
Without dCTP	0.12
Without dGTP	<0.10

A preparation of R-MLV was provided by the Viral Resources Program of the National Cancer Institute. The virus had been purified from the plasma of infected Swiss mice by differential centrifugation. The preparation had a titre of $10^{4.88}$ spleen enlarging doses (50 per cent end point) per ml. Before use the preparation was centrifuged at 105,000*g* for 30 min and the pellet was suspended in 0.137 M NaCl–0.003 M KCl–0.01 M phosphate buffer (*p*H 7.4)–0.6 mM EDTA (PBS–EDTA) at 1/20 of the initial volume. The concentrated virus suspension contained 3.1 mg/ml. of protein. The assay mixture contained, in 0.1 ml., 5 μmoles Tris-HCl (*p*H 8.3) at 37° C, 0.6 μmole magnesium acetate, 6 μmoles NaCl, 2 μmoles dithiothreitol, 0.08 μmole each of dATP, dCTP and dGTP, 0.001 μmole [^{3}H-*methyl*]–TTP (708 c.p.m. per pmole) (New England Nuclear) and 15 μg viral protein. The reaction mixture was incubated for 45 min at 37° C. The acid-insoluble radioactivity in the sample was then determined by addition of sodium pyrophosphate, carrier yeast RNA and trichloroacetic acid followed by filtration through a membrane filter and counting in a scintillation spectrometer, all as previously described[12]. The radioactivity of an unincubated sample was subtracted from each value (less than 7 per cent of the incorporation in the complete reaction mixture).

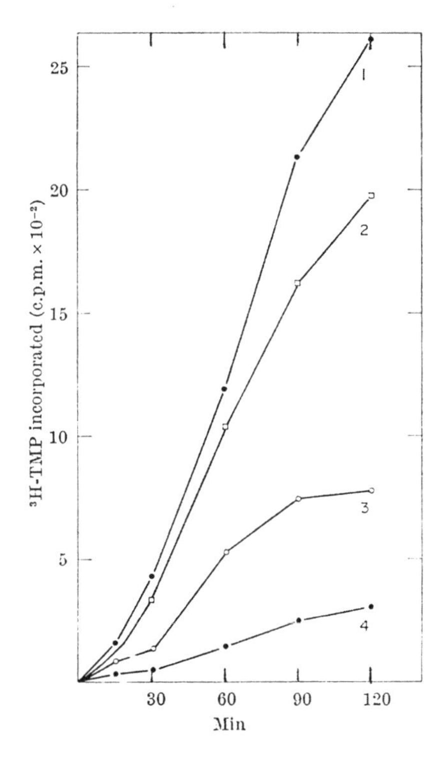

Fig. 1. Incorporation of radioactivity from ^{3}H-TTP by the R-MLV DNA polymerase in the presence and absence of ribonuclease. A 1.5-fold standard reaction mixture was prepared with 30 μg of viral protein and ^{3}H–TTP (specific activity 950 c.p.m. per pmole). At various times, 20 μl. aliquots were added to 0.5 ml. of non-radioactive 0.1 M sodium pyrophosphate and acid insoluble radioactivity was determined[12]. For the preincubated samples, 0.06 ml. of H_2O and 0.01 ml. of R-MLV (30 μg of protein) were incubated with or without 10 μg of pancreatic ribonuclease at 22° C for 20 min, chilled and brought to 0.15 ml. with a concentrated mixture of the components of the assay system. Curve 1, no treatment; curve 2, preincubated; curve 3, 10 μg ribonuclease added to the reaction mixture; curve 4, preincubated with 10 μg ribonuclease.

per cent 'Triton N-101' (a non-ionic detergent) also markedly depressed activity.

Characterization of the product

The nature of the reaction product was investigated by determining its sensitivity to various treatments. The product could be rendered acid-soluble by either pancreatic deoxyribonuclease or micrococcal nuclease but was unaffected by pancreatic ribonuclease or by alkaline hydrolysis

Table 2 Characterization of the polymerase product

Expt.	Treatment	Acid-insoluble radioactivity	Percentage undigested product
1	Untreated	1,425	(100)
	20 μg deoxyribonuclease	125	9
	20 μg micrococcal nuclease	69	5
	20 μg ribonuclease	1,361	96
2	Untreated	1,644	(100)
	NaOH hydrolysed	1,684	100

For experiment 1, 93 μg cf viral protein was incubated for 2 h in a reaction mixture twice the size of that described in Table 1, with ^{3}H-TTP having a specific activity of 1,133 c.p.m. per pmole. A 50 μl. portion of the reaction mixture was diluted to 5 ml. with 10 mM $MgCl_2$ and 0.5 ml. aliquots were incubated for 1.5 h at 37° C with the indicated enzymes. (The sample with micrococcal nuclease also contained 5 mM $CaCl_2$.) The samples were then chilled, precipitated with trichloroacetic acid and radioactivity was counted. For experiment 2, two standard reaction mixtures were incubated for 45 min at 37° C, then to one sample was added 0.1 ml. of 1 M NaOH and it was boiled for 5 min. It was then chilled and both samples were precipitated with trichloroacetic acid and counted. In a separate experiment (unpublished) it was shown that the alkaline hydrolysis conditions would completely degrade the RNA product of the VSV virion polymerase.

(Table 2). The product therefore has the properties of DNA. If 50 μg/ml. of deoxyribonuclease was added to a reaction mixture there was no loss of acid-insoluble product. The product is therefore protected from the enzymc, probably by the envelope of the virion, although merely diluting the reaction mixture into 10 mM $MgCl_2$ enables the product to be digested by deoxyribonuclease (Table 2).

Localization of the enzyme and its template

To investigate whether the DNA polymerase and its template were associated with the virions, a R-MLV suspension was centrifuged to equilibrium in a 15–50 per cent sucrose gradient and fractions of the gradient were assayed for DNA polymerase activity. Most of the activity was found at the position of the visible band of virions (Fig. 2). The density at this band was 1.16 g/cm^3, in agreement with the known density of the virions[14]. The polymerase and its template therefore seem to be constituents of the virion.

The template is RNA

Virions of the RNA tumour viruses contain RNA but no DNA[15,16]. The template for the virion DNA polymerase is therefore probably the viral

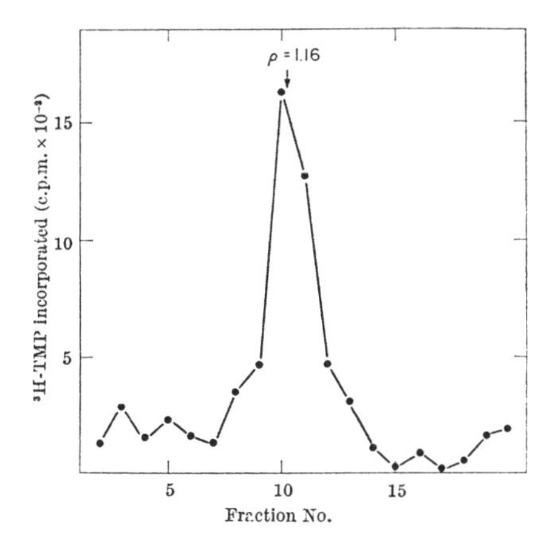

Fig. 2. Localization of DNA polymerase activity in R-MLV by isopycnic centrifugation. A preparation of R-MLV containing 150 μg of protein in 50 μl. was layered over a linear 5.2 ml. gradient of 15–50 per cent sucrose in PBS–EDTA. After centrifugation for 2 h at 60,000 r.p.m. in the Spinco 'SW65' rotor, 0.27 ml. fractions of the gradient were collected and 0.1 ml. portions of each fraction were incubated for 60 min in a standard reaction mixture. The acid-precipitable radioactivity was then collected and counted. The density of each fraction was determined from its refractive index. The arrow indicates the position of a sharp, visible band of light-scattering material which occurred at a density of 1.16.

RNA. To substantiate further that RNA is the template, the effect of ribonuclease on the reaction was investigated. When 50 μg/ml. of pancreatic ribonuclease was included in the reaction mixture, there was a 50 per cent inhibition of activity during the first hour and more than 80 per cent inhibition during the second hour of incubation (Fig. 1, curve 3). If the virions were preincubated with the enzyme in water at 22° C and the components of the reaction mixture were then added, an earlier and more extensive inhibition was evident (Fig. 1, curve 4). Preincubation in water without ribonuclease caused only a slight inactivation of the virion polymerase activity (Fig. 1, curve 2). Increasing the concentration of ribonuclease during preincubation could inhibit more than 95 per cent of the DNA polymerase activity (Table 3). To ensure that the inhibition by ribonuclease was attributable to the enzymic activity of the added protein, two other basic proteins were preincubated with the virions. Only ribonuclease was able to inhibit the reaction (Table 3). These experiments substantiate the idea that RNA is the template for the reaction. Hybridization experiments are in progress to determine if the DNA is complementary in base sequence to the viral RNA.

Table 3 Effect of ribonuclease on the DNA polymerase activity of Rauscher mouse leukaemia virus

Conditions	pmoles ^{3}H-TMP incorporation
No preincubation	2.50
Preincubated with no addition	2.20
Preincubated with 20 μg/ml. ribonuclease	0.69
Preincubated with 50 μg/ml. ribonuclease	0.31
Preincubated with 200 μg/ml. ribonuclease	0.08
Preincubated with no addition	3.69
Preincubated with 50 μg/ml. ribonuclease	0.52
Preincubated with 50 μg/ml. lysozyme	3.67
Preincubated with 50 μg/ml. cytochrome *c*	3.97

In experiment 1, for the preincubation, 15 μg of viral protein in 5 μl. of solution was added to 45 μl. of water at 4° C containing the indicated amounts of enzyme. After incubation for 30 min at 22° C, the samples were chilled and 50 μl. of a 2-fold concentrated standard reaction mixture was added. The samples were then incubated at 37° C for 45 min and acid-insoluble radioactivity was measured. In experiment 2, the same procedure was followed, except that the preincubation was for 20 min at 22° C and the 37° C incubation was for 60 min.

Ability of the enzyme to incorporate ribonucleotides

The deoxyribonucleotide incorporation measured in these experiments could be the result of an RNA polymerase activity in the virion which can polymerize deoxyribonucleotides when they are provided in the reaction mixture. The VSV RNA polymerase and the R-MLV DNA polymerase were therefore compared. The VSV RNA polymerase incorporated only ribonucleotides. At its *p*H optimum of 7.3 (my unpublished observation), in the presence of the four common ribonucleoside triphosphates, the enzyme incorporated ^{3}H-GMP extensively[12]. At this *p*H, however, in the presence of the four deoxyribonucleoside triphosphates, no ^{3}H-TMP incorporation was demonstrable (Table 4). Furthermore, replacement of even a single ribonucleotide by its homologous deoxyribonucleotide led to no detectable synthesis (my unpublished observation). At *p*H 8.3, the optimum for the R-MLV DNA polymerase, the VSV polymerase catalysed much less ribonucleotide incorporation and no significant deoxyribonucleotide incorporation could be detected.

The R-MLV polymerase incorporated only deoxyribonucleotides. At *p*H 8.3, ^{3}H-TMP incorporation was readily demonstrable but replacement of dATP by ATP completely prevented synthesis (Table 4). Furthermore, no significant incorporation of ^{3}H-GMP could be found in the presence of the four ribonucleotides. At *p*H 7.3, the R-MLV polymerase was also

Table 4 Comparison of nucleotide incorporation by vesicular stomatitis virus and Rauscher mouse leukaemia virus

		Incorporation in 45 min (pmoles)	
Precursor	*p*H	Vesicular stomatitis virus	Mouse leukaemia virus
^{3}H-TTP	8.3	<0.01	2.3
^{3}H-TTP (omit dATP)	8.3	N.D.	0.06
^{3}H-TTP (omit dATP; plus ATP)	8.3	N.D.	0.08
^{3}H-GTP	8.3	0.43	<0.03
^{3}H-GTP	7.3	3.7	<0.03

When ^{3}H-TTP was the precursor, standard reaction conditions were used (see Table 1). When ^{3}H-GTP was the precursor, the reaction mixture contained, in 0.1 ml., 5 μmoles Tris-HCl (*p*H as indicated), 0.6 μmoles magnesium acetate, 0.3 μmoles mercaptoethanol, 9 μmoles NaCl, 0.08 μmole each of ATP, CTP, UTP; and 0.001 μmole ^{3}H-GTP (1,040 c.p.m. per pmole). All VSV assays included 0.1 per cent 'Triton N-101' (ref. 12) and 2–5 μg of viral protein. The R-MLV assays contained 15 μg of viral protein.

inactive with ribonucleotides. The polymerase in the R-MLV virions is therefore highly specific for deoxyribonucleotides.

DNA polymerase in Rous sarcoma virus

A preparation of the Prague strain of Rous sarcoma virus was assayed for DNA polymerase activity (Table 5). Incorporation of radioactivity from ^{3}H-TTP was demonstrable and the activity was severely reduced by omission of either Mg^{2+} or dATP from the reaction mixture. RNA-dependent DNA polymerase is therefore probably a constituent of all RNA tumour viruses.

Table 5 Properties of the Rous sarcoma virus DNA polymerase

Reaction system	pmoles ^{3}H-TMP incorporated in 120 min
Complete	2.06
Without magnesium acetate	0.12
Without dATP	0.19

A preparation of the Prague strain (sub-group C) of Rous sarcoma virus[16] having a titre of 5×10^7 focus forming units per ml. was provided by Dr Peter Vogt. The virus was purified from tissue culture fluid by differential centrifugation. Before use the preparation was centrifuged and the pellet dissolved in 1/10 of the initial volume as described for the R-MLV preparation. For each assay 15 μl. of the concentrated Rous sarcoma virus preparation was assayed in a standard reaction mixture by incubation for 2 h. An unincubated control sample had radioactivity corresponding to 0.14 pmole which was subtracted from the experimental values.

These experiments indicate that the virions of Rauscher mouse leukaemia virus and Rous sarcoma virus contain a DNA polymerase. The inhibition of its activity by ribonuclease suggests that the enzyme is an RNA-dependent DNA polymerase. It seems probable that all RNA tumour viruses have such an activity. The existence of this enzyme strongly supports the earlier suggestions[1–7] that genetically specific DNA synthesis is an early event in the replication cycle of the RNA tumour viruses and that DNA is the template for viral RNA synthesis. Whether the viral DNA ("provirus")[2] is integrated into the host genome or remains as a free template for RNA synthesis will require further study. It will also be necessary to determine whether the host DNA-dependent RNA polymerase or a virus-specific enzyme catalyses the synthesis of viral RNA from the DNA.

I thank Drs G. Todaro, F. Rauscher and R. Holdenreid for their assistance in providing the mouse leukaemia virus. This work was supported by grants from the US Public Health Service and the American Cancer Society and was carried out during the tenure of an American Society Faculty Research Award.

Department of Biology, Massachusetts Institute of Technology, Cambridge, Massachusetts 02139

Received June 2, 1970 [Published 27 June]

References

1. Green, M., *Ann. Rev. Biochem.*, **39** (1970, in the press).
2. Temin, H. M., *Virology*, **23**, 486 (1964).
3. Bader, J. P., *Virology*, **22**, 462 (1964).
4. Vigler, P., and Golde, A., *Virology*, **23**, 511 (1964).
5. Duesberg, P. H., and Vogt, P. K., *Proc. US Nat. Acad. Sci.*, **64**, 939 (1969).
6. Temin. H. M., in *Biology of Large RNA Viruses* (edit. by Barry, R., and Mahy, B.) (Academic Press, London, 1970).
7. Temin, H. M., *Virology*, **20**, 577 (1963).
8. Kates, J. R., and McAuslan, B. R., *Proc. US Nat. Acad. Sci.*, **58**, 134 (1967).
9. Munyon, W., Paoletti, E., and Grace, J. T. J., *Proc. US Nat. Acad. Sci.*, **58**, 2280 (1967).
10. Shatkin, A. J., and Sipe, J. D., *Proc. US Nat. Acad. Sci.*, **61**, 1462 (1968).
11. Borsa, J., and Graham, A. F., *Biochem. Biophys. Res. Commun.*, **33**, 895 (1968).
12. Baltimore, D., Huang, A. S., and Stampfer, M., *Proc. US Nat. Acad. Sci.* **66** (1970, in the press).
13. Temin, H. M., and Mizutani, S., *Nature*, **226**, 1211 (1970) (following article).
14. O'Conner, T. E., Rauscher, F. J., and Zeigel, R. F., *Science*, **144**, 1144 (1964).
15. Crawford, L. V., and Crawford, E. M., *Virology*, **13**, 227 (1961).
16. Duesberg, P., and Robinson, W. S., *Proc. US Nat. Acad. Sci.*, **55**, 219 (1966).
17. Duff, R. G., and Vogt, P. K., *Virology*, **39**, 18 (1969).

RNA-dependent DNA polymerase in virions of Rous sarcoma virus

Howard M. Temin and Satoshi Mizutani

Infection of sensitive cells by RNA sarcoma viruses requires the synthesis of new DNA different from that synthesized in the *S*-phase of the cell cycle (refs. 1, 2 and unpublished results of D. Boettiger and H. M. T.); production of RNA tumour viruses is sensitive to actinomycin D[3,4]; and cells transformed by RNA tumour viruses have new DNA which hybridizes with viral RNA[5,6]. These are the basic observations essential to the DNA provirus hypothesis—replication of RNA tumour viruses takes place through a DNA intermediate, not through an RNA intermediate as does the replication of other RNA viruses[7].

Formation of the provirus is normal in stationary chicken cells exposed to Rous sarcoma virus (RSV), even in the presence of 0.5 μg/ml. cycloheximide (our unpublished results). This finding, together with the discovery of polymerases in virions of vaccinia virus and of reovirus[8–11], suggested that an enzyme that would synthesize DNA from an RNA template might be present in virions of RSV. We now report data supporting the existence of such an enzyme, and we learn that David Baltimore has independently discovered a similar enzyme in virions of Rauscher leukaemia virus[12].

The sources of virus and methods of concentration have been described[13]. All preparations were carried out in sterile conditions. Concentrated virus was placed on a layer of 15 per cent sucrose and centrifuged at 25,000 r.p.m. for 1 h in the 'SW 25.1' rotor of the Spinco ultra-centrifuge on to a cushion of 60 per cent sucrose. The virus band was collected from the interphase and further purified by equilibrium sucrose density gradient centrifugation[14]. Virus further purified by sucrose velocity density gradient centrifugation gave the same results.

The polymerase assay consisted of 0.125 μmoles each of dATP, dCTP, and dGTP (Calbiochem) (in 0.02 M Tris-HCl buffer at *p*H 8.0, containing 0.33 M EDTA and 1.7 mM 2-mercaptoethanol); 1.25 μmoles of $MgCl_2$ and 2.5 μmoles of KCl; 2.5 μg phosphoenolpyruvate (Calbiochem); 10 μg pyruvate kinase (Calbiochem); 2.5 μCi of ^{3}H-TTP (Schwarz) (12 Ci/mmole); and 0.025 ml. of enzyme (10^8 focus forming units of disrupted Schmidt-Ruppin virus, $A_{280\ \mathrm{nm}}$ = 0.30) in a total volume of 0.125 ml. Incubation was at 40° C for 1 h. 0.025 ml. of the reaction mixture was withdrawn and assayed for acid-insoluble counts by the method of Furlong[15].

Table 1 Activation of enzyme

System	^{3}H-TTP incorporated (d.p.m.)
No virions	0
Non-disrupted virions	255
Virions disrupted with 'Nonidet'	
At 0° + DTT	6,730
At 0° − DTT	4,420
At 40° + DTT	5,000
At 40° − DTT	425

Purified virions untreated or incubated for 5 min at 0° C or 40° C with 0.25 per cent 'Nonidet P–40' (Shell Chemical Co.) with 0 or 1 per cent dithiothreitol (DTT) (Sigma) were assayed in the standard polymerase assay.

To observe full activity of the enzyme, it was necessary to treat the virions with a non-ionic detergent (Tables 1 and 4). If the treatment was at 40° C the presence of dithiothreitol (DTT) was necessary to recover activity. In most preparations of virions, however, there was some activity: 5–20 per cent of the disrupted virions, in the absence of detergent treatment, which probably represents disrupted virions in the preparation. It is known that virions of RNA tumour viruses are easily disrupted[16,17], so that the activity is probably present in the nucleoid of the virion.

The kinetics of incorporation with disrupted virions are shown in Fig. 1. Incorporation is rapid for 1 h. Other experiments show that incorporation continues at about the same rate for the second hour. Preheating disrupted virus at 80° C prevents any incorporation, and so does pretreatment of disrupted virus with crystalline trypsin.

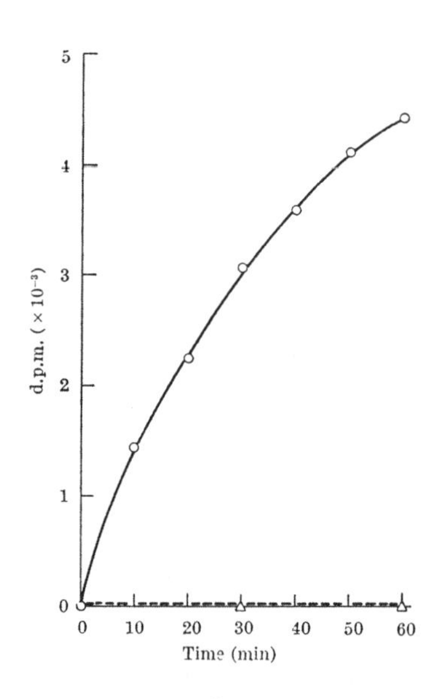

Fig. 1. Kinetics of incorporation. Virus treated with 'Nonidet' and dithiothreitol at 0° C and incubated at 37° C (○—○) or 80° C (△---△) for 10 min was assayed in a standard polymerase assay. ○, Unheated; △, heated.

Table 2 Requirements for enzyme activity

System	^{3}H-TTP incorporated (d.p.m.)
Complete	5,675
Without $MgCl_2$	186
Without $MgCl_2$, with $MnCl_2$	5,570
Without $MgCl_2$, with $CaCl_2$	18
Without dATP	897
Without dCTP	1,780
Without dGTP	2,190

Virus treated with 'Nonidet' and dithiothreitol at 0° C was incubated in the standard polymerase assay with the substitutions listed.

Fig. 2 demonstrates that there is an absolute requirement for $MgCl_2$, 10 mM being the optimum concentration. The data in Table 2 show that $MnCl_2$ can substitute for $MgCl_2$ in the polymerase assay, but $CaCl_2$ cannot. Other experiments show that a monovalent cation is not required for activity, although 20 mM KCl causes a 15 per cent stimulation. Higher concentrations of KCl are inhibitory: 60 per cent inhibition was observed at 80 mM.

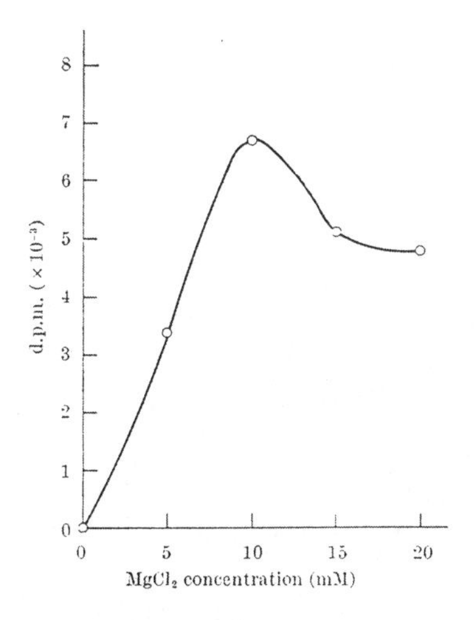

Fig. 2. $MgCl_2$ requirement. Virus treated with 'Nonidet' and dithiothreitol at 0° C was incubated in the standard polymerase assay with different concentrations of $MgCl_2$.

When the amount of disrupted virions present in the polymerase assay was varied, the amount of incorporation varied with second-order kinetics. When incubation was carried out at different temperatures, a broad optimum between 40° C and 50° C was found. (The high temperature of this optimum may relate to the fact that the normal host of the virus is the chicken.) When incubation was carried out at different *p*Hs, a broad optimum at *p*H 8–9.5 was found.

Table 2 demonstrates that all four deoxyribonucleotide triphosphates are required for full activity, but some activity was present when

Table 3 RNA dependence of polymerase activity

Treatment	^{3}H-TTP incorporated (d.p.m.)
Non-treated disrupted virions	9,110
Disrupted virions preincubated with ribonuclease A (50 μg/ml.) at 20° C for 1 h	2,650
Disrupted virions preincubated with ribonuclease A (1 mg/ml.) at 0° C for 1 h	137
Disrupted virions preincubated with lysozyme (50 μg/ml.) at 0° C for 1 h	9,650

Disrupted virions were incubated with ribonuclease A (Worthington) which was heated at 80° C for 10 min, or with lysozyme at the indicated concentration in the specified conditions, and a standard polymerase assay was performed.

only three deoxyribonucleotide triphosphates were added and 10–20 per cent of full activity was still present with only two deoxyribonucleotide triphosphates. The activity in the presence of three deoxyribonucleotide triphosphates is probably the result of the presence of deoxyribonucleotide triphosphates in the virion. Other host components are known to be incorporated in the virion of RNA tumour viruses[18,19].

The data in Table 3 demonstrate that incorporation of thymidine triphosphate was more than 99 per cent abolished if the virions were pretreated at 0° with 1 mg ribonuclease per ml. Treatment with 50 μg/ml. ribonuclease at 20° C did not prevent all incorporation of thymidine triphosphate, which suggests that the RNA of the virion may be masked by protein. (Lysozyme was added as a control for non-specific binding of ribonuclease to DNA.) Because the ribonuclease was heated for 10 min at 80° C or 100° C before use to destroy deoxyribonuclease it seems that intact RNA is necessary for incorporation of thymidine triphosphate.

To determine whether the enzyme is present in supernatants of normal cells or in RNA leukaemia viruses, the experiment of Table 4 was performed. Normal cell supernatant did not contain activity even after treatment with 'Nonidet'. Virions of avian myeloblastosis virus (AMV) contained activity that was increased ten-fold by treatment with 'Nonidet'.

The nature of the product of the polymerase assay was investigated by treating portions with deoxyribonuclease, ribonuclease or KOH. About 80 per cent of the product was made acid soluble by treatment with deoxyribonuclease, and the product was resistant to ribonuclease and KOH (Table 5).

To determine if the polymerase might also make RNA, disrupted virions were incubated with the four ribonucleotide triphosphates, including ^{3}H-UTP (Schwarz, 3.2 Ci/mmole). With either $MgCl_2$ or $MnCl_2$ in the incubation mixture, no incorporation was detected. In a parallel incuba-

Table 4 Source of polymerase

Source	^{3}H-TTP incorporated (d.p.m.)
Virions of SRV	1,410
Disrupted virions of SRV	5,675
Virions of AMV	1,875
Disrupted virions of AMV	12,850
Disrupted pellet from supernatant of uninfected cells	0

Virions of Schmidt-Ruppin virus (SRV) were prepared as before (experiment of Table 2). Virions of avian myeloblastosis virus (AMV) and a pellet from uninfected cells were prepared by differential centrifugation. All disrupted preparations were treated with 'Nonidet' and dithiothreitol at 0° C and assayed in a standard polymerase assay. The material used per tube was originally from 45 ml. of culture fluid for SRV, 20 ml. for AMV, and 20 ml. for uninfected cells.

tion with deoxyribonucleotide triphosphates, 12,200 d.p.m. of ^{3}H-TTP was incorporated.

These results demonstrate that there is a new polymerase inside the virions of RNA tumour viruses. It is not present in supernatants of normal cells but is present in virions of avian sarcoma and leukaemia RNA tumour viruses. The polymerase seems to catalyse the incorporation of deoxyribonucleotide triphosphates into DNA from an RNA template. Work is being performed to characterize further the reaction and the product. If the present results and Baltimore's results[12] with Rauscher leukaemia virus are upheld, they will constitute strong evidence that the DNA provirus hypothesis is correct and that RNA tumour viruses have a DNA genome when they are in cells and an RNA genome when they are in virions. This result would have strong implications for theories of viral carcinogenesis and, possibly, for theories of information transfer in other biological systems[20].

Table 5 Nature of product

Treatment	Residual acid-insoluble ^{3}H-TTP (d.p.m.)	
	Experiment A	Experiment B
Buffer	10,200	8,350
Deoxyribonuclease	697	1,520
Ribonuclease	10,900	7,200
KOH	—	8,250

A standard polymerase assay was performed with 'Nonidet' treated virions. The product was incubated in buffer or 0.3 M KOH at 37° C for 20 h or with (*A*) 1 mg/ml. or (*B*) 50 μg/ml. of deoxyribonuclease I (Worthington), or with 1 mg/ml. of ribonuclease A (Worthington) for 1 h at 37° C, and portions were removed and tested for acid-insoluble counts.

This work was supported by a US Public Health Service research grant from the National Cancer Institute. H. M. T. holds a research career development award from the National Cancer Institute.

McArdle Laboratory for Cancer Research, University of Wisconsin, Madison, Wisconsin 53706

Received June 15, 1970 [Published 27 June]

References

1. Temin, H. M., *Cancer Res.*, **28**, 1835 (1968).
2. Murray, R. K., and Temin, H. M., *Intern. J. Cancer* (in the press).
3. Temin, H. M., *Virology*, **20**, 577 (1963).
4. Baluda, M. B., and Nayak, D. P., *J. Virol.*, **4**, 554 (1969).
5. Temin, H. M., *Proc. US Nat. Acad. Sci.*, **52**, 323 (1964).
6. Baluda, M. B., and Nayak, D. P., in *Biology of Large RNA Viruses* (edit. by Barry, R., and Mahy, B.) (Academic Press, London, 1970).
7. Temin, H. M., *Nat. Cancer Inst. Monog.*, **17**, 557 (1964).
8. Kates, J. R., and McAuslan, B. R., *Proc. US Nat. Acad. Sci.*, **57**, 314 (1967).
9. Munyon, W., Paoletti, E., and Grace, J. T., *Proc. US Nat. Acad. Sci.*, **58**, 2280 (1967).
10. Borsa, J., and Graham, A. F., *Biochem. Biophys. Res. Commun.*, **33**, 895 (1968).
11. Shatkin, A. J., and Sipe, J. D., *Proc. US Nat. Acad. Sci.*, **61**, 1462 (1968).
12. Baltimore, D., *Nature*, **226**, 1209 (1970) (preceding article).
13. Altaner, C., and Temin, H. M., *Virology*, **40**, 118 (1970).
14. Robinson, W. S., Pitkanen, A., and Rubin, H., *Proc. US Nat. Acad. Sci.*, **54**, 137 (1965).
15. Furlong, N. B., *Meth. Cancer Res.*, **3**, 27 (1967).
16. Vogt, P. K., *Adv. Virus. Res.*, 11, 293 (1965).
17. Bauer, H., and Schafer, W., *Virology*, **29**, 494 (1966).
18. Bauer, H., *Z. Naturforsch.*, 21b, 453 (1966).
19. Erikson, R. L., *Virology*, **37**, 124 (1969).
20. Temin, H. M., *Persp. Biol. Med.* (in the press).

s, and microaphanitic, and further on a geo-
o micrgriranular, microlamellar, and micro-
recently, Kloss[12] reported that, unlike macro-
tz crystals, microcrystalline quartz crystals
no sharp inversion point, and the inversion
an interval of nearly 50° C. Our work suggests
describe the product of powdering quartz as
quartz, which has an X-ray structure corre-
rtz but is not normally detectable by d.t.a.
ut at high heating rat . In additi this
quartz contains chemisor $_2$O
fessor R. A. Howie, Moret oc . N
Miss P. S. Osborn, and th en

G. S. M. Mo
H. E. Ro

Laboratory,
Mechanical Engineering,
Strand,

10, 1972.

., and Ritchie, P. D., *J. Appl. Chem.*, **3**, 182 (1953).
F., *J. Amer. Chem. Soc.*, **30**, 1120 (1908).
and Tuttle, O. F., *Amer. J. Sci.*, Bowen vol., 203

oc. Roy. Soc., **101**, A, 509 and 640 (1922).
, and Merwin, H. E., *J. Wash. Acad. Sci.*, **14**, 117

ans. Brit. Ceram. Soc., **23**, 211 (1924).
., and Ritchie, P. D., *J. Appl. Chem.*, **2**, 42 (1952).
., and Ritchie, P. D., *J. Appl. Chem.*, **3**, 187 (1953).
al., *Bull. Nat. Inst. Indust. Jap.*, **4**, 1 (1960).
em. Soc. Jap., **34**, 1491 (1961).
, *The Phases of Silica*, 218 (Rutgers University

, W., *Contr. Mineral. Petrol.*, **36**, 1 (1972).

rmation by Induced Local ns: Examples Employing Magnetic Resonance

bject may be defined as a graphical representa-
l distribution of one or more of its properties.
usually requires that the object interact with a
on field characterized by a wavelength compar-
r than the smallest features to be distinguished,
of interaction may be restricted and a resolved

on the wavelength of the field may be removed,
of image generated, by taking advantage of
teractions. In the presence of a second field
interaction of the object with the first field to a
the resolution becomes independent of wave-
tead a function of the ratio of the normal width
n to the shift produced by a gradient in the
ecause the interaction may be regarded as a
two fields by the object, I propose that image
is technique be known as zeugmatography,
ζευγμα, "that which is used for joining".

Assuming uniform signal strength across the regio
transmitter–receiver coil, the signal in the presen
gradient represents a one-dimensional projection
content of the object, integrated over planes perp
the gradient direction, as a function of the gradi
ate (Fig. 1). One method of constructing a two-
projected image of the object, as represented by it
tent, is to combine several projections, obtained
the object about an axis perpendicular to the gradie
Fig ting the gradient about the o
e methods for reconstructio
fro r projec –5. Fig. 2 was generated by a
si to tha rdon and Herman[4], applied
je s, spaced a ig. 1, so as to construct a 2
x. The rep ation shown was produced
in con erpolated between the ma
and clearly reveals the locations and dimensions
columns of H_2O. In the second experiment, one ca
tained pure H_2O, and the other contained a 0.19 m
of $MnSO_4$ in H_2O. At low radio-frequency powe
mgauss) the two capillaries gave nearly identical in

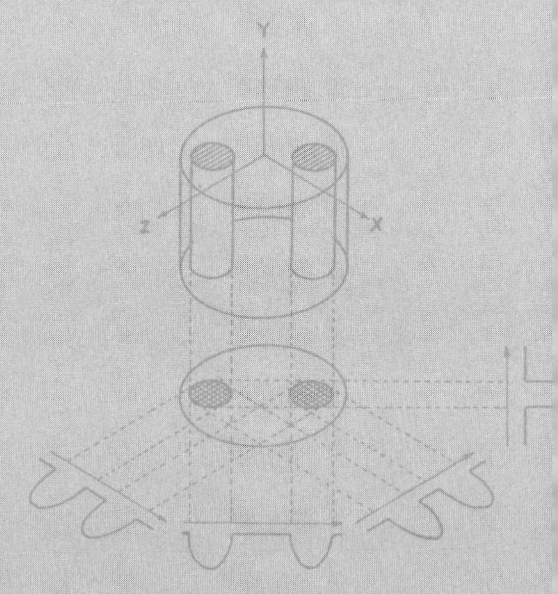

Fig. 1 Relationship between a three-dimensional objec
dimensional projection along the Y-axis, and four o
sional projections at 45° intervals in the XZ-plane. T
indicate the gradient directions.

THIRTEEN

Images of body and brain

Marcus E. Raichle

The magnetic properties of atomic nuclei, which provide the basis for magnetic resonance imaging (MRI), were demonstrated in the 1940s. Creating images, however, awaited the seminal work of Paul Lauterbur, who in 1973 described how these properties might be used to produce detailed pictures of the body and brain. Hospitals and medical clinics now routinely use MRI for diagnosis and for planning treatments. Essentially the same technique also provides a way of probing the mysteries of the human mind.

Before the early 1970s, physicians had only a limited set of tools with which to obtain information about the living human body, and diagnose disease and institute appropriate treatments. Various X-ray techniques provided excellent pictures of the bones, as they do today. But images of the soft tissues of organs such as the brain and spinal cord, and the joints of the arms and legs, were poor and often obtained at the price of great discomfort and risk to the patient. That changed dramatically in the early 1970s when Godfrey Hounsfield, a self-taught engineering genius working for the EMI company in London, introduced X-ray computed tomography[1]—CT scanning as it is now called.

In devising CT, Hounsfield had arrived at a practical way of creating cross-sectional "tomographic" images of an intact object. The images were obtained by passing a narrow beam of X-rays through the object at many different angles, recording the intensity of the emerging beam at each angle, and reconstructing the cross-section of the object from this information. Hounsfield's invention had an immediate impact: it not only changed the practice of medicine but also set the stage for the introduction of other imaging techniques. For this work, he and Allan Cormack, who had independently developed the theoretical basis for CT,[2,3] were awarded the Nobel Prize in Physiology or Medicine in 1979.

Among those other imaging techniques was one that had its roots in the study of the magnetic properties of atomic nuclei. Here we must return to the 1940s and look at some physics.

In a series of experiments, first by I. I. Rabi, and later by Edward Purcell and colleagues, and by Felix Bloch, the basic principles of nuclear magnetic resonance, NMR, were established.[4] Here again, Nobel Prizes were won, but in physics (Rabi in 1944, and Purcell and Bloch in 1952). NMR arises because protons and neutrons, the two constituents of atomic nuclei, each possess an inherent angular momentum or "spin." Although a full understanding of such concepts as angular momentum and spin requires a knowledge of quantum mechanics, one can think of particles with spin as behaving like tiny bar magnets, or "magnetic dipoles," which can have the north pole pointing up or down. In nuclei containing an even number of protons and neutrons, the up and down spins cancel out, giving such nuclei a zero spin. But nuclei with an odd number of protons and neutrons, such as hydrogen, which contains a single proton, have a nonzero spin, and accordingly behave as magnetic dipoles. These are the nuclei that can exhibit NMR; they are said to have a "magnetic moment."

When atoms with such nuclei are placed in a strong, static magnetic field, a small percentage of the dipoles align with the field. They also circle ("precess") about the axis of the field, just as a spinning top may circle slowly about a vertical axis. The frequency of this precession, known as the Larmor frequency, depends on both the particular atomic nucleus and the strength of the static magnetic field. Left undisturbed in the static magnetic field, atomic nuclei capable of NMR produce no externally detectable signal. But if the nuclei are subjected to radio-frequency pulses of energy at their Larmor frequency, energy is absorbed and then reemitted as the NMR signal. A systematically varying voltage at the Larmor frequency is induced in a receiver instrument, which can be characterized in several ways by its change in magnitude over time. Because these time-dependent changes in voltage are a function of the local environment surrounding the NMR nuclei, deductions can be made about the composition and structure of the object under study.

During the 1950s and 1960s, NMR was known primarily as a powerful tool for investigating the molecular structure of compounds. It could be applied only to a small sample, typically less than one cubic centimeter in volume, which was placed in very powerful magnets. Samples were set spinning to minimize the undesirable consequences of spatial differences—inhomogeneities—in the magnetic field. No attention was paid to the spatial variation of the signals within the sample. Indeed, vigorous attempts were made to eliminate such variations by the construction of magnets with extremely uniform fields, and by placing little magnetic "shims" into the magnets along with the sample. These shims could be adjusted to correct for the effect of spatial inhomogeneity induced in the sample by small local variations in the magnetic field. Otherwise the result was an NMR signal containing spurious, unwanted frequencies ("lumps and bumps").

Although the possibility that NMR might provide medically valuable information was appreciated at that time,[4] it was not apparent how images of the sort obtained by CT could be created from the information-rich NMR signals. It was the insight of Paul Lauterbur that provided the answer—an insight that, once announced in his *Nature* paper of 1973,[5] left many feeling "Why didn't I think of that?"

In the fall of 1971, Lauterbur was the acting president of a small, struggling Pennsylvania company called NMR Specialties, as well as a faculty member at the State University of New York at Stony Brook. Lauterbur had a sophisticated understanding of NMR and a hands-on feel for the manipulation of its signals. Observing the work of a graduate student, Leon Saryan, he became intrigued by the possibility of using deliberately introduced magnetic-field inhomogeneity to investigate the internal structure of three-dimensional objects. His published demonstration[5] of this idea illustrates the concept beautifully.

Imagine looking at a test object containing two small tubes of water. Water—H_2O—produces a strong NMR signal because of its high concentration of NMR-detectable hydrogen atoms, whose nuclei contain a single proton. One of the tubes is situated to your left and one to your right. In a static magnetic field that is not uniform but increases linearly from left to right, the voltage that is detected exhibits two different frequencies (fig. 13.1). The NMR voltage signal from the left tube of water will have a lower frequency than the one on the right. This is a straightforward prediction of the Larmor equation, which says that the Larmor frequency is proportional to the strength of the magnetic field. If you know the exact relationship of the magnetic field gradient to the object you are studying, you now have a one-dimensional "shadow" of the distribution of its water content.

Lauterbur's next challenge was to expand the one-dimensional projection to two and then three dimensions. To acquire the necessary data he simply rotated the magnetic field gradient about the object in a stepwise fashion, obtaining views through the object at different angles. He proceeded to

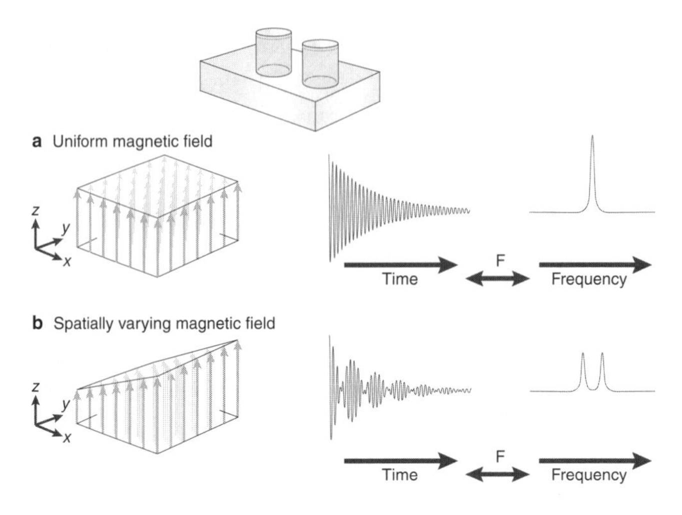

Fig. 13.1. The effect of the magnetic field on the signal obtained with nuclear magnetic resonance (NMR). If NMR is applied to a volume containing two tubes of water (top) the resulting signal is very different when the magnetic field in which the tubes reside is uniform (a) as compared to spatially varying (b). In the case of a uniform magnetic field (a), a time-varying voltage is observed, which consists of a single frequency and decays rapidly over time. When the magnetic field is caused to vary, for example along the *x* dimension of volume (b), a time-varying voltage is again observed but in this instance it consists of two frequencies because of the difference in position of the two water-containing objects relative to the spatially varying magnetic field. By knowing the strength of the magnetic field along the *x* dimension of the volume it is possible, as Lauterbur demonstrated,[5] to locate the projected position of the objects. By viewing the object under similar conditions from many different angles it becomes possible to establish precisely the location of the objects within the volume.

develop a mathematical strategy that allowed this information to be used to form an image of the distribution of the object's water content. Unknown to him, similar mathematical formalisms had been developed some years before by people such as Cormack,[2] and these were being employed in the first CT scanners. Be that as it may, Lauterbur's genius lay in recognizing the possibilities of using the magnetic field gradient to encode spatial information.

Lauterbur proposed that his technique be known as "zeugmatography," from the Greek for "that which is used for joining." Although this colorful and descriptive name did not endure, the technique itself has remained absolutely central to image encoding and reconstruction in MRI. Some may also wonder why the name magnetic resonance imaging (MRI), and not nuclear magnetic resonance (NMR), was adopted. The reason was largely because of concern that patients might be worried about a "nuclear" technique when, in fact, no radioactivity was involved. Almost a decade passed between Lauterbur's discovery and the emergence of clinically useful images, with many people contributing.[4] Notable among them was Peter Mansfield, working at the University of Nottingham in Britain, who also used a field-gradient scheme and in 1976 developed a rapid-scanning MRI method called echo-planar imaging.

Although several biologically important nuclei can be studied with MRI techniques, the hydrogen proton forms the basis of most clinical images. This is because of its abundance in the body, in water (all of the soft tissues have a high water content), and its high sensitivity to MRI. As the technique of MRI was refined during the 1980s and 1990s, its capacity to provide detailed images of the soft tissues of the body, and its ability to complement the CT approach, became increasingly apparent (fig. 13.2).

In the United States alone, well over twelve million MRI scans are now done annually. About two-thirds of clinical studies concentrate on the head and the spinal cord, regions that are particularly difficult to image with other techniques. Brain diseases of especial interest include cancers, multiple sclerosis, stroke, Alzheimer's disease, and conditions characterized by excess bleeding or fluid. In the spine, CT remains best for investigating the

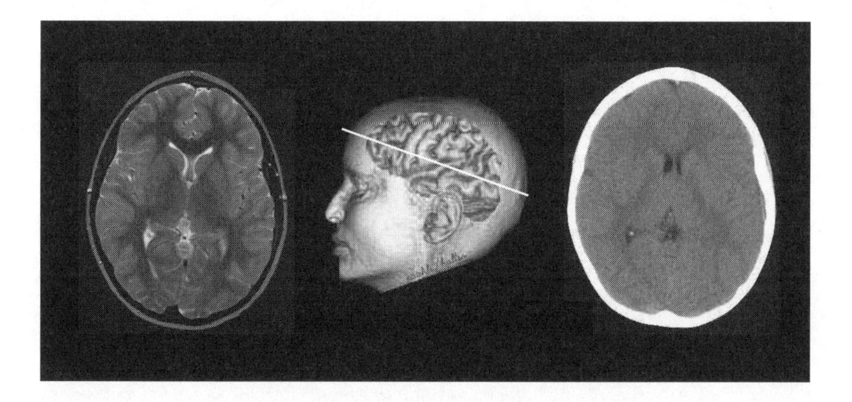

Fig. 13.2. Comparison of computed tomography (CT) and magnetic resonance imaging (MRI). On the left is an image of the brain obtained with MRI in a normal adult; on the right is an X-ray CT image from the same individual for comparison. In the center is a three-dimensional reconstruction of a head obtained with MRI, superimposed upon which is a three-dimensional lateral view of the brain. The white line represents the approximate orientation of the MRI and CT images. Bone is seen clearly by CT, but produces almost no signal in MRI. In contrast, MRI reveals much greater anatomical detail within the soft tissues of the brain. It is this that makes MRI so useful in the clinic. The center image also shows the three-dimensional nature of the MRI data, which allows representation of the data in many different anatomical views.

bony structures. But MRI excels for examining soft tissues, including the spinal cord itself and nerve roots. The musculoskeletal system, particularly the soft tissues of the joints (ligaments and cartilages), is also an important arena for clinical MRI.

Over the past ten years, a further application has come to prominence: the use of MRI to study the function of the living human brain, as well as its structure. This is my own area of research. Here, developments hinged on the finding that local variations in brain activity are associated with changes in oxygen content.[6,7] Seiji Ogawa and colleagues[8] went on to show that MRI could detect these changes, and an avalanche of information has followed. We can now see what happens, moment to moment, within the human brain.

It is a safe bet that more exciting things lie ahead. For example, the movement (diffusion) of water along the fiber tracks of the brain can be detected with MRI.[9] These tracks represent the "wires" connecting regions within the brain, and the brain with the body. With the MRI information it is now possible to study the wiring diagram of the living human brain. Interestingly enough, Lauterbur[5] anticipated such a development, including the MRI measurement of diffusion coefficients which forms the basis of the fiber-tracking technique.

The invention of the microscope and the telescope opened up unexpectedly vast domains of scientific discovery. A similar opportunity has been created in the study of the living human body by the introduction of CT and MRI, and another imaging technique called positron emission tomography (PET). Thanks to Lauterbur and the other pioneers who laid the groundwork and developed these techniques, we have at hand tools with the potential to provide unparalleled insights into the anatomy, biochemistry, metabolism, and function of the organs within the body.

References

1. Hounsfield, G. N. Computerized transverse axial scanning (tomography): Part I. Description of system. *Brit. J. Radiol.* **46,** 1016–1022 (1973).
2. Cormack, A. M. Representation of a function by its line integrals, with some radiological physics. *J. Appl. Phys.* **34,** 2722–2727 (1963).
3. Cormack, A. M. Reconstruction of densities from their projections, with radiological applications. *Phys. Med. Biol.* **18,** 195–207 (1973).
4. Kevles, B. H. *Naked to the Bone: Medical Imaging in the Twentieth Century* (Rutgers University Press, 1997).
5. Lauterbur, P. C. Image formation by induced local interactions: examples employing nuclear magnetic resonance. *Nature* **242,** 190–191 (1973).
6. Fox, P. T. & Raichle, M. E. Regional uncoupling of cerebral blood flow and oxygen metabo-

lism during focal physiological activation: A positron emission tomography study. *J. Cereb. Blood Flow Metab.* **5,** S177 (1985).
7. Fox, P. T., Raichle, M. E., Mintun, M. A. & Dence, C. Nonoxidative glucose consumption during focal physiologic neural activity. *Science* **241,** 462–464 (1988).
8. Ogawa, S., Lee, T. M., Kay, A. R. & Tank, D. W. Brain magnetic resonance imaging with contrast dependent on blood oxygenation. *Proc. Natl Acad. Sci. USA* **87,** 9868–9872 (1990).
9. Conturo, T. E. *et al.* Tracking neuronal fiber pathways in the living human brain. *Proc. Natl Acad. Sci. USA* **96,** 10422–10427 (1999).

Further reading

Kevles, B. H. *Naked to the Bone: Medical Imaging in the Twentieth Century* (Rutgers University Press, 1997).

Nobel e-Museum. *The Nobel Prize in Physiology or Medicine 1979* (http://www.nobel.se/medicine/laureates/1979/). (Also physics prizes in 1944 and 1952.)

Raichle, M. E. "A brief history of human functional brain mapping." In *Brain Mapping: The Systems* (eds Toga, A. W. & Mazziotta, J. C.). 33–75 (Academic, San Diego, 2000).

1973

Image formation by induced local interactions: examples employing nuclear magnetic resonance

P. C. Lauterbur

An image of an object may be defined as a graphical representation of the spatial distribution of one or more of its properties. Image formation usually requires that the object interact with a matter or radiation field characterized by a wavelength comparable to or smaller than the smallest features to be distinguished, so that the region of interaction may be restricted and a resolved image generated.

This limitation on the wavelength of the field may be removed, and a new class of image generated, by taking advantage of induced local interactions. In the presence of a second field that restricts the interaction of the object with the first field to a limited region, the resolution becomes independent of wavelength, and is instead a function of the ratio of the normal width of the interaction to the shift produced by a gradient in the second field. Because the interaction may be regarded as a coupling of the two fields by the object, I propose that image formation by this technique be known as zeugmatography, from the Greek ζευγμα, "that which is used for joining."

The nature of the technique may be clarified by describing two simple examples. Nuclear magnetic resonance (NMR) zeugmatography was performed with 60 MHz (5 m) radiation and a static magnetic field gradient corresponding, for proton resonance, to about 700 Hz cm^{-1}. The test object consisted of two 1 mm inside diameter thin-walled glass capillaries of H_2O attached to the inside wall of a 4.2 mm inside diameter glass tube of D_2O. In the first experiment, both capillaries contained pure water. The proton resonance line width, in the absence of the transverse field gradient, was about 5 Hz. Assuming uniform signal strength across the

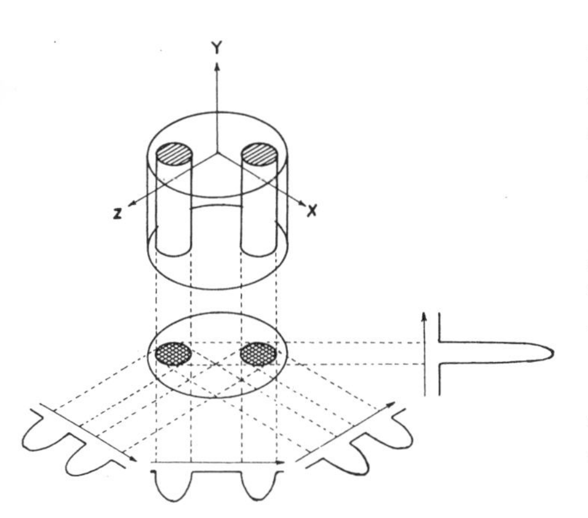

Fig. 1 Relationship between a three-dimensional object, its two-dimensional projection along the Y-axis, and four one-dimensional projections at 45° intervals in the XZ-plane. The arrows indicate the gradient directions.

region within the transmitter-receiver coil, the signal in the presence of a field gradient represents a one-dimensional projection of the H_2O content of the object, integrated over planes perpendicular to the gradient direction, as a function of the gradient coordinate (Fig. 1). One method of constructing a two-dimensional projected image of the object, as represented by its H_2O content, is to combine several projections, obtained by rotating the object about an axis perpendicular to the gradient direction (or, as in Fig. 1, rotating the gradient about the object), using one of the available methods for reconstruction of objects from their projections[1–5]. Fig. 2 was generated by an algorithm, similar to that of Gordon and Herman[4], applied to four projections, spaced as in Fig. 1, so as to construct a 20 × 20 image matrix. The representation shown was produced by shading within contours interpolated between the matrix points, and clearly reveals the locations and dimensions of the two columns of H_2O. In the second experiment, one capillary contained pure H_2O, and the other contained a 0.19 mM solution of $MnSO_4$ in H_2O. At low radio-frequency power (about 0.2 mgauss) the two capillaries gave nearly identical images in the zeugmatogram (Fig. 3*a*). At a higher power level (about 1.6 mgauss), the pure water sample gave much more saturated signals than the sample whose spin-lattice relaxation time T_1 had been shortened by the addition of the paramagnetic Mn^{2+} ions, and its zeugmatographic image vanished at the contour level used in Fig. 3*b*. The sample region with long T_1 may be selectively emphasized (Fig. 3*c*) by constructing a difference zeugmatogram

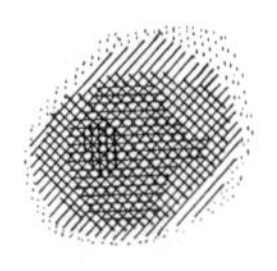

Fig. 2 Proton nuclear magnetic resonance zeugmatogram of the object described in the text, using four relative orientations of object and gradients as diagrammed in Fig. 1.

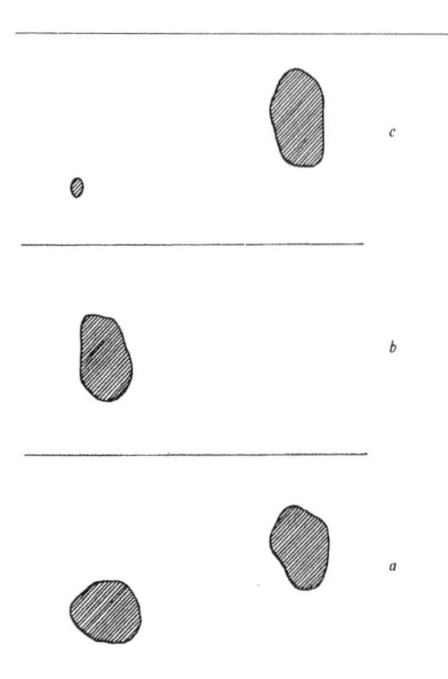

Fig. 3 Proton nuclear magnetic resonance zeugmatograms of an object containing regions with different relaxation times. *a*, Low power; *b*, high power, *c*, difference between *a* and *b*.

from those taken at different radiofrequency powers.

Applications of this technique to the study of various inhomogeneous objects, not necessarily restricted in size to those commonly studied by magnetic resonance spectroscopy, may be anticipated. The experiments outlined above demonstrate the ability of the technique to generate pictures of the distributions of stable isotopes, such as H and D, within an object. In the second experiment, relative intensities in an image were made to depend upon relative nuclear relaxation times. The variations in water contents and proton relaxation times among biological tissues should permit the generation, with field gradients large compared to internal magnetic inhomogeneities, of useful zeugmatographic images from the rather sharp water resonances of organisms, selectively picturing the various soft structures and tissues. A possible application of considerable interest at this time would be to the *in vivo* study of malignant tumours, which have been shown to give proton nuclear magnetic resonance signals with much longer water spin-lattice relaxation times than those in the corresponding normal tissues[6].

The basic zeugmatographic principle may be employed in many different ways, using a scanning technique, as described above, or transient methods. Variations on the experiment, to be described later, permit the generation of two- or three-dimensional images displaying chemical compositions, diffusion coefficients and other properties of objects measurable by spectroscopic techniques. Although applications employing nuclear magnetic resonance in liquid or liquid-like systems are simple and attractive because of the ease with which field gradients large enough to shift the narrow resonances by many line widths may be generated, NMR zeugmatography of solids, electron spin resonance zeugmatography, and analogous experiments in other regions of the spectrum should also be possible. Zeugmatographic techniques should find many useful applications in studies of the internal structures, states, and compositions of microscopic objects.

Department of Chemistry, State University of New York at Stony Brook, Stony Brook, New York 11790

Received October 30, 1972; revised January 8, 1973 [Published 16 March]

References

1. Bracewell, R. N., and Riddle, A. C., *Astrophys. J.*, **150**, 427 (1967).
2. Vainshtein, B. K., *Soviet Physics—Crystallography*, **15**, 781 (1971).
3. Ramachandran, G. N., and Lakshminarayan, A. V., *Proc. US Nat. Acad. Sci.*, **68**, 2236 (1971).
4. Gordon, R., and Herman, G. T., *Comm. Assoc. Comput. Mach.*, **14**, 759 (1971).
5. Klug, A., and Crowther, R. A., *Nature*, **238**, 435 (1972).
6. Weisman, I. D., Bennett, L. H., Maxwell, Sr., L. R., Woods, M. W., and Burk, D., *Science*, **178**, 1288 (1972).

er 17, 1973.
jöberg, O., and Möller, G., *Transplantn. Rev.*,
., and Janossy, G., *Transplantn. Rev.*, **11**, 87
Transplantn. Rev., **11**, 178 (1972).
J., Stutman, O., and Good, A., *Sem. Hæmat.*,
nd Greaves, M. F., *Clin. exp. Immun.*, **10**,
nd Vassalli, P., *Eur. J.* ... **3**, ...
., Bauminger, S., and Jan... G., ...
37 (1972).
derson, J., Pohlit, H., an... ...be...
13, 89 (1973).
Rosenthal, A., and Paul, W. E., *J. Im*...
., and Bauminger, S., *Nature new* ... **235**,
Edelman, G. M., Moller, G., and Sjöberg, O.,
n., **2**, 233 (1972).
Clin. exp. Immun., **11**, 551 (1972).
Roitt, I. M., *Nature new Biol.*, **241**, 254 (1973).
, and Phillips, B., *J. exp. Med.*, **138**, 64 (1973).
, *Transplantn. Rev.*, **11**, 39 (1972).
nd Greaves, M. F., *Clin. exp. Immun.*, **9**, 483
itor), *Transplantn. Rev.*, **16** (1973).
, *Current Titles in Immunology: Transplanta-*
rgy, **1**, 193 (1973).
, and Brown, G., *J. Immun.*, **112**, 420 (1974).
ell, H., *Clin. exp. Immun.*, **14**, 171 (1973).
., Strober, S., Herzenberg, L. A., and De Pam-
J. Immun. (in the press).
, K. J., Park, B. H., Biggar, W. D., and Good,
. *Invest.*, **52**, 919 (1973).
Rosen, F. S., Filler, R. M., Janeway, C. A.,
B., and Kay, H. E. M., *Lancet*, **ii**, 1210 (1968).
Waksman, B., *J. exp. Med.*, **136**, 143 (1972).
Moller, G., and Sjöberg, O., *Eur. J. Immun.*,
nd Unanue, E. R., *J. Immun.*, **109**, 1022 (1972).
munology, **19**, 583 (1970).
reaves, M. F., Doenhoff, M. J., and Snajdr, J.,
mmun., **14**, 581 (1973).
Greaves, M. F., *Eur. J. Immun.* (in the press).
., and Brown, G., *Nature new Biol.*, **246**, 116
Wedner, H. J., Parker, C. W., *Immun. Commun.*,
Simpson, E., and Herzenberg, L. A., *Eur. J.*
345 (1973).
d Fudenberg, H. H., *Int. Archs. Allergy appl.*
18 (1968).

of *in vitro* T cell-mediated in lymphocytic chorio- within a syngeneic or neic system

ments[1–3] indicate that cooperation between lymphocytes (T cells) and antibody-forming (B cells) is restricted by the H-2 gene com- tivity *in vivo* operates only when T cells and least one set of H-2 antigenic specificities. sented here that the interaction of cytotoxic ther somatic cells budding[4–5] lymphocytic

Table 1 Cytotoxic activity of spleen cells from vario mice injected i.c. 7 d previously with 300 LD_{50}* of virus for monolayers of LCM-infected or normal mouse L cells.

Experiment	Mouse strain	H-2 type	% ^{51}Cr Infected
1	CBA/H	*k*	65.1 ± 3.3
	Balb/C	*d*	17.9 ± 0.9
	C57Bl	*b*	22.7 ± 1.4
	C...H × C57Bl	*k/b*	56.1 ± 0.5
	... × Balb/C	*b/d*	24.8 ± 2.4
	... or +/+		42.8 ± 2.0
	nu...		23.3 ± 0.6
	...	*k*	85.5 ± 3.1
	...	*k*	71.2 ± 1.6
	DI...2	*d*	24.5 ± 1.2
3	C...H	*k*	77.9 ± 2.7
	C3H/HeJ	*k*	77.8 ± 0.8

* Other mice were injected with 2 × 10^6 LD_{50}, b specific release were invariably lower due to the high d paralysis[8,20] associated with viscerotropic (WE3) LCM

† % ^{51}Cr release by normal spleen cells on infe ranged from: (experiment 1) 17.1 ± 0.3 to 20.0 ± 0.7; 2) 20.0 ± 1.4 to 25.3 ± 0.7; (experiment 3) 27.2 ± 2.0.

infected L cells by CBA/H immune spleen cell shown to be a property of specifically sensitise derived lymphocytes, which act in the absenc macrophages and substances secreted into the large[6–8].

Various strains of mice were injected intracereb with 300 mouse LD_{50} of WE3 LCM virus. Mice we 7 d later when, in CBA/H mice, maximal cytoto is found in lymphoid tissue[6,7]. Only spleen pr from mice sharing at least one set of H-2 antige cities with the target monolayer caused high leve 50%) of specific lysis (Table 1). Spleen cells control (nu/+ or +/+) mice, derived locally and Balb/C stock (Dr J. B. Smith, personal comm were less active and lymphocytes from histoir mice caused minimal specific release (< 5%) of ^{51}

Spleen preparations from mice immunised 10 previously were also assayed, as Marker and Vo reported that maximal cytotoxicity of C3H lymp L cells infected with the Traub strain of LCM v at 11 d after inoculation. High levels of specific were again recognised only in the histocompati (Table 2) activity declining, as has been shown from a peak on day 7.

Demonstration of reciprocal exclusion of cy essential to establish that mice possessing other antigenic specificities are capable of generating cells. Comparisons were thus made using similar t allogeneic mouse strains. Peritoneal macrophage tained[10] from normal Balb/C and CBA/H mic in plastic tissue culture trays and infected[11] with virus. Specific lysis was restricted to the synge

Table 2 % ^{51}Cr release* from infected C3H L cells o spleen cells from mice sampled at 7, 10 and 13 d after inoculation with 2,000 LD_{50} of WE3 LCM vir

FOURTEEN

Journey to the T cell

Jonathan C. Howard

Immunization against viral infection has saved countless lives. But how it works long remained a mystery. In 1974, a brief paper opened the way to an understanding of how a class of white blood cells known as T cells kill virus-infected cells in the body, and so prevent the spread of viruses. This cell-mediated immunity complements the antibody response to infection.

During the era I am going to describe, the last train to Cambridge from London's Liverpool Street station was a miserable affair. Even if the driver turned up at all, it was stop-go all the way; on one occasion, just before Cambridge, we even went backward. But we reached our destination in the end. The journey from Rolf Zinkernagel and Peter Doherty's famous but highly technical paper[1] of 1974 to an understanding of what it meant reminds me of that terrible ride. But the world eventually achieved that understanding and, in 1996, Zinkernagel and Doherty received the Nobel Prize in Physiology or Medicine. Twenty-two years is a long time to wait nowadays. Why did it take so long?

The theme is acquired immunity. This is the lifelong resistance we can muster against many infectious diseases, in which the immune "memory" of one infection results in a much more effective response against a second such infection. Many of these diseases were killers. Now, through immunization programs, they are becoming rarities. Smallpox has been eradicated worldwide; polio will be the next to go. But until recently these great advances were made in deep ignorance of fundamental immune mechanisms.

The discovery of antibodies at the end of the nineteenth century[2] revealed one mechanism. Antibodies are proteins made by the immunized or infected animal that bind specifically to infectious agents or their toxic products (antigens) and inactivate them. Antibodies clearly saved lives. But how were they made? In particular, how could an animal make antibodies specific for virtually any antigen?[3] This magical power of producing an infinite repertoire of antibodies seemed to imply an infinite repertoire of genes to code for them. And if an infinite repertoire of antibodies could be made, why did not some attack the body's own tissues—that is, why is the body "self-tolerant"?

By the 1950s the immune system was definitively localized to the lymphocytes, abundant white blood cells. In 1957, MacFarlane Burnet, an Australian virologist, produced the grandest of immunological theories to reconcile the antibody and self-tolerance problems. Burnet's "clonal selection theory"[4] proposed that the enormous repertoire of antibodies was made by an equally enormous repertoire of lymphocytes, each potentially able to make antibody of just one specificity and carrying representatives of that antibody on its surface as receptors. When the receptor bound to a specific triggering molecule, say an infecting bacterium or virus, the cell would be activated and divide repeatedly to make more cells with the same specificity. These cells would then release large quantities of the soluble form of the antibody into the bloodstream to destroy the invaders and clear up the infection (fig. 14.1).

Burnet proposed that the lymphocyte repertoire was generated by a genetic process, then of unknown nature, occurring during the development of lymphocytes. Self-tolerance, he argued, was achieved during a "sensitive" phase in the embryo when triggering of specific receptors would result in selective death of the lymphocytes concerned. So all lymphocytes whose receptors recognized "self" molecules would be eliminated and the animal would be self-tolerant. Failures of this elimination process could be responsible for autoimmune disease. This sure-footed scheme, with the underlying genetics now understood, remains essentially what we believe today. Burnet had to wait only until 1960 for his Nobel Prize.

But something was missing, for it seemed that another immune mechanism existed—one that required direct cell-cell contact to do its destructive work, rather than using soluble proteins such as antibodies. Most conspicuously, this mechanism showed itself in the recognition and destruction of foreign-tissue transplants. It seemed that individuals of a species were genetically distinct at a cellular level, so that cells or tissues transplanted from one to another were rejected by the immune system of the recipient. One genetic system seemed to be of overwhelming importance in every species

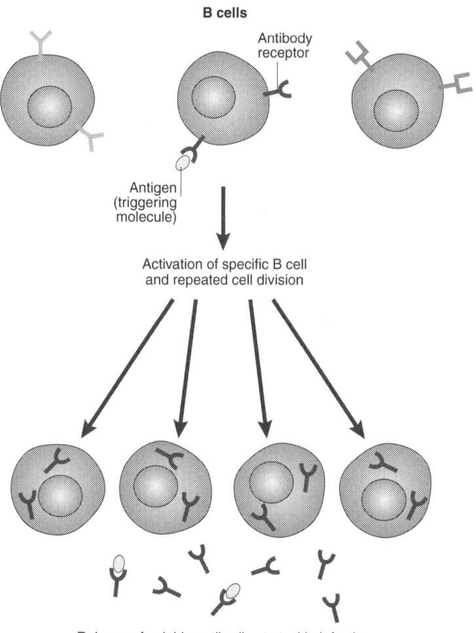

Fig. 14.1. The antibody response to infection. This response involves the type of lymphocytes known as B cells. When the antibody receptor on the cell surface is triggered by a specific antigen, for instance an infectious agent or part of it, a specific type of B cell is activated. Others do not respond. The activated B cell then divides repeatedly to produce large numbers of identical cells; this is Burnet's "clonal selection." These cells release large quantities of soluble antibodies into the bloodstream to tackle the infection.

tested. This was a complex of genes coding for cell-surface proteins, which were unusual in that virtually every individual in the population appeared to carry distinct forms of the genes (or alleles, as such variant forms are known). As a result, every person carried a distinctive set of these proteins on their cell surfaces. Because this gene complex controlled such powerful tissue compatibility factors, it became known as the major histocompatibility complex, or MHC. The recognition and destruction of cells with foreign MHC proteins on their surface could be achieved in cell culture, when lymphocytes from one individual were cultured with cells from another individual: the lymphocytes could be seen to bind to target cells and kill them in a few hours.

Were the destructive cells responsible for tissue rejection members of the known lymphocyte repertoire, but using the surface form of the antibody instead of the secreted, soluble form? It seemed the obvious explanation, but it was wrong. Indeed by this time, around 1970, it had become clear that there are two different kinds of lymphocyte. They all looked the same, but only about half of them, called B cells because in mammals they mature in the bone marrow, appeared to be responsible for the antibody response.

The other half, known as T cells because their development depended on the thymus gland, carried no known receptor. Yet it seemed that T cells were dedicated to the specific recognition of foreign structures on cell surfaces, as witnessed by the recognition and destruction of cells differing at the MHC part of their genome. Furthermore, T cells that behaved very like the ones that destroyed foreign tissues could also be found in animals infected with certain viruses: these T cells would kill only virus-infected cells, and only cells infected with the same virus that caused the immunity in the first place. The properties of T cells promised to open a new door to understanding immunity against viruses. And so they did.

What Zinkernagel and Doherty[1] showed was that T cells from a mouse immunized against a virus could kill cells infected with the same virus, but only—and here's the main point—if these cells expressed the same MHC type as the infected animal. Cells infected with the same virus, but taken from a mouse carrying different alleles at the MHC, could not be killed. The killing was said to be "restricted" by the MHC. In some way, recognition and killing of the target cell had two requirements: the right virus component, the antigen, and the right sequence of a gene or genes from the infected animal itself, the MHC genes. So was the T cell recognizing two different things? If so, was it doing it with two receptors, one for the virus antigen and one for the MHC? Or was a single T-cell receptor recognizing one complex, made of structures from the MHC and from the virus?

By 1976 the two-receptor model had gained the ascendancy. The general structure of the antigen-binding site on B cells was by then known. If the T-cell receptor was built along the same lines, as was generally assumed (and as turned out to be correct), it would simply be too small to be able to recognize structural features of both virus antigen and MHC protein. By 1980, around the time when my train was going backward, all of the textbooks showed T cells with two distinct receptors sticking out of the cell membrane, one for the virus and one for the MHC. The simplicity of Burnet's scheme for B cells—one cell, one antibody—was rapidly abandoned for T cells because of the difficulty of conceiving the molecular nature of what the T cells were recognizing on virus-infected cells.

There were, of course, occasional dissenters. In a brilliant yet little quoted review in 1981, Polly Matzinger[5] held out against the tide and argued for the existence of a single T-cell receptor. Her analysis proved to be extraordinarily correct. By this time other experiments, most notably the intricate and beautiful work of John Kappler and Pippa Marrack,[6] were creating appalling difficulties for the two-receptor model. The end for that model came in 1983 or 1984, with the conclusive identification of the T-cell receptor. There was just one kind on each cell, bound to the surface, and antibody-like in its structure and diversity.

If there was only one kind of receptor on the T cell, how was dual specificity for antigen and MHC achieved? The answer came all in a rush in 1987, with the determination of the three-dimensional structure of an MHC molecule by Pamela Bjorkman and Don Wiley.[7,8] I still remember the revelation of seeing Bjorkman describe the structure at a talk just before it appeared in *Nature.* The MHC molecule is itself a receptor. The site that recognizes and binds parts of other molecules is a deep cleft just long enough to hold short peptides—fragments of virus protein that proved to constitute the antigen for T-cell recognition. All the famous genetic diversity of the MHC centers on the specification of the structure of the binding cleft; the products of different alleles have binding clefts with differently shaped and positioned pockets and therefore preferentially bind different virus peptides. A complicated intracellular machinery chops up viral proteins into peptides and assembles those that happen to fit into newly synthesized MHC proteins on their way to the cell surface. There the complex of MHC and viral antigen awaits the arrival of a T cell of the appropriate specificity, which will recognize and kill the virus-infected cell (fig. 14.2).

What did Zinkernagel and Doherty really observe in 1974? Only after this long journey do we know. Infection of a mouse with live virus stimulated an immune response, resulting in the development of a population of active, virus-specific T cells able to recognize and kill virus-infected cells. But these killer T cells were not directly stimulated by the virus, or even by a viral protein. Their receptors were stimulated by recognizing self MHC molecules, on the surface of infected cells, which contained fragments of viral proteins that had been degraded inside the infected cell. When the active killer cells were confronted *in vitro* with more virus-infected cells carrying the same MHC molecules, they again recognized the same viral peptides in the molecular complex and initiated their killing program. But when Zinkernagel and Doherty confronted them with virus-infected cells with different MHC molecules on them, different viral peptides were loaded into a different MHC binding cleft, the new complexes failed to stimulate the receptor and no killing reaction was initiated.

T-cell immunity is indeed profoundly different from B-cell immunity. Knowledge of it has transformed the prospects for designing new and effective vaccines. To get into Zinkernagel and Doherty's little paper, and

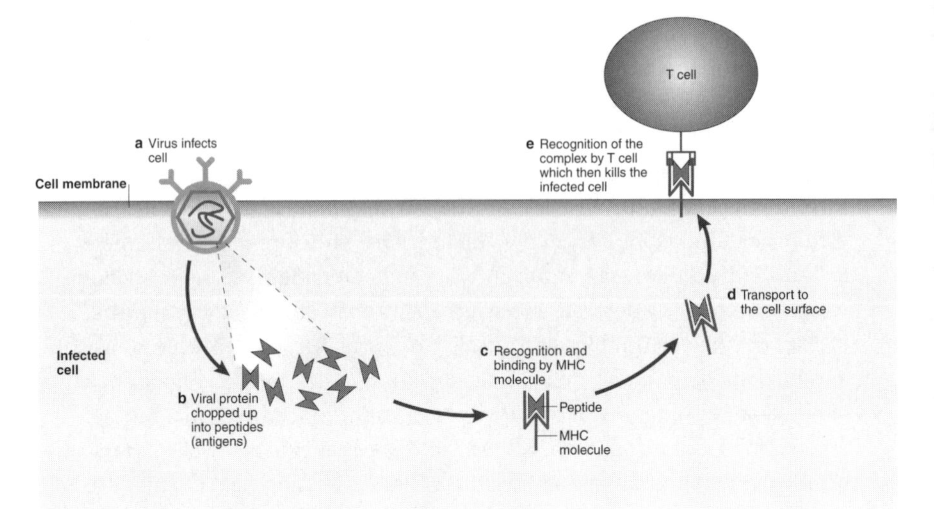

Fig. 14.2. The cell-mediated (T-cell) response to infection. The purpose of this response is to kill virus-infected cells, to prevent virus spread, and it depends on the "presentation" of virus peptides (antigens) at the cell surface by the machinery of the major histocompatibility complex (MHC). a, A virus infects a cell. b, As part of its defense response, the cell chops up some viral proteins. c, The resulting peptides can be recognized with exquisite specificity by an MHC molecule and captured in its binding cleft. d, The complex of MHC molecule and viral antigen is transported to the cell surface. e, At the cell surface, the complex is recognized by a T cell, which goes on to kill the infected cell.

then out again to our present understanding, was an arduous but satisfying ride.

References

1. Zinkernagel, R. M. & Doherty, P. C. Restriction of *in vitro* T cell-mediated cytotoxicity in lymphocytic choriomeningitis within a syngeneic or semiallogeneic system. *Nature* **248,** 701–702 (1974).
2. Behring, E. v. & Kitasato, S. Ueber das Zustandekommen der Diphterie-Immunität und der Tetanus-Immunität bei Thieren. *Dt. Med. Wschr.* **49,** 1113–1114 (1890).
3. Landsteiner, K. *The Specificity of Serological Reactions,* revised edition (Harvard University Press, 1944).
4. Burnet, F. M. *The Clonal Selection Theory of Acquired Immunity* (Cambridge University Press, 1959).
5. Matzinger, P. A one-receptor view of T-cell behaviour. *Nature* **292,** 497–501 (1981).
6. Kappler, J. W. *et al.* Antigen-inducible, H-2-restricted, interleukin-2-producing T cell hybridomas. Lack of independent antigen and H-2 recognition. *J. Exp. Med.* **153,** 1198–1214 (1981).
7. Bjorkman, P. J. *et al.* The foreign antigen binding site and T cell recognition regions of class I histocompatibility antigens. *Nature* **329,** 512–518 (1987).

8. Bjorkman, P. J. *et al.* Structure of the human class I histocompatibility antigen HLA-A2. *Nature* **329,** 506–512 (1987).

Further reading

Nobel e-Museum. *The Nobel Prize in Physiology or Medicine 1996* (http://www.nobel.se/medicine/laureates/1996/). (Also physiology or medicine prize in 1960.)

Rudolph, M. G., Luz, J. G. & Wilson, I. A. Structural and thermodynamic correlates of T cell signaling. *Annu. Rev. Biophys. Biomol. Struct.* **31,** 121–149 (2002).

1974

Restriction of *in vitro* T cell-mediated cytotoxicity in lymphocytic choriomeningitis within a syngeneic or semiallogeneic system

R. M. Zinkernagel and P. C. Doherty

Recent experiments[1–3] indicate that cooperation between thymus derived lymphocytes (T cells) and antibody-forming cell precursors (B cells) is restricted by the H-2 gene complex. Helper activity *in vivo* operates only when T cells and B cells share at least one set of H-2 antigenic specificities. Evidence is presented here that the interaction of cytotoxic T cells with other somatic cells budding[4–5] lymphocytic choriomeningitis (LCM) virus is similarly restricted.

Both the cytotoxic assay used and the characteristics of the cells involved have been described previously[6–8]. Briefly, monolayers of C3H mouse fibroblasts (L cells) are grown in plastic tissue culture trays, infected with a high multiplicity of the WE3 strain of LCM virus and the cells labelled with ^{51}Cr and overlaid (40:1) with the spleen cell preparation to be tested. Supernatants are removed between 15 and 16 h later and % ^{51}Cr release calculated[7]. Results are expressed as mean ± s.e.m. for four replicates. Cytolysis of infected L cells by CBA/H immune spleen cells has been shown to be a property of specifically sensitised thymus-derived lymphocytes, which act in the absence of both macrophages and substances secreted into the medium at large[6–8].

Various strains of mice were injected intracerebrally (i.c.) with 300 mouse LD_{50} of WE3 LCM virus. Mice were sampled 7 d later when, in CBA/H mice, maximal cytotoxic activity is found in lymphoid tissue[6,7]. Only spleen preparations[6] from mice sharing at least one set of H-2 antigenic specificities with the target monolayer caused high levels (40% to 50%) of specific lysis (Table 1). Spleen cells from nude control (nu/+ or

Table 1 Cytotoxic activity of spleen cells from various strains of mice injected i.c. 7 d previously with 300 LD_{50}* of WE3 LCM virus for monolayers of LCM-infected or normal C3H (H-2^k) mouse L cells.

Experiment	Mouse strain	H-2 type	% ^{51}Cr release†	
			Infected	Normal
1	CBA/H	*k*	65.1 ± 3.3	17.2 ± 0.7
	Balb/C	*d*	17.9 ± 0.9	17.2 ± 0.6
	C57Bl	*b*	22.7 ± 1.4	19.8 ± 0.9
	CBA/H × C57Bl	*k/b*	56.1 ± 0.5	16.7 ± 0.3
	C57Bl × Balb/C	*b/d*	24.8 ± 2.4	19.8 ± 0.9
	nu/+ or +/+		42.8 ± 2.0	21.9 ± 0.7
	nu/nu		23.3 ± 0.6	20.0 ± 1.4
2	CBA/H	*k*	85.5 ± 3.1	20.9 ± 1.2
	AKR	*k*	71.2 ± 1.6	18.6 ± 1.2
	DBA/2	*d*	24.5 ± 1.2	21.7 ± 1.7
3	CBA/H	*k*	77.9 ± 2.7	25.7 ± 1.3
	C3H/HeJ	*k*	77.8 ± 0.8	24.5 ± 1.5

* Other mice were injected with 2 × 10^6 LD_{50}, but levels of specific release were invariably lower due to the high dose immune paralysis[8,20] associated with viscerotropic (WE3) LCM virus.

† % ^{51}Cr release by normal spleen cells on infected targets ranged from: (experiment 1) 17.1 ± 0.3 to 20.0 ± 0.7; (experiment 2) 20.0 ± 1.4 to 25.3 ± 0.7; (experiment 3) 27.2 ± 2.0.

+/+) mice, derived locally from CBA and Balb/C stock (Dr J. B. Smith, personal communication), were less active and lymphocytes from histoincompatible mice caused minimal specific release (<5%) of ^{51}Cr.

Spleen preparations from mice immunised 10 and 13 d previously were also assayed, as Marker and Volkert[9] have reported that maximal cytotoxicity of C3H lymphocytes for L cells infected with the Traub strain of LCM virus occurs at 11 d after inoculation. High levels of specific ^{51}Cr release were again recognised only in the histocompatible system (Table 2) activity declining, as has been shown previously[7], from a peak on day 7.

Demonstration of reciprocal exclusion of cytolysis was essential to establish that mice possessing other than H-2^k antigenic specificities are capable of generating cytotoxic T cells. Comparisons were thus made using similar targets from allogeneic mouse strains. Peritoneal macrophages were obtained[10] from normal Balb/C and CBA/H mice, cultured in plastic tissue culture trays and infected[11] with WE3 LCM virus. Specific lysis was restricted to the syngeneic system (Table 3). Comparable levels of specific ^{51}Cr release from isologous infected macrophages were caused

Table 2 % ^{51}Cr release* from infected C3H L cells overlaid with spleen cells from mice sampled at 7, 10 and 13 d after intravenous inoculation with 2,000 LD_{50} of WE3 LCM virus.

	Days after inoculation		
Mouse strain	7	10	13
CBA/H	72.0 ± 2.0	66.4 ± 1.4	27.5 ± 0.5
Balb/C	26.1 ± 0.7	28.0 ± 1.6	22.7 ± 1.8
C57Bl	27.3 ± 1.1	24.3 ± 1.8	24.0 ± 0.4

* Levels of ^{51}Cr release due to overlaying normal L cells with immune spleen cells, infected L cells with control spleen cells or with medium alone ranged from 17.1 ± 0.4 to 24.0 ± 1.4. Other mice were injected with 2 × 10^6 LD_{50}, but levels of specific release were invariably lower.

by Balb/C and CBA/H spleen cells (overlaid at 20:1) from mice infected at the same time with the same dose of LCM virus, whereas histoincompatible macrophages were not damaged. Lysis was completely abrogated by treatment with AKR anti-θ ascitic fluid and guinea pig complement, but not by normal AKR ascitic fluid and complement. Though levels of

Table 3 % ^{51}Cr release from normal and infected peritoneal macrophages by spleen cells from control mice and from mice injected i.c. with 300 LD_{50} of WE3 LCM virus 7 d previously.

			% ^{51}Cr release from macrophages			
			Experiment 1		Experiment 2	
Spleen cells		Macrophage source	Infected	Normal	Infected	Normal
Balb/C	Immune	Balb/C	61.8 ± 4.2^c	27.6 ± 1.9^e	77.5 ± 4.2^d	47.0 ± 3.5^d
	Anti-θ*		ND	ND	40.6 ± 2.5^e	ND
	N ascitic*		ND	ND	90.0 ± 2.7	ND
	Control		42.0 ± 4.8^a	40.5 ± 5.2^a	49.6 ± 2.5	43.5 ± 1.6
CBA/H	Immune		42.7 ± 6.7^a	33.7 ± 5.4^a	32.9 ± 3.0^a	48.6 ± 3.9^a
	Control		28.0 ± 4.1	40.5 ± 5.2	46.5 ± 3.7	39.7 ± 4.3
CBA/H	Immune	CBA/H	69.1 ± 2.8^e	30.9 ± 3.4^e	72.5 ± 5.2^d	40.0 ± 2.9^d
	Anti-θ		ND	ND	44.0 ± 2.5^d	ND
	N ascitic		ND	ND	74.3 ± 8.4	ND
	Control		34.2 ± 1.1	35.1 ± 3.7	46.5 ± 3.6	44.4 ± 6.2
Balb/C	Immune		46.2 ± 3.3^a	30.4 ± 3.8^b	44.0 ± 2.9^a	41.0 ± 2.4^a
	Control		34.9 ± 5.7	33.7 ± 5.6	40.5 ± 2.5	41.0 ± 2.4

* Treated with AKR anti-θ (C3H) ascitic fluid and guinea pig complement, or normal AKR ascitic fluid and guinea pig complement.

a, b, c, d, e, Differences by Student's *t* test between values for immune spleen cells treated with anti-θ ascitic fluid or normal ascitic fluid, immune and control spleen cells overlaid on infected macrophages (infected column), or immune spleen cells overlaid on infected and normal macrophages (normal column). *a*, $P > 0.05$; *b*, $P < 0.05$; *c*, $P < 0.02$; *d*, $P < 0.01$; *e*, $P < 0.001$.

ND, Not done.

^{51}Cr release from macrophages were more variable than for L cells, probably because of inconsistencies in target cell concentrations and the higher background of non-specific lysis, the effect was both highly significant and repeatable.

The ability of T cells to cause lysis across an allogeneic barrier, when lymphocytes are sensitised to antigens specified by the H-2 gene complex, is well established[12,13]. This also applies to L cells (H-2^k), which are readily lysed by spleen cells from C57Bl (H-2^b) mice stimulated in mixed lymphocyte culture by CBA/H (H-2^k) but not by Balb/C (H-2^d) lymph node cells. Sufficiently close association for lysis to occur is possible if the T cells are sensitised to alloantigens present on the surface of the target. Interaction between immune lymphocytes and cells expressing antigens expressed by LCM virus is, however, apparently confined to a histocompatible system, perhaps because it is only in this situation that the necessary intimacy of contact is achieved.

This restriction may possibly be overcome by previous treatment of the target population with trypsin. Balb/C (H-2^d) immune spleen cells will lyse recently trypsinised L cells (H-2^k) infected with the E-350 strain of LCM virus, the effect being completely abrogated by treatment with a rabbit anti-mouse brain serum cytotoxic for T cells[14]. Our experiments with previously trypsinised WE3-infected L cells have, to date, given equivocal results because of the high background of non-specific ^{51}Cr release.

An alternative possibility that must be considered in LCM is that the process of virus maturation[5–6] through the cell membrane causes changes in self components, which are recognised only within the syngeneic or semi-allogeneic system. There is ample evidence[15,16], for instance, that concentrations of H-2 antigens in the cell membrane are decreased in cells productively infected with budding viruses. The cytotoxic T cell may thus be recognising altered self, the implication being that LCM is essentially an autoimmune phenomenon.

These results impose a possible constraint on attempts to demonstrate cytotoxic T cells in infections of man and domestic mammals, where histocompatible cell lines and inbred strains are not available. Perhaps isologous macrophages or lectin-transformed peripheral blood leukocytes may prove suitable targets in at least some disease states. Restriction of cell-mediated cytotoxicity within a syngeneic or semiallogeneic system may prove a reliable index of T cell involvement in species where the θ marker is not available, lysis across this barrier indicating an antibody-associated process[17–19].

Dr Zinkernagel is supported by the Schweizerische Stiftung Fuer Biologisch-Medizinische Stipendien.

Department of Microbiology, John Curtin School of Medical Research, Australian National University, Canberra

Received December 10, 1973 [Published 19 April 1974]

Reference

1. Kindred, B., and Shreffler, D. C., *J. immun.*, **109**, 940 (1972).
2. Katz, D. H., Hamaoka, T., and Benacerraf, B., *J. exp. Med.*, **137**, 1405 (1973).
3. Katz, D. H., Hamaoka, T., Dorf, M. E., and Benacerraf, B., *Proc. natn. Acad. Sci., U.S.A.*, **70**, 2624 (1973).
4. Abelson, H. T., Smith, G. H., Hoffman, H. A., and Rowe, W. P., *J. natn. Cancer Inst.*, **42**, 497 (1969).
5. Kajima, M., and Majde, J., *Naturwissenschaften*, **57**, 93 (1970).
6. Zinkernagel, R. M., and Doherty, P. C., *J. exp. Med.*, **138**, 1266 (1973).
7. Doherty, P. C., Zinkernagel, R. M., and Ramshaw, I. A., *J. Immun.* (in the press).
8. Doherty, P. C., and Zinkernagel, R. M., *Transpln. Rev.*, **18** (in the press).
9. Marker, O., and Volkert, M., *J. exp. Med.*, **137**, 1511 (1973).
10. Blanden, R. V., Mackaness, G. B., and Collins, F. M., *J. exp. Med.*, **124**, 585 (1966).
11. Mims, C. A., and Subrahmanyan, T. P., *J. Path. Bact.*, **91**, 403 (1966).
12. Cerottini, J.-C., Nordin, A. A., and Brunner, K. T., *Nature*, **228**, 1308 (1970).
13. Cerottini, J.-C., and Brunner, K. T., *Adv. Immun.* (in the press).
14. Cole, G. A., Prendergast, R. A., and Henney, C. S., *Fedn. Proc.*, **32**, 964 (1973).
15. Hecht, T. T., and Summers, D. F., *J. Virol.*, **10**, 578 (1972).
16. Lengerova, A., *Adv. Cancer Res.*, **16**, 235 (1972).
17. Perlmann, P., Perlmann, H., and Wigzell, H., *Transpln. Rev.*, **13**, 91 (1972).
18. MacLennan, I. C. M., *Transpln. Rev.*, **13**, 67 (1972).
19. Steele, R. W., Hensen, S. A., Vincent. M. M., Fuccillo, D. A., and Bellanti, J. A., *J. Immun.*, **110**, 1502 (1973).
20. Hotchin, J., *Monogr. Virol.*, **3**, 1 (1971).

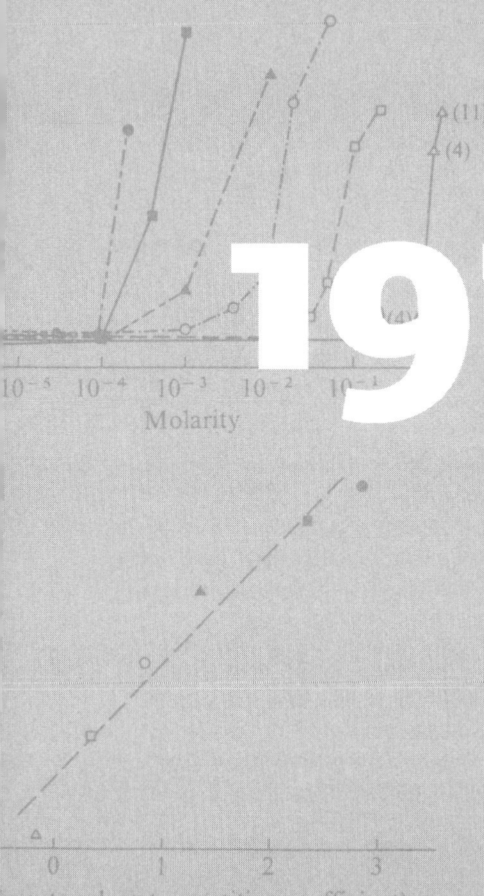

ncy of aliphatic alcohols in increasing the rate P decay. ●, C_8; ■, C_7; ▲, C_5; ○, C_4; □, C_3; e of the carbon chain is indicated by the number C_8 is *n*-octanol). An aliphatic alcohol (in fortified ter) was perfused for 30 min, beginning 30 min train; and then a train was given in the presence In contrast with ethanol, the higher alcohols plitudes of all e.p.s.p.s. PTP still decayed, however, ponential time course with a rate constant which ermined. Each data point represents the average from three different preparations, except for the number of preparations at each data point is that the effect of each alcohol increases markedly ange of concentration. *b*, The potency of the effect it alcohols (measured as the concentration, *C*, ase the rate of decay of PTP fourfold) correlates ilicity of the alcohol (as measured by the *n*-octanol: coefficient, *P*). The data points were obtained nterpolating the concentration of each alcohol ve a rate constant of PTP decay of 20×10^{-3} s^{-1}.) against ln*P* gives a straight line with the equation ln*P*+1.66. Other biological systems susceptible nilarly give straight lines in these coordinates[6]. Symbols as in (*a*).

ynaptic membrane organisation (for example, e number or effectiveness of "vesicle attach- In this latter case, the relaxation of this isation would correspond to the decay of d be directly affected by the fluidity of com-

San Diego, and by a grant from the National Alcohol Abuse and Alcoholism.

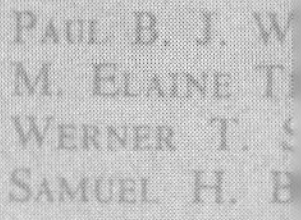
Paul B. J. W
M. Elaine T
Werner T. S
Samuel H. B

Department ...hiatry,
Veterans ...inistration Hospital,
...eg... ...nia 92161, and
D... ...m... ...iatry,
S...ol o... ...dici...
...ersit... ...nia, San Diego,
... Jolla, Ca... ...a 92093

Received January 2; accepted March 2, 1976.

1 Gage, P. W., and Hubbard, J. I., *J. Physiol., Lond.*, 184, 353–375
2 Schlapfer, W. T., Tremblay, J. P., Woodson, P. B. J., and *Brain Res.* (in the press).
3 Schlapfer, W. T., Woodson, P. B. J., Smith, G. A., Tremblay, J. S. H., *Nature*, 258, 623–625 (1975).
4 Paterson, S. J., *et al.*, *Biochim. biophys. Acta*, 266, 597–602 (197
5 Schlapfer, W. T., Woodson, P. B. J., Tremblay, J. P., and *Brain Res.*, 76, 267–280 (1974).
6 Hansch, D., and Anderson, S. M., *J. med. Chem.*, 10, 745–753 (
7 Rosenthal, J., *J. Physiol., Lond.*, 203, 121–133 (1969).
8 Wilson, D. F., and Skirboll, L. R., *Am. J. Physiol.*, 227, 92–95 (
9 Pfenninger, K. H., and Rovainen, C. M., *Brain Res.*, 72, 1–23 (

Single-channel currents recorded fron membrane of denervated frog muscle

THE ionic channel associated with the acetylch receptor at the neuromuscular junction of ske fibres is probably the best described channel i membranes. Nevertheless, the properties of channels are still unknown, as previous studies cerned with average population properties. N conductance fluctuations occurring in the preser were analysed to provide estimates for single c ductance and mean open times[1–3]. The value however, depended on assumptions about the s elementary conductance contribution—for exa the elementary contribution is a square pulse- Clearly, it would be of great interest to refine te conductance measurement in order to resol changes in conductance which are expected to single channels open or close. This has not be so far because of excessive extraneous backgr We report on a more sensitive method of c measurement, which, in appropriate conditions, crete changes in conductance that show ma features that have been postulated for single ioni

The key to the high resolution in the present lies in limiting the membrane area from which measured to a small patch, and thereby decrea ground membrane noise. This is achieved b closely the tip of a glass pipette, 3–5 μm in diam the muscle surface, thus isolating electrically a of membrane (Fig. 1). This method has been a viously in various modifications and mostly pipette tips to muscle[4], molluscan neurones[5,6], axon[7]. The pipette, which has fire-polished edg

FIFTEEN

Molecular switches for "animal electricity"

Fred J. Sigworth

The electric currents that flow across cell membranes carry information in our brain, trigger our heartbeat, and allow us to sense our environment. The currents are carried by ions, and are controlled by ion-channel proteins that are embedded in a membrane. The operation of single ion channels became clear in electrical recordings reported in 1976. The technique involved, known as the "patch clamp," for the first time allowed investigation of the behavior of single protein molecules, and has become a standard method in biology.

In the early 1970s, two young German neuroscientists were investigating the electrical events in synapses—the sites of contact between nerve cells, or between nerve and muscle cells. The scientists were Erwin Neher and Bert Sakmann, and that interest led them to develop an unprecedentedly sophisticated technique for measuring the current flow across cell membranes. These currents influence the function of all cells but are central to the action of nerve and muscle in particular. Neher and Sakmann's technique was described in a *Nature* paper[1] of 1976: it was an experimental tour de force, the first of several studies that in 1991 won them the Nobel Prize in Physiology or Medicine.

To see the setting for this work, we must go back to 1902 and to Julius Bernstein,[2] who formulated possible explanations for the "animal electricity" that can be measured in nerve and muscle cells. In one hypothesis he postulated that the cells are bounded by a thin membrane that, under some conditions, is permeable to the positively charged potassium ions that are plentiful inside cells. Because the potassium concentration outside cells is relatively low, the diffusion of ions across the membrane would set up an electrical potential difference. This "potassium-ion battery" explained the steady flow of electric currents from the cut ends of nerve or muscle fibers. When the cells were stimulated to evoke impulses, however, the currents became transiently much smaller, as if the membrane permeability suddenly and reversibly changed to become less ion-selective. This change in membrane permeability would, according to Bernstein's hypothesis, give rise to the transient reversals of electrical potential across membranes that are known as "action potentials" and characterize the activity of nerve and muscle cells.

Direct evidence for Bernstein's hypothesis, and a detailed understanding of the action potential, came in the 1940s. Alan Hodgkin and Andrew Huxley[3] showed that the nerve-cell membrane undergoes voltage-dependent changes in ion permeability. The ion gradients across the membrane are the power source for the process (we now know that these gradients—whether of sodium, potassium, or calcium ions—are maintained by "molecular pumps"). Hodgkin and Huxley found that a voltage change increases the sodium permeability, resulting in a larger voltage change; this regenerative amplification produces the action potential, which spreads along nerve or muscle fibers. For this work Hodgkin and Huxley received a share in the Nobel Prize in Physiology or Medicine in 1963.

Meanwhile, it became clear that something similar happens at synapses. During the 1950s and 1960s, Bernard Katz—also, in 1970, a Nobel laureate—and his colleagues worked intensively on the nerve-muscle synapse, the neuromuscular junction. Here, it turned out, membrane permeability is altered in response to a neurotransmitter substance called acetylcholine. Small packets of acetylcholine are released by the nerve ending and attach to the muscle membrane; they change the permeability of that membrane, trigger an action potential and so induce muscle contraction.

But what was the molecular nature of these permeability changes? What special properties of biological membranes make them sensitive to changes in voltage or to molecules of a neurotransmitter? These behaviors are now known as voltage-gating and ligand-gating of ion channels, a ligand being a molecule that binds to another, larger molecule.

Hints to the nature of the permeability changes came from various sources. One was work with neurotoxins. A snake toxin, alpha-bungarotoxin, causes paralysis by blocking the transmission at the neuromuscular junction. Very few molecules of bungarotoxin are needed to do this, implying that the blockage sites are discrete and few in number. The proteins to which bungarotoxin binds were biochemically isolated and purified, and were found to be ligand-gated ion channels that respond to acetylcholine; in other words, they were the acetylcholine receptors.

Another hint came from recordings of membrane currents, which showed an increase in electrical "noise" when the permeability mechanisms were activated, for example through the application of acetylcholine to a muscle fiber. These current fluctuations could be explained if the ionic current were carried by independent channels that open and close randomly. Analysis of the fluctuations yielded estimates of the conductance of single channels, and of the mean open time of the channels underlying the acetylcholine-activated current.

A third hint came from experiments with artificial cell membranes. In life, cell membranes are sheets composed of two layers of lipid (fat) molecules. Artificial membranes formed from purified lipid constituents have only low conductivity for ions. When the membrane is "doped" with certain molecules such as gramicidin A, however, the conductivity rises dramatically. At very low concentrations of gramicidin, discrete bursts of ionic conductivity were evident. These were interpreted as the contributions of single ion channels formed by gramicidin. They corresponded to currents that were difficult to measure (in the range of a picoampere, a million millionth of an ampere) but which, nevertheless, consist of the transport of millions of ions per second.

So we come to Neher and Sakmann and the challenge of measuring single-channel currents in a living cell. Before embarking on the experiments described in their 1976 paper,[1] Sakmann had worked with Katz on the frog neuromuscular junction and the acetylcholine sensitivity of the muscle membrane. Neher had worked with glass pipettes to make recordings of currents in snail neurons, and had also studied the currents carried by gramicidin A channels. Both researchers were experimental virtuosi—together they were uniquely prepared to carry out the elaborate experiments that pioneered the patch-clamp technique.

The basic technique is depicted in figure 1 of Neher and Sakmann's paper (p. 225). A single muscle cell was impaled by two microelectrodes, which are connected to an amplifier that provides electronic feedback. The effect is to hold the membrane potential "clamped" at a preset value, with reference to a bath electrode located outside the cell. Another electrode, consisting of a blunt micropipette filled with saline solution and connected to an amplifier, collected for measurement the current flowing through a tiny piece—a patch—of membrane beneath the pipette tip. It was thought that there would be only a few ion channels in the membrane patch, because of its small area and the low density of channels expected in the extrasynaptic region of the muscle fiber. Even so, the resolution of the recording was limited by thermal noise: the random motion of ions in the solution between the pipette tip and the cell membrane produces a fluctuating current that can easily obscure the single-channel currents. Neher and Sakmann tackled the problem by treating the muscle fiber with enzymes. This produced a "clean" membrane that the patch pipette could approach very closely, increasing the electrical resistance of the glass-membrane seal and decreasing the thermal noise.

The technique yielded what was arguably the first biophysical measurement in a living cell. Inside the patch pipette was a low concentration of acetylcholine or a related molecule of the same cholinergic chemical family. In the jargon of the trade, these activating molecules are called agonists. In this case, the agonist induced the acetylcholine receptor channels to "chatter"—open and close rapidly—and so allowed current to flow.

Roughly speaking, each pulse of current represented the residence time of an agonist molecule on the receptor—explaining, for example, why one type of cholinergic agonist (suberyldicholine, which binds tightly) yielded longer channel-open times than another (carbamylcholine, which binds weakly). The waiting time between pulses, on the other hand, reflected the time between the departure of one agonist molecule and the arrival of another. The randomness of the open and closed durations of the channel reflected the randomness of agonist binding and unbinding, and the randomness of the transitions between the open and closed states of the acetylcholine receptor. These molecular events could be observed only because the acetylcholine receptor acts as its own reporter, revealing its state through switching current on and off.

The first patch-clamp recordings were technically demanding in the extreme, and the technique did not become routine until after further advances had been made. One memorable Monday morning in early 1980, a few weeks after I had joined his laboratory, Erwin Neher cheerfully announced that my research project was going to be much easier than I had thought. Over the weekend he had discovered that under the right conditions a very tight seal, with a huge electrical resistance, formed between a patch pipette and a cell membrane. This seal greatly reduced the background noise, which was good news to me: my project was to record the small and very short-lived currents in voltage-gated sodium channels. The tight seal relaxed the requirements on the shape of the pipettes, and allowed those (like myself) who were not experimental virtuosi also to make excellent recordings.

I started work on improving the amplifiers, to exploit the increased sensitivity of the tight-seal recording technique. In the neighboring laboratory, Sakmann and Owen Hamill, working with redesigned pipettes, discovered that they could excise a bit of membrane from a cell, yielding an inside-out patch of membrane spanning the pipette tip. Meanwhile Alain Marty and Neher found that a new "whole-cell" configuration not only still allowed electrical recording but also permitted exchange of molecules from the cell's cytoplasm with the contents of the patch pipette. (A decade later, this technique was used to sample messenger RNA from neurons after making electrical recordings.) These approaches, described in a 1981 paper[4] and shown here in fig. 15.1, opened up many possibilities for studying molecular mes-

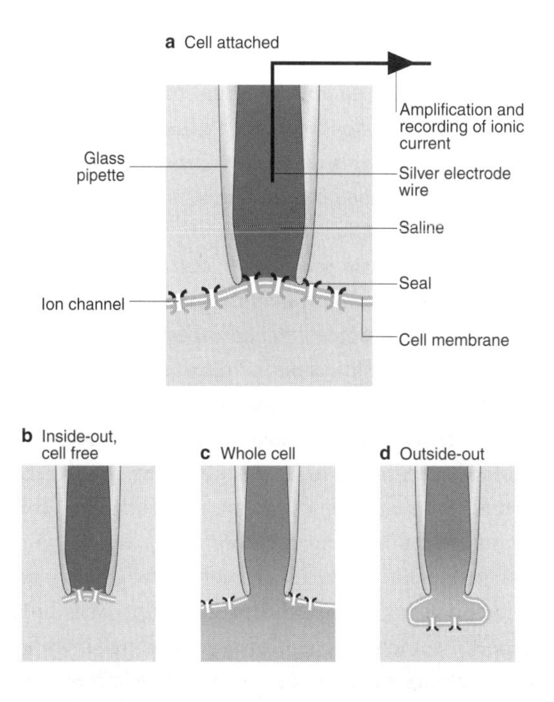

Fig. 15.1. Forms of the modern patch-clamp technique. a, The basic "cell-attached" recording configuration, in which the tip of a saline-filled glass pipette is sealed against the cell membrane and collects ionic current from a small membrane patch. b, Membrane patches can be excised from cells, allowing inside-out, "cell-free" experiments to be performed, for example in the investigation of channels activated by intracellular ligands. c, Alternatively, the patch can be ruptured after the seal is formed, providing direct access from the patch pipette to the inside of the cell and "whole-cell" recording. d, When the pipette is withdrawn from the cell, the membrane pinches off, forming an "outside-out" patch. To give an idea of the scale, the opening of the tip of a typical pipette used for patch-clamping is one micrometer across. A small cell is about ten micrometers across. (Figure modified from fig. 4-55 of *Molecular Biology of the Cell,* by Bruce Alberts *et al.,* 3d ed., Garland, New York, 1994.)

senger systems, and their effects on ionic currents, inside a cell as well as outside it.

The routine recording of single ion-channel currents allowed the various channel types to be distinguished simply by the sizes of their current pulses and by their patterns of openings and closings. Hints of the great diversity of ion channels were already becoming apparent in the 1980s, as several varieties of potassium and calcium channel currents were seen in patches from individual cells. And the mechanisms underlying touch, hearing, sight, and smell became clearer with the discovery of new types of channels that are gated by mechanical forces or by chemical signals from within the cells.

As for the acetylcholine receptor, the mechanism that couples the binding of agonists to the opening of the channel was worked out in great detail in the 1980s and 1990s. The fine structure of channel openings and closings, now recorded at high resolution, was interpreted using the mathematics of random processes. In this application of chemical kinetics at the single-molecule level, the binding of two agonist molecules and the channel open-closed transitions could all be described from the statistics of channel "chatterings" as seen in the electrical recordings.

Erwin Neher and Bert Sakmann went on to tackle bigger questions. Neher applied the high resolution of patch recordings to detect the small electrical changes that accompany secretion of substances from a cell following fusion between membrane-bound sacs inside the cell and the surface membrane. His new technique has opened a wide window on neurotransmitter release and other secretory processes. Sakmann returned to the study of synaptic transmission in the brain. In investigating the activity of neurotransmitter receptors and the modulation of synapse strength, he has used patch pipettes to record from the cell body and individual projections of nerve cells.

Thanks to techniques descended from Neher and Sakmann's work, the functional properties of ion channels are now quite well understood. It has emerged that ion channels are essential in a wide variety of physiological processes and that channel malfunction underlies some diseases of muscle and heart rhythm. Ion channels are involved in the causes of stroke and certain headache and neurodegenerative disorders, and their inactivity or overactivity also underlies diseases such as cystic fibrosis, and types of hypertension and diabetes.

What remains a challenge is working out the molecular structures of the proteins from which channels are made. What forms the "gate" which switches the ion current on and off? In the case of the acetylcholine receptor, approaches using mutations of the channel genes, chemical labelling, and

X-ray and electron crystallography are converging on an answer involving the physical occlusion of a narrow part of the pore.[5,6] And what about the selective permeability to potassium that gives rise to "animal electricity" in the first place? In 1998 MacKinnon and colleagues[7] determined by X-ray diffraction the structure of a potassium channel known as KcsA. In this structure an elegant, multi-stage ion selectivity filter can be seen.[8] It is potassium channels akin to KcsA, sparsely distributed in a cell membrane, that produce the selective permeability that over a century ago Bernstein first proposed as the origin of the electrical polarization of nerve and muscle cells.

References

1. Neher, E. & Sakmann, B. Single-channel currents recorded from membrane of denervated frog muscle fibres. *Nature* **260,** 799–802 (1976).
2. Bernstein, J. Untersuchungen zur Thermodynamik der bioelektrischen Ströme. *Pflügers Arch.* **92,** 521–562 (1902).
3. Hodgkin, A. L. & Huxley, A. F. A quantitative description of membrane current and its application to conduction and excitation in nerve. *J. Physiol. (Lond.)* **90,** 211–232 (1952).
4. Hamill, O. P., Marty, A., Neher, E., Sakmann, B. & Sigworth, F. J. Improved patch-clamp techniques for high-resolution current recording from cells and cell-free membrane patches. *Pflügers Arch.* **391,** 85–100 (1981).
5. Karlin, A. Emerging structure of the nicotinic acetylcholine receptors. *Nature Rev. Neurosci.* **3,** 102–114 (2002).
6. Unwin, N. Nicotinic acetylcholine receptors and the structural basis of fast synaptic transmission. *Phil. Trans. R. Soc. Lond. B* **355,** 1813–1829 (2000).
7. Doyle, D. A. *et al.* The structure of the potassium channel. Molecular basis of K^+ conduction and selectivity. *Science* **280,** 69–77 (1998).
8. Zhou, Y., Morais-Cabral, J. H., Kaufman, A. & MacKinnon, R. Chemistry of ion coordination and hydration revealed by a K^+ channel–Fab complex at 2.0 Å resolution. *Nature* **414,** 43–48 (2001).

Further reading

Hille, B. *Ionic Channels of Excitable Membranes* (Sinauer, Sunderland, MA, 2001).

Nicholls, J. G., Wallace, B. G., Martin, A. R. & Fuchs, P. A. *From Neuron to Brain* (Sinauer, Sunderland, MA, 2001).

Nobel e-Museum. *The Nobel Prize in Physiology or Medicine 1991* (http://www.nobel.se/medicine/laureates/1991/). (Also physiology or medicine prizes in 1963 and 1970.)

Sakmann, B. & Neher, E. (eds) *Single Channel Recording* (Plenum, New York, 1996).

1976

Single-channel currents recorded from membrane of denervated frog muscle fibres

Erwin Neher and Bert Sakmann

The ionic channel associated with the acetylcholine (ACh) receptor at the neuromuscular junction of skeletal muscle fibres is probably the best described channel in biological membranes. Nevertheless, the properties of individual channels are still unknown, as previous studies were concerned with average population properties. Macroscopic conductance fluctuations occurring in the presence of ACh were analysed to provide estimates for single channel conductance and mean open times[1–3]. The values obtained, however, depended on assumptions about the shape of the elementary conductance contribution—for example, that the elementary contribution is a square pulse-like event[2]. Clearly, it would be of great interest to refine techniques of conductance measurement in order to resolve discrete changes in conductance which are expected to occur when single channels open or close. This has not been possible so far because of excessive extraneous background noise. We report on a more sensitive method of conductance measurement, which, in appropriate conditions, reveals discrete changes in conductance that show many of the features that have been postulated for single ionic channels.

The key to the high resolution in the present experiments lies in limiting the membrane area from which current is measured to a small patch, and thereby decreasing background membrane noise. This is achieved by applying closely the tip of a glass pipette, 3–5 μm in diameter, on to the muscle surface, thus isolating electrically a small patch of membrane (Fig. 1). This method has been applied previously in various modifications and mostly with larger pipette tips to muscle[4], molluscan neu-

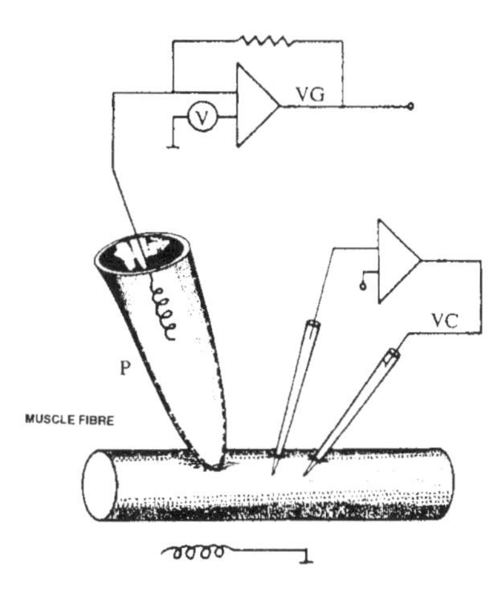

Fig. 1 Schematic circuit diagram for current recording from a patch of membrane with an extracellular pipette. VC, Standard two-microelectrode voltage clamp circuit to set locally the membrane potential of the fibre to a fixed value. P, Pipette, fire polished, with 3–5 μm diameter opening, containing Ringer's solution and agonist at concentrations between 2×10^{-7} and 6×10^{-5} M. d. c. resistance of the pipette: 2–5 MΩ. The pipette tip applied closely on to the muscle fibre within 200 μm of the intracellular clamp electrodes. VG, Virtual ground circuit, using a Function Modules Model 380K operational amplifier and a 500 MΩ feedback resistor to measure membrane current. The amplifier is mounted together with a shielded pipette holder on a motor-driven micromanipulator. V, Bucking potential and test signal for balancing of pipette leakage and measuring pipette resistance.

rones[5,6], and squid axon[7]. The pipette, which has fire-polished edges, is filled with Ringer's solution and contains the cholinergic agonist at micromolar concentrations. Its interior is connected to the input of a virtual-ground circuit, which clamps the potential inside the pipette to ground and at the same time measures current flowing through the pipette, that is, through the patch of membrane covered by the pipette opening. The interior of the muscle fibre is clamped locally to a fixed value by a conventional two-microelectrode clamp[8]. Thus, voltage-clamp conditions are secured across the patch of membrane under investigation. Since current densities involved are very small, a simple virtual ground inside the pipette is preferable to more complicated arrangements for stabilising potential described previously[6].

The dominant source of background noise in these measurements was the leakage shunt under the pipette rim between membrane and glass. It was constantly monitored by measuring the electrical conductance between pipette interior and bath. Discrete conductance changes could be resolved only when the conductance between pipette interior and bath decreased by a factor of four or more after contact between pipette and membrane. To minimise the leakage conductance, the muscle was treated with collagenase and protease[9]. This enzyme treatment digested connective tissue and the basement membrane, thereby enabling closer contact between glass and membrane. At the same time, however, it made the membrane fragile and more sensitive to damage by the approaching pipette. It did not, however, change the ACh sensitivity of the fibre or alter the properties of ACh-induced conductance fluctuations (E.N. and B.S., unpublished).

All experiments were carried out on the extrasynaptic region of denervated hypersensitive muscle fibres. The uniform ACh sensitivity found over most of the surface of these fibres greatly enhanced the probability of the occurrence of agonist-induced conductance changes at the membrane patch under investigation. Extrasynaptic ACh channels of denervated muscle fibres have mean open times which are about three to five times longer than those of endplate channels[1,10–12]. The longer duration facilitated the detection of conductance changes. Additional measures were taken which are known to either increase the size of the elementary current pulse or prolong its duration: the membrane was hyperpolarised up to −120 mV; suberyldicholine (SubCh) was used as an agonist in most of the experiments; the preparation was cooled to 6–8 °C.

Figure 2 shows a current recording taken in the conditions outlined above. Current can be seen to switch repeatedly between different levels. The discrete changes are interpreted as the result of opening and closing of individual channels. This interpretation is based on the very close similarity to single-channel recordings obtained in artificial membrane systems[13]. The preparation under study is, however, subject to a number of additional sources of artefact. Therefore it is necessary to prove that the recorded events do show the properties which are assigned to ionic chan-

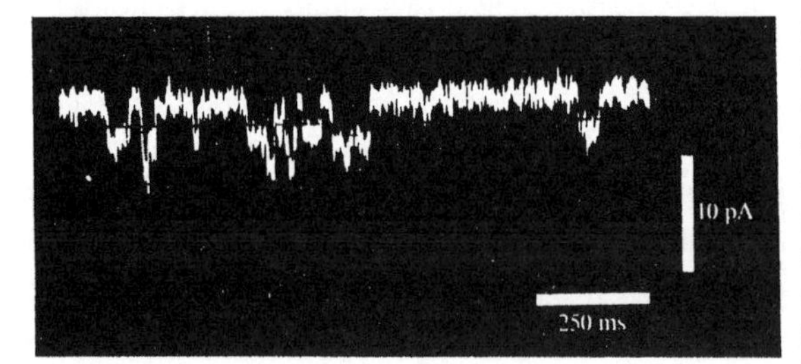

Fig. 2 Oscilloscope recording of current through a patch of membrane of approximately 10 μm². Downward deflection of the trace represents inward current. The pipette contained 2 × 10⁻⁷ M SubCh in Ringer's solution. The experiment was carried out with a denervated hypersensitive frog cutaneus pectoris (*Rana pipiens*) muscle in normal frog Ringer's solution. The record was filtered at a bandwidth of 200 Hz. Membrane potential: −120 mV. Temperature: 8 °C.

nels of the cholinergic system. These are: a correlation with the degree of hypersensitivity of the muscle membrane; an amplitude dependent on membrane potential as predicted by noise analysis; a mean length or channel open time, which should depend on voltage in a characteristic manner[2]; pharmacological specificity with different mean open times for different cholinergic agonists[14,15]. The experiments bore out all of the above-mentioned points as outlined below.

The frequency of occurrence of single blips depended on the sensitivity of the patch under investigation. A plot of the number of current pulses per second against the iontophoretically measured sensitivity of the membrane region, determined either immediately before or after the pipette experiment, revealed a distinct correlation between both quantities with a correlation coefficient of 0.91 for a linear regression (Fig. 3*b*). Student's *t* test assigned a significance better than 0.1% to the relationship.

To estimate the size of the current pulses amplitude histograms were calculated from the current recordings (Fig. 3*c*). The histograms show a prominent peak of gaussian shape around zero deviation, the width of which is a measure of the high frequency background noise of the current trace. Multiple, equally spaced peaks at larger deviations represent the probabilities that either one, two, or three channels are open simultaneously. The peak separation gives the amplitude of the single-channel contribution, which was 3.4 pA for the histogram shown in Fig. 3*c*. This was obtained from an experiment at -120 mV membrane potential. A similar histogram from the same muscle fibre obtained at -80 mV yields a current pulse amplitude of 2.2 pA. These two values extrapolate to an equilibrium potential of -7 mV. Channel conductance is estimated as 28 pmhos in this case. It scattered from fibre to fibre with a mean value of 22.4 ± 0.3 pmhos (mean $\pm$ s.e.; number of determinations $= 27$). This value is somewhat lower than the one derived from noise analysis at normal endplates, which is 28.6 pmhos for SubCh[15]. Higher order peaks in the histograms are not merely scaled images of the zero order peak. They tend to be smeared out due to non-uniformity of current pulse amplitudes. We cannot decide at present whether this is a real feature of the channels or a measurement artefact. Such an effect could arise if not all of the channels are located ideally in the central region of the pipette opening. Current contributions from peripherally located channels would only partially be picked up by the pipette. This source of error is also likely to lead to an underestimate of channel size if the pipette seal is not optimal.

Temporal analysis of the current records was carried out partly by measurement of individual channel length and averaging 40–50 mea-

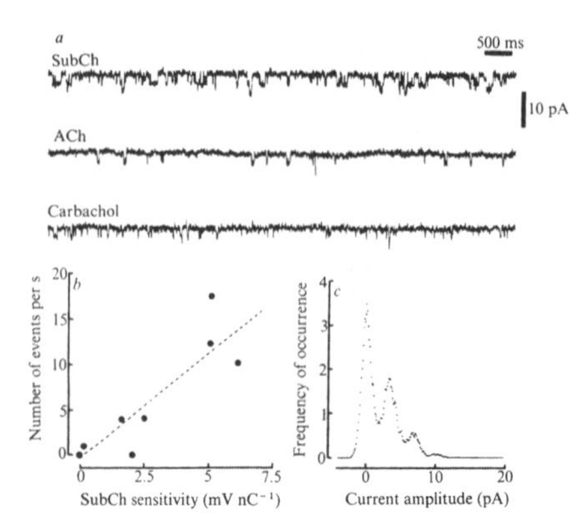

Fig. 3 Characterisation of single-channel currents. *a*, Comparison of current recordings obtained with different cholinergic agonists at concentrations of 2×10^{-7} M(SubCh), 2×10^{-6} M(ACh), and 6×10^{-5} M(carbachol). Downward deflection of the trace represents inward current. Three different experiments. Pen records replayed from analogue tape at a bandwidth of 100 Hz. All experiments at −120 mV membrane potential; 8 °C. *b*, Number of current blips per second is plotted against iontophoretic sensitivity of the membrane region under investigation. Sensitivity was determined by 100-ms iontophoretic pulses delivered from a pipette filled with 1 M SubCh (40 MΩ pipette resistance, 10 nA bucking current, sensitivity measured at resting potential). Pooled data from eight experiments. Broken line represents linear regression. *c*, Amplitude histogram of membrane current. Current traces were digitalised and baselines fitted by eye to data records, each 4 s in length. Frequency of occurrence of deviations from the baseline is shown in arbitrary units. Histograms were calculated on a PDP-11 computer; 2–8 histograms were averaged to obtain curves like the one shown. 8 °C; −120 mV membrane potential.

surements, and partly by calculation of the power spectrum of the current recordings. In the latter case, the cutoff frequency f_0 of the Lorentzian spectrum yielded an estimate of mean channel open time τ (or pulse duration) through the relationship $\tau = 1/(2\pi f_0)$. Values of mean open times obtained by the two methods were consistent within ± 30%. For SubCh as an agonist and at a temperature of 8 °C it was 45 ± 3 ms ($n = 11$) at −120 mV and 28 ± 3 ms ($n = 14$) at −80 mV. These values are approximately three times longer than the corresponding mean open times of endplate channels derived from noise analysis[15]. Note, however, that lengthening of channel durations by factors of three to five at extrajunctional sites with respect to endplate values has been measured independently by conventional noise analysis[12]. The voltage dependence of the values given above corresponds to an *e*-fold change per 80 mV, which is within the range of published values[2].

Channel open times were different when different cholinergic agonists were used (Fig. 3*a*). For −120 mV and 8 °C, mean channel open time was 45 ± 3 ms ($n = 11$) for SubCh, 26 ± 5 ms ($n = 4$) for ACh, and 11 ± 2 ms ($n = 3$) for carbachol. This sequence reflects the well known relationship between the open times of channels induced by these drugs at normal endplates[14,15] and at extrasynaptic membrane of hypersensitive fibres[12].

The results obtained so far, especially the pharmacological specificity, lead us to conclude that the observed conductance changes are indeed recordings of single-channel currents. They are consistent with the conclusions drawn from statistical analysis of endplate current fluctuations, and show that current contributions of individual channels are of the form of square pulses. In addition, analysis of areas under the peaks of histograms like Fig. 3*c* indicates that in our experimental conditions opening of individual channels is statistically independent, since the probabilities of zero, one, or two channels being open simultaneously follow—within the limits of experimental resolution—a Poisson distribution.

Recordings of single-channel currents finally resolves the third level of quantification in the process of neuromuscular transmission after the discovery of endplate currents and miniature endplate currents. It should facilitate discrimination between factors influencing the properties of single channels and agents creating or modifying different populations of channels.

We thank J. H. Steinbach for help with some experiments. Supported by a USPHS grant to Dr C. F. Stevens, and a stipend of the Max-Planck-Gesellschaft.

E.N.: *Yale University School of Medicine, Department of Physiology, New Haven, Connecticut 06510;* B.S.: *Max-Planck-Institut für Biophysikalische Chemie, 3400 Göttingen, Am Fassberg, West Germany*

Received January 26; accepted March 1, 1976 [Published 29 April]

References

1. Katz, B., and Miledi, R., *J. Physiol., Lond.*, **224**, 665–699 (1972).
2. Anderson, C. R., and Stevens, C. F., *J. Physiol., Lond.*, **235**, 655–691 (1973).
3. Ben Haim, D., Dreyer, F., and Peper, K., *Pflügers Arch. ges. Physiol.*, **355**, 19–26 (1975).
4. Strickholm, A., *J. gen. Physiol.*, **44**, 1073–1087 (1961).
5. Frank, K., and Tauc, L., in *The Cellular Function of Membrane Transport* (edit by Hoffman, J.) (Prentice Hall, Englewood Cliffs, New Jersey, 1963).
6. Neher, E., and Lux, H. D., *Pflügers Arch. ges. Physiol.*, **311**, 272–277 (1969).
7. Fishman, H. M., *Proc. natn. Acad. Sci. U.S.A.*, **70**, 876–879 (1973).
8. Takeuchi, A., and Takeuchi, N., *J. Neurophysiol.*, **22**, 395–411 (1959).
9. Betz, W., and Sakmann, B., *J. Physiol., Lond.*, **230**, 673–688 (1973).
10. Neher, E., and Sakmann, B., *Pflügers Arch. ges. Physiol.*, **355**, R63 (1975).
11. Dreyer, F., Walther, Ch., and Peper, K., *Pflügers Arch. ges. Physiol.*, **359**, R71 (1975).
12. Neher, E., and Sakmann, B., *J. Physiol., Lond.* (in the press).
13. Hladky, S. B., and Haydon, D. A., *Nature*, **225**, 451–453 (1970).
14. Katz, B., and Miledi, R., *J. Physiol., Lond.*, **230**, 707–717 (1973).
15. Colquhoun, D., Dionne, V. E., Steinbach, J. H., and Stevens, C. F., *Nature*, **253**, 204–206 (1975).

cles

1977

otide sequ ce f b t i ge
4 DNA

G. M. Air*, B. G. Barrell, N   Brow A. R. C   son, J. C. Fiddes,
ison III‡, P. M. Slocombe§ & M. Smith¹

of Molecular Biology, Hills Road, Cambridge CB2 2QH, UK

*ce for the genome of bacteriophage ΦX174
ly 5,375 nucleotides has been determined
and simple 'plus and minus' method. The
fies many of the features responsible for the
he proteins of the nine known genes of the
ding initiation and termination sites for the
NAs. Two pairs of genes are coded by the
DNA using different reading frames.*

f bacteriophage ΦX174 is a single-stranded,
of approximately 5,400 nucleotides coding for
eins. The order of these genes, as determined by
nes[2-4], is *A–B–C–D–E–J–F–G–H*. Genes *F*, *G*
structural proteins of the virus capsid, and gene
sequence work) codes for a small basic protein
of the virion. Gene *A* is required for double-
replication and single-strand synthesis. Genes
re involved in the production of viral single-
however, the exact function of these gene
clear as they may either be involved directly in
or be required for DNA packaging, which is
gle-strand production. Gene *E* is responsible for

cleotide sequences established in ΦX were
ts[5-7] obtained by the Burton and Petersen[8]
ocedure. The longer tracts could be obtained
ces of up to 10 nucleotides were obtained. More
ell[9] has improved the hydrazinolysis method to
r purine tracts. These results are included in the
n Fig. 1.

e ΦX sequences were obtained using partial
chniques, particularly with endonuclease IV
). Ziff *et al.*[12,13] used this enzyme in condi-
hydrolysis to obtain fragments 50–200 nucleo-
ch were purified as ^{32}P-labelled material by
on polyacrylamide gels. The fragments came
egion of the genome and the sequence of a 48-
fragment (band 6, positions 1,047–1,094) was
g mainly further degradation with endonuclease
exonuclease digestions.

strand DNA of ΦX has the same sequence as the ml
certain conditions, will bind ribosomes so that
fragment can be isolated and sequenced. Only on
was found. By comparison with the amino acid seq
was found that this ribosome binding site sequence c
initiation of the gene *G* protein[15] (positions 2,362–

At this stage sequencing techniques using prim
with DNA polymerase were being developed[16] a
synthesised a decanucleotide with a sequence comp
part of the ribosome binding site. This was used t
the intercistronic region between the *F* and *G* genes,
polymerase and ^{32}P-labelled triphosphates[18]. The r
tion technique[16] facilitated the sequence determin
labelled DNA produced. This decanucleotide-pri
was also used to develop the plus and minus metho
synthetic primers are, however, difficult to prep
DNA fragments generated by restriction enzym
readily available these have been used for most
reported here.

Another approach to DNA sequencing is to ma
copy using RNA polymerase with α-^{32}P-labelled
phates and then to determine the RNA sequen
established methods. Blackburn[19,20] used this a
intact single-stranded ΦX and on fragments c
digestion with endonuclease IV or with restricti
Sedat *et al.*[21] were extending their studies on the
nuclease IV fragments and their results, taken in
with the transcription of the DNA fragments[20],
sequence of the F protein[22], and the plus and mi
results, made it possible to deduce a sequence of 28
(positions 1,016–1,296, Fig. 1) within the *F* ge
scription of *Hin*dII fragment 10, amino acid sequenc
G gene, and the plus and minus method using *Hin*d
2 and 10 as primers, gave a sequence of 195
(positions 2,387–2,582, Fig. 1) at the N terminus
(ref. 24).

The 'plus and minus' method

Further work on the ΦX sequence has been done
the plus and minus method primed with restrictio

SIXTEEN

DNA sequencing: the silent revolution

Peter Little

The complete DNA sequence of any organism's genome—not least our own—provides unprecedented knowledge because nothing is left out. Traditionally, biological research has been based on incomplete information. So the possession of complete lists of genes and proteins will open up fresh vistas in biology and wholly new ways of studying ourselves. A landmark paper of 1977 not only established the central techniques for DNA sequencing, it also pioneered this innovation in thinking.

The historian's task would be much easier if revolutions were invariably accompanied by the crash of falling palace gates or some other noisy event. Certainly, given the incessant media coverage and the pronouncements of heads of state, future historians should have no difficulty in identifying the years 2000 and 2001 as revolutionary in the endeavor to define the DNA sequence of the human genome. This period saw the announcement and subsequent publication of the largely complete sequence. But few biological revolutions are like this; some even go unrecognized by the revolutionaries themselves. Such, I suspect, was the case when in 1977 Fred Sanger and his colleagues published their paper[1] "Nucleotide sequence of bacteriophage ΦX174 DNA."

I doubt that, at the time, anyone realized how far-reaching the ramifications of the paper would be (not least, as I shall mention, in one less-than-obvious way). I also doubt that many modern molecular biologists have any idea what an intractable problem DNA sequencing then seemed. DNA is fundamentally a very boring molecule, combining relative lack of chemical reactivity with massive size and numbing repetition. DNA is a coiled ladder whose "rungs" consist of just four chemicals, called bases and given the simple abbreviations of A, G, C, and T. The beautiful structure of DNA announced by James Watson and Francis Crick[2] in 1953 (see chapter 6) showed that the two "side rails" of DNA are perfectly parallel, because each rung is exactly the same length, and is made up of pairs of bases that obey a simple rule: base A always pairs with base T, and base G with base C (fig. 16.1).

The beauty of its structure emerges only when DNA is split into two down the long axis of its ladder, as happens when a cell divides and the DNA has to be copied during the process known as replication. Although every rung is broken in half, each half rung remains attached to its rail and the rule of A with T and G with C means it is a simple matter to rebuild the missing half of the DNA. This feature immediately suggested how DNA could be faithfully copied and passed from cell to cell, and from generation to generation.

In the twenty years following Watson and Crick's paper came elucidation of how DNA stores information and the breaking of the "genetic code." The truth was marvelously simple: it is the order of the bases, read in groups of three down the DNA, that is the code. Through an intermediary molecule, messenger RNA, each group of three—a "codon"—encodes one or another of the twenty amino acids that, strung together, make up the proteins in

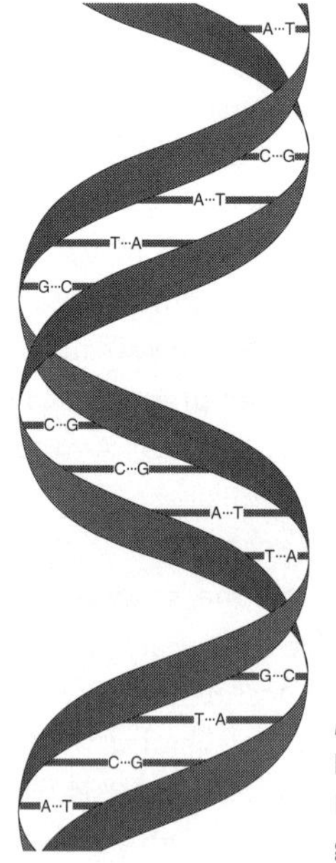

Fig. 16.1. The famous double helix of DNA, showing the rule of base pairing: A with T, and G with C. This principle means that DNA can be faithfully copied and passed from cell to cell, and from generation to generation.

our bodies. Once a cell has a way of making specific proteins, all of life follows because every living organism is made of protein and the products of protein-catalyzed reactions.

The challenge this understanding posed was simple: if we could read the codons in DNA by identifying the chemical composition of the rungs and the order or sequence of the bases (hence the term "DNA sequencing"), then we could figure out the amino acids in any protein encoded in DNA. Sanger had already worked on a similar problem of establishing the order of amino acids in a protein chain, that of insulin. The technique involved using a range of chemical reactions to break proteins at the location of different amino acids, and then combining the information gleaned from different reactions: for this Sanger won his first Nobel Prize, that in chemistry in 1958. Sanger's prize was well deserved because this was the first proof that a protein had a specific amino-acid sequence: the implication was that there had to be some mechanism within a cell to store the sequence, and of course we now know that this is the very essence of the genetic code.

Many people had thought about using similar approaches for DNA sequencing. But the difficulty is that DNA is too simple—there are just the four bases, and so chemical attack breaks up the DNA too indiscriminately. (In fact such an approach was finally made to work by Allan Maxam and Walter Gilbert,[3] for which Gilbert was to share the 1980 Nobel Prize in Chemistry.) However, Sanger and his colleagues developed an ingenious enzymatic method based on making copies of DNA using purified DNA and DNA polymerases, the enzymes that copy DNA strands during replication. Sanger realized that leaving out one of the four bases in the reaction would cause the copying reaction to stop at that point, with the length of the copied strand revealing the location of the omitted base. This became the basis of the "plus minus" method that he and Alan Coulson[4] published in 1975. It was a difficult technique that depended on carefully reducing the amount of one of the four nucleotides in a DNA-synthesis reaction, thus causing the synthesis to stop, randomly, at the nucleotide that was present at low amounts in the reaction; the length of the resulting mixture of DNA molecules indicated the location of all, for example, A bases. Similar experiments in parallel yielded the locations of C, G, and T bases.

This was one breakthrough: until then, establishing the sequence of bases in the DNA rungs had foundered on the relative chemical inactivity of DNA. Some success had been achieved using chemical degradation; or by making complete or partial copies of DNA into the more chemically reactive RNA;[5] or even by the heroic route of tricking the protein-synthesis machinery of cells into making proteins[6] directly from DNA, rather than from

RNA, and then sequencing the resulting proteins! Difficult though it was, the "plus minus" method was a huge improvement on all of these baroque techniques.

Sanger's problems did not end with availability of the enzymatic techniques for sequencing DNA, because these techniques did not work over more than a few hundred rungs at a time. This was clearly not sufficient to tackle the 5,375 DNA rungs of ΦX174, a bacteriophage (a virus that infects bacteria). The second breakthrough was putting the sequencing technology together with the idea of breaking the DNA into overlapping fragments using enzymes that could break DNA at specific sequences. These enzymes are called restriction enzymes, and their discovery won Werner Arber, Daniel Nathans, and Hamilton Smith the 1978 Nobel Prize in Physiology or Medicine.

The important principle in using restriction enzymes was to break the DNA molecules into fragments that could be fully sequenced. More importantly still, using fragments made from different enzymes would result in the same region of the original being sequenced several times; computers could readily identify the overlaps and join the sequences to generate the whole DNA sequence. This was the second key element in producing the complete DNA sequence of ΦX174 (fig. 16.2).

With hindsight, the revolution seems easy to spot. This was the first paper to report the complete DNA content of an organism. And the paper

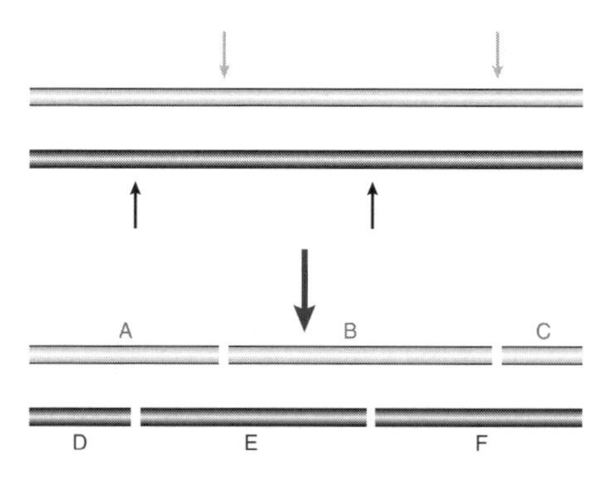

Fig. 16.2. Restriction enzymes and DNA sequencing. ΦX174 DNA is 5,375 base pairs in length, much too long to be analyzed in a single sequencing reaction. But if the DNA is cut with a restriction enzyme (top arrows) it breaks into smaller fragments (A, B, and C) that can be sequenced, at least in part. If a second restriction enzyme is used to cut the DNA at different points (middle arrows), fragments D, E, and F are produced and can also be sequenced. If the sequences derived from all six fragments A to F are then combined, the result is the whole DNA sequence. In practice, Sanger *et al.*[1] had to use more than two enzymes, and ΦX174 DNA is actually a circle, but that makes no difference to the concepts outlined here.

seems neatly to mark the beginning of DNA sequencing work that has moved in complexity from viruses to bacteria to humans, and from the 5,375 DNA rungs of ΦX174 to the enormous three billion of human DNA. True, Sanger[7] had still to go on to develop the "di-deoxy" method of sequencing that was at the heart of the human genome project and other sequencing endeavors. Di-deoxy sequencing uses blocked nucleotides to terminate DNA synthesis randomly, so that by mixing blocked and nonblocked nucleotides synthesis of a DNA chain can be terminated at, for example, every A base. The specificity of this reaction and its great simplicity compared to the delicate plus minus method, combined with the technology to separate DNA fragments that differ by only one base in length, lies at the very heart of later sequencing projects. But the completed ΦX174 sequence combined the practical and conceptual framework for future research, and the work featured prominently in the wonderful decision to award Sanger a share in another Nobel Prize (in chemistry, in 1980).

We now know the complete DNA sequence of over one hundred organisms other than ourselves, ranging from the fruit fly to primitive bacteria, and including human pathogens and organisms central to experimental research. At the time of writing, the sequences of more than 550 genomes are in the process of being determined. The resulting information is being brought to bear on such questions as the evolutionary relationships of bacteria, how certain bacteria cause disease, and the mechanisms of molecular evolution. The genomes of human pathogens that have now been sequenced, such as those that cause malaria, will provide a plethora of new targets for drug treatments. In short, genome sequencing powers whole swaths of biology and will underpin biological research in the decades and centuries to come.

But was the description of the methods for genome sequencing the truly revolutionary content of Sanger's paper? I think not. Rather its greater significance lies in an entirely new concept that had nothing directly to do with sequencing technology itself. It is found in just one phrase, buried in the middle of the paper (p. 246), "In considering the sequence of ΦX174 as a functional unit. . . ." This statement could be made only because Sanger in principle knew everything that there was to know about the genes of ΦX174 and he could be certain of this because the DNA sequence was complete. Until this moment, research scientists had always accepted that they must necessarily be ignorant of parts of the biological problems that they studied; they could never have imagined that a day would come when they might have the entire "parts list" of proteins that make up a complicated organism such as ourselves. We did not realize—nor, without hindsight, could we have realized—the implications that would flow from the idea of having complete knowledge about an organism, and that less than a quarter century later we would be able to apply such thinking to ourselves and to our own genetic composition.

Having complete knowledge of the genes of any organism opens wholly new ways to do biology. We now speak of analyzing the "transcriptome" (the complete messenger RNA content of cells) and the "proteome" (analogously, the complete protein complement). These studies have already led to new methods of classifying cancers,[8] and of identifying coordinately controlled genes and protein functions,[9] which are most likely the targets of the next generation of therapeutic drugs. They have also led to new ways of analyzing evolutionary relationships of organisms, and even to the discovery that between 250 and 350 genes are all that are required to make a simple living organism.[10]

This is the real revolution. We are presented with the idea that living organisms are organized through networks of subtle interactions where the behavior of the whole cannot readily be predicted from that of its components. But in parallel we have developed the knowledge to study the whole instead of its parts. The biological problem we face is in understanding such networks, and the complete knowledge of our protein composition that is emerging from genome studies will be the key contributor to that understanding.

This is the future of biology and the challenge is clear: to develop the analytical and technical abilities to study the isolated systems that we have discovered over the past years within a complete biological context. Now we can follow where Sanger and his colleagues led. One day someone will write in a paper, "In considering the sequence of the human being as a functional unit. . . ." That will be the sound of crashing palace gates.

References

1. Sanger, F. *et al.* Nucleotide sequence of bacteriophage ΦX174 DNA. *Nature* **265,** 687–695 (1977).
2. Watson, J. D. & Crick, F. H. C. A structure for deoxyribose nucleic acid. *Nature* **171,** 737–738 (1953).
3. Maxam, A. M. & Gilbert, W. A new method for sequencing DNA. *Proc. Natl Acad. Sci. USA* **74,** 560–564 (1977).
4. Sanger, F. & Coulson, A. R. A rapid method for determining sequences in DNA by primed synthesis with DNA polymerase. *J. Mol. Biol.* **94,** 441–448 (1975).
5. Sanger, F. *et al.* Use of DNA polymerase I primed by a synthetic oligonucleotide to determine a nucleotide sequence in phage f1 DNA. *Proc. Natl Acad. Sci. USA* **70,** 1209–1213 (1973).

6. Robertson, H. D. *et al.* Isolation and sequence analysis of a ribosome-protected fragment from bacteriophage ΦX174 DNA. *Nature New Biol.* **241,** 38–40 (1973).
7. Sanger, F., Nicklen, S. & Coulson, A. R. DNA sequencing with chain-terminating inhibitors. *Proc. Natl Acad. Sci. USA* **74,** 5463–5467 (1977).
8. Golub, T. R. *et al.* Molecular classification of cancer: class discovery and class prediction by gene expression monitoring. *Science* **286,** 531–537 (1999).
9. Bassett, D. E., Eisen, M. B. & Boguski, M. S. Gene expression informatics—it's all in your mind. *Nature Genet.* **21** (suppl.) 51–55 (1999).
10. Hutchison, C. A. *et al.* Global transposon mutagenesis and a minimal *Mycoplasma* genome. *Science* **286,** 2165–2169 (1999).

Further reading

Sequences of the human genome were published on 15 and 16 February 2001, in *Nature* and *Science* respectively. Gaps of difficult-to-sequence stretches remained, however, and work has continued to provide a fully complete sequence. See also http://www.nature.com/genomics/human/ and http://www.sciencemag.org/content/vol291/issue5507/.

Davies, K. *Cracking the Genome: Inside the Race to Unlock Human DNA* (Free Press, New York, 2000).

Nobel e-museum. *The Nobel Prize in Chemistry 1980* (http://www.nobel.se/chemistry/laureates/1980/). (Also chemistry prize in 1958 and physiology or medicine prize in 1978.)

Ridley, M. *Genome: The Autobiography of a Species in 23 Chapters* (HarperCollins, London, 2000).

1977

Nucleotide sequence of bacteriophage ΦX174 DNA

F. Sanger, G. M. Air, B. G. Barrell,
N. L. Brown, A. R. Coulson, J. C. Fiddes, C. A. Hutchison III,
P. M. Slocombe, and M. Smith

A DNA sequence for the genome of bacteriophage ΦX174 of approximately 5,375 nucleotides has been determined using the rapid and simple 'plus and minus' method. The sequence identifies many of the features responsible for the production of the proteins of the nine known genes of the organism, including initiation and termination sites for the proteins and RNAs. Two pairs of genes are coded by the same region of DNA using different reading frames.

The genome of bacteriophage ΦX174 is a single-stranded, circular DNA of approximately 5,400 nucleotides coding for nine known proteins. The order of these genes, as determined by genetic techniques[2–4], is *A–B–C–D–E–J–F–G–H*. Genes *F, G* and *H* code for structural proteins of the virus capsid, and gene *J* (as defined by sequence work) codes for a small basic protein that is also part of the virion. Gene *A* is required for double-stranded DNA replication and single-strand synthesis. Genes *B, C* and *D* are involved in the production of viral single-stranded DNA: however, the exact function of these gene products is not clear as they may either be involved directly in DNA synthesis or be required for DNA packaging, which is coupled with single-strand production. Gene *E* is responsible for lysis of the host.

The first nucleotide sequences established in ΦX were pyrimidine tracts[5–7] obtained by the Burton and Petersen[8] depurination procedure. The longer tracts could be obtained pure and sequences of up to 10 nucleotides were obtained. More recently Chadwell[9] has improved the hydrazinolysis method to obtain the longer purine tracts. These results are included in the sequence given in Fig. 1.

More extensive ΦX sequences were obtained using partial degradation techniques, particularly with endonuclease IV (refs 10 and 11). Ziff *et al.*[12,13] used this enzyme in conditions of partial hydrolysis to obtain fragments 50–200 nucleotides long which were purified as ^{32}P-labelled material by electrophoresis on polyacrylamide gels. The fragments came from the same region of the genome and the sequence of a 48-nucleotide long fragment (band 6, positions 1,047–1,094) was determined using mainly further degradation with endonuclease IV and partial exonuclease digestions.

Another 50-nucleotide long fragment was obtained by Robertson *et al.*[14] as a ribosome binding site. The viral (or plus) strand DNA of ΦX has the same sequence as the mRNA and, in certain conditions, will bind ribosomes so that a protected fragment can be isolated and sequenced. Only one major site was found. By comparison with the amino acid sequence data it was found that this ribosome binding site sequence coded for the initiation of the gene *G* protein[15] (positions 2,362–2,413).

At this stage sequencing techniques using primed synthesis with DNA polymerase were being developed[16] and Schott[17] synthesised a decanucleotide with a sequence complementary to part of the ribosome binding site. This was used to prime into the intercistronic region between the *F* and *G* genes, using DNA polymerase and ^{32}P-labelled triphosphates[18]. The ribo-substitution technique[16] facilitated the sequence determination of the labelled DNA produced. This decanucleotide-primed system was also used to develop the plus and minus method[1]. Suitable synthetic primers are, however, difficult to prepare and as DNA fragments generated by restriction enzymes are more readily available these have been used for most of the work reported here.

Another approach to DNA sequencing is to make an RNA copy using RNA polymerase with α-^{32}P-labelled ribotriphosphates and then to determine the RNA sequence by more established methods. Blackburn[19,20] used this approach on intact single-stranded ΦX and on fragments obtained by digestion with endonuclease IV or with restriction enzymes. Sedat *et al.*[21] were extending their studies on the larger endonuclease IV fragments and their results, taken in conjunction with the transcription of the DNA fragments[20], amino acid sequence of the F protein[22], and the plus and minus method results, made it possible to deduce a sequence of 281 nucleotides (positions 1,016–1,296, Fig. 1) within the *F* gene[23].

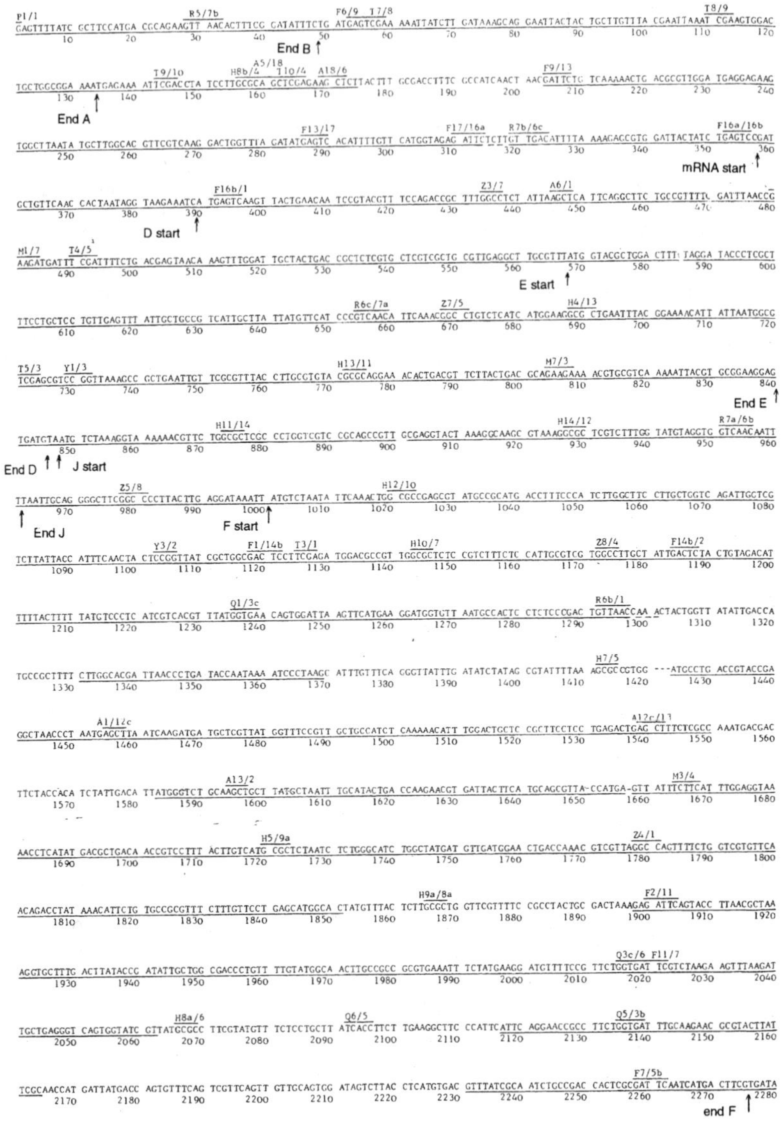

```
P1/1                                  R5/7b                      F6/9  T7/8                                                                                          T8/9
GACTTTTATC GCTTCCATGA CGCAGAAGTT AACACTTTCG GATATTTCTG ATGAGTCGAA AAATTATCTT GATAAAGCAG GAATTACTAC TGCTTGTTTA CGAATTAAAT CGAAGTGGAC
        10         20         30         40         50         60         70         80         90        100        110        120
                                              End B ↑

                                      A5/18
                      T9/10     H8b/4   T10/4  A18/6                                F9/13
TGCTGGCGGA AAATGAGAAA ATTCGACCTA TCCTTGCGCA GCTCGAGAAG CTCTTACTTT GCGACCTTTC GCCATCAACT AACGATTCTG TCAAAAACTG ACGCGTTGGA TGAGGAGAAG
       130  ↑     140        150        160        170        180        190        200        210        220        230        240
     End A

                                             F13/17                      F17/16a   R7b/6c                                  F16a/16b
TGGCTTAATA TGCTTGGCAC GTTCGTCAAG GACTGGTTTA GATATGAGTC ACATTTTGTT CATGGTAGAG ATTCTCTTGT TGACATTTTA AAAGACCGTG GATTACTATC TGAGTCCGAT
       250        260        270        280        290        300        310        320        330        340        350        360
                                                                                                                         mRNA start ↑

                                 F16b/1                                      Z3/7       A6/1
GCTGTTCAAC CACTAATAGG TAAGAAATCA TGAGTCAAGT TACTGAACAA TCCGTACGTT TCCAGACCGC TTTGGCCTCT ATTAAGCTCA TTCAGGCTTC TGCCGTTTTG GATTTAACCG
       370        380        390        400        410        420        430        440        450        460        470        480
                     D start ↑

M1/7    T4/5
AAGATGATTT CGATTTTCTG ACGAGTAACA AAGTTTGGAT TGCTACTGAC CGCTCTCGTG CTCGTCGCTG CGTTGAGGCT TGCGTTTATG GTACGCTGGA CTTTGTAGGA TACCCTCGCT
       490        500        510        520        530        540        550        560        570        580        590        600
                                                                                      E start ↑

                                                       R6c/7a     Z7/5                  H4/13
TTCCTGCTCC TGTTGAGTTT ATTGCTGCCG TCATTGCTTA TTATGTTCAT CCCGTCAACA TTCAAACGGC CTGTCTCATC ATGGAAGGCG CTGAATTTAC GGAAAACATT ATTAATGGCG
       610        620        630        640        650        660        670        680        690        700        710        720

T5/3    Y1/3                                           H13/11                           M7/3
TCGAGCGTCC GGTTAAAGCC GCTGAATTGT TCGCGTTTAC CTTGCGTGTA CGCGCAGGAA ACACTGACGT TCTTACTGAC GCAGAAGAAA ACGTGCGTCA AAAATTACGT GCGGAAGGAG
       730        740        750        760        770        780        790        800        810        820        830        840
                                                                                                                             End E ↑

                                 H11/14                                                 H14/12                           R7a/6b
TGATGTAATG TCTAAAGGTA AAAAACGTTC TGGCGCTCGC CCTGGTCGTC CGCAGCCGTT GCGAGGTACT AAAGGCAAGC GTAAAGGCGC TCGTCTTTGG TATGTAGGTG GTCAACAATT
       850        860        870        880        890        900        910        920        930        940        950        960
End D ↑ ↑ J start

                 Z5/8                                        H12/10
TTAATTGCAG GGGCTTCGGC CCCTTACTTG AGGATAAATT ATGTCTAATA TTCAAACTGG CGCCGAGCGT ATGCCGCATG ACCTTTCCCA TCTTGGCTTC CTTGCTGGTC AGATTGCTCG
↑      970        980        990       1000 ↑     1010       1020       1030       1040       1050       1060       1070       1080
 End J                              F start

                      Y3/2       F1/14b  T3/1             H10/7                             Z8/4       F14b/2
TCTTATTACC ATTTCAACTA CTCCGGTTAT CGCTGGCGAC TCCTTCGAGA TGGACGCCGT TGGCGCTCTC CGTCTTTCTC CATTGCGTCG TGGCCTTGCT ATTGACTCTA CTGTAGACAT
      1090       1100       1110       1120       1130       1140       1150       1160       1170       1180       1190       1200

                                 Q1/3c                                                            R6b/1
TTTTACTTTT TATGTCCCTC ATCGTCACGT TTATGGTGAA CAGTGGATTA AGTTCATGAA GGATGGTGTT AATGCCACTC CTCTCCCGAC TGTTAACCAA ACTACTGGTT ATATTGACCA
      1210       1220       1230       1240       1250       1260       1270       1280       1290       1300       1310       1320

                                                                                                  H7/5
TGCCGCTTTT CTTGGCACGA TTAACCCTGA TACCAATAAA ATCCCTAAGC ATTTGTTTCA GGGTTATTTG ATATCTATAG CGTATTTTAA AGCGCCGTGG ---ATGCCTG ACCGTACCGA
      1330       1340       1350       1360       1370       1380       1390       1400       1410       1420       1430       1440

           A1/12c                                                                                            A12c/13
GGCTAACCCT AATGAGCTTA ATCAAGATGA TGCTCGTTAT GGTTTCCGTT GCTGCCATCT CAAAAACATT TGGACTGCTC CGCTTCCTCC TGAGACTGAG CTTTCTCGCC AAATGACGAC
      1450       1460       1470       1480       1490       1500       1510       1520       1530       1540       1550       1560

                                 A13/2                                                                               M3/4
TTCTACCACA TCTATTGACA TTATGGGTCT GCAAGCTGCT TATGCTAATT TGCATACTGA CCAAGAACGT GATTACTTCA TGCAGCGTTA-CCATGA-GTT ATTTCTTCAT TTGGAGGTAA
      1570       1580       1590       1600       1610       1620       1630       1640       1650       1660       1670       1680

                                 H5/9a                                                           Z4/1
AACCTCATAT GACGCTGACA ACCGTCCTTT ACTTGTCATG CGCTCTAATC TCTGGGCATC TGGCTATGAT GTTGATGGAA CTGACCAAAC GTCGTTAGGC CAGTTTTCTG GTCGTGTTCA
      1690       1700       1710       1720       1730       1740       1750       1760       1770       1780       1790       1800

                                                                  H9a/8a                                F2/11
ACAGACCTAT AAACATTCTG TGCCGCGTTT CTTTGTTCCT GAGCATGGCA CTATGTTTAC TCTTGCGCTG GTTCGTTTTC CGCCTACTGC GACTAAAGAG ATTCAGTACC TTAACGCTAA
      1810       1820       1830       1840       1850       1860       1870       1880       1890       1900       1910       1920

                                                                                                   Q3c/6 F11/7
AGGTGCTTTG ACTTATACCG ATATTGCTGG CGACCCTGTT TTGTATGGCA ACTTGCCGCC GCGTGAAATT TCTATGAAGG ATGTTTTCCG TTCTGGTGAT TCGTCTAAGA AGTTTAAGAT
      1930       1940       1950       1960       1970       1980       1990       2000       2010       2020       2030       2040

                      H8a/6                      Q6/5                                              Q5/3b
TGCTGAGGGT CAGTGGTATC GTTATGCGCC TTCGTATGTT TCTCCTGCTT ATCACCTTCT TGAAGGCTTC CCATTCATTC AGGAACCGCC TTCTGGTGAT TTGCAAGAAC GCGTACTTAT
      2050       2060       2070       2080       2090       2100       2110       2120       2130       2140       2150       2160

                                                                                                      F7/5b
TCGCAACCAT GATTATGACC AGTGTTTCAG TCGTTCAGTT GTTGCAGTGG ATAGTCTTAC CTCATGTGAC GTTTATCGCA ATCTGCCGAC CACTCGCGAT TCAATCATGA CTTCGTGATA
      2170       2180       2190       2200       2210       2220       2230       2240       2250       2260       2270     ↑ 2280
                                                                                                                          end F
```

Fig. 1 For full description, see p. 244.

R1/9 H6/3
AAAGATTGAG TGTGAGGTTA TAACCGAAGC GGTAAAAATT TTAATTTTTG CCGCTGAGGG GTTGACCAAG CGAAGCGCGG TAGGTTTTCT GCTTAGGAGT TTAATCATGT TTCAGACTTT
2290 2300 2310 2320 2330 2340 2350 2360 2370 2380 2390 2400
G start ↑

A2/16 A16/15a M4/10 A15a/3 R9/10
TATTTCTCGC CACAATTCAA ACTTTTTTTC TGATAAGCTG GTTCTCACTT CTGTTACTCC AGCTTCTTCG GCACCTGTTT TACAGACACC TAAAGCTACA TCGTCAACGT TATATTTTGA
2410 2420 2430 2440 2450 2460 2470 2480 2490 2500 2510 2520

M10/9 R10/2
TAGTTTGACG GTTAATGCTG GTAATGGTGG TTTTCTTCAT TGCATTCAGA TGGATACATC TGTCAACGCC GCTAATCAGG TTGTTTCAGT TGGTGCTGAT ATTGCTTTTG ATGCCGACCC
2530 2540 2550 2560 2570 2580 2590 2600 2610 2620 2630 2640

F5b/8 M9/2
TAAATTTTTT GCCTGTTTGG TTCGCTTTGA GTCTTCTTCG GTTCCGACTA CCCTCCCGAC TGCCTATGAT GTTTATCCTT TGGATGGTCG CCATGATGGT GGTTATTATA CCGTCAAGGA
2650 2660 2670 2680 2690 2700 2710 2720 2730 2740 2750 2760

Y2/5 F8/4
CTGTGTGACT ATTGACGTCC TTCCCCGTAC GCCGGGCAAT AACGTCTACG TTGGTTTCAT GGTTTGGTCT AACTTTACCG CTACTAAATG CCGCGGATTG GTTTCGCTGA ATCAGGTTAT
2770 2780 2790 2800 2810 2820 2830 2840 2850 2860 2870 2880

Q3b/4 H3/2
TAAAGAGATT ATTTGTCTCC AGCCACTTAA GTGAGGTGAT TTATGTTTGG TGCTATTGCT GGCGGTATTG CTTCTGCTCT TGCTGGTGGC GCCATGTCTA AATTGTTTGG AGGCGGTCAA
2890 2900 2910 2920 2930 2940 2950 2960 2970 2980 2990 3000
End G ↑ H start ↑

Y5/4 Q4/7 Q7/2
AAAGCCGCCT CCGGTGGCAT TCAAGGTGAT GTGCTTGCTA CCGATAACAA TACTGTAGGC ATGGGTGATG CTGGTATTAA ATCTGCCATT CAAGGCTCTA ATGTTCCTAA CCCTGATGAG
3010 3020 3030 3040 3050 3060 3070 3080 3090 3100 3110 3120

Z1/2 A3/9
GCCGCCCCTA GTTTTGTTTC GTGTGCTATT GCTAAAGCTG GTAAAGGACT TCTTGAAGGT ACGTTGCAGG CTGGCACTTC TGCCGTTTCT GATAAGTTGC TTGATTTGGT TGGACTTGGT
3130 3140 3150 3160 3170 3180 3190 3200 3210 3220 3230 3240

A9/12d R2/6a
GCCAAGTCTG CCGCTGATAA AGGAAAGGAT ACTCGTGATT ATCTTGCTGC TGCATTTCCT GAGCTTAATG CTTGGGAGCG TGCTGGTGCT GATGCTTCCT CTGCTGGTAT GGTTGACGCC
3250 3260 3270 3280 3290 3300 3310 3320 3330 3340 3350 3360

Y4/1 F4/14a A12d/7c F14a/12
GGATTTGAGA ATCAAAAAGA GCTTACTAAA ATGCAACTGG ACAATCAGAA AGAGATTGCC GAGATGCAAA ATGAGACTCA AAAAGAGATT GCTGGCATTC AGTCGGCGAC TTCACGCCAG
3370 3380 3390 3400 3410 3420 3430 3440 3450 3460 3470 3480

F12/10
AATACGAAAG ACCAGGTATA TGCACAAAAT GAGATGCTTG CTTATTC-AC AGAAGGAGTC TACTGCTGCG TTGCGTCTAT TATGGAAAAC ACCAATCTTT CCAAGCAACA GCAGGTTTCC
3490 3500 3510 3520 3530 3540 3550 3560 3570 3580 3590 3600

H2/9b A7c/8 F10/15 R6a/4
GAGATTATGC GCCAAATGCT TACTCAAGCT CAAACGGCTG GTCAGTATTT TACCAATGAC CAAATCAAAG AAATGACTCG CAAGGTTAGT GCTGAGGTTG ACTTAGTTCA TCAGCAAACG
3610 3620 3630 3640 3650 3660 3670 3680 3690 3700 3710 3720

F15/5c M2/5 H9b/1 A8/14
CAGAATCAGC GGTATGGCTC TTCTCATATT GGCGCTACTG CAAAGGATAT TTCTAATGTC GTCACTGATG CTGCTTCTGG TGTGGTTGAT ATTTTTCATG GTATTGATAA AGCTGTTGCC
3730 3740 3750 3760 3770 3780 3790 3800 3810 3820 3830 3840

A14/7b
GATACTTGGA ACAATTTCTG GAAAGACGGT AAAGCTGATG GTATTGGCTC TAATTTGTCT AGGAAATAAC CGTCAGGATT GACACCCTCC CAATTGTATG TTTTCATGCC TCCAAATCTT
3850 3860 3870 3880 3890 3900 3910 3920 3930 3940 3950 3960
End H ↑ mRNA start ↑

T1/6
GGAGGCTTTT TTATGGTTCG TTCTTATTAC CCTTCTGAAT GTCACGCTGA TTATTTTGAC TTTGAGCGTA TCGAGGCTCT TAAACCTGCT ATTGAGGCTT GTGGCATTTC TACTCTTTCT
3970 3980 3990 4000 4010 4020 4030 4040 4050 4060 4070 4080
A start ↑

A7b/7a M5/8 F5c/3 T6/2 Q2/3a R4/3 Z2/6b
CAATCCCCAA TGCTTGGCTT CCATAAGCAG ATGGATAACC GCATCAAGCT CTTGGAAGAG ATTCTGTCTT TTCGTATGCA GGGCGTTGAG TTCGATAATG GTGATATGTA TGTTGACGGC
4090 4100 4110 4120 4130 4140 4150 4160 4170 4180 4190 4200

CATAAGGCTG CTTCTGACGT TCGTGATGAG TTTGTATCTG TTACTGAGAA GTTAATGGAT GAATTGGCAC AATGCTACAA TGTGCTCCCC CAACTTGATA TTAATAACAC TATAGACCAC
4210 4220 4230 4240 4250 4260 4270 4280 4290 4300 4310 4320

M8/6 A7a/4
CGCCCCGAAG GGGACGAAAA ATGGTTTTTA GAGAACGAGA AGACGGTTAC GCAGTTTTGC AAGCTGGCTG CTGAACGCCC TCTTAAGGAT ATTCGCGATG AGTATAATTA CCCCAAAAAG
4330 4340 4350 4360 4370 4380 4390 4400 4410 4420 4430 4440

Z6b/6a
AAAGGTATTA AGGATGAGTG TTCAAGATTG CTGGAGGCCT CCACTAAGAT ATCGCGTAGA GGCTTTGCTA TTCAGCGTTT GATGAATGCA ATGCGACAGG CTCATGCTGA TGGTTGGTTT
4450 4460 4470 4480 4490 4500 4510 4520 4530 4540 4550 4560

ATCGTTTTTG ACACTCTCAC GTTGGCTGAC GACCGATTAG AGGCGTTTTA TGATAATCCC AATGCTTTGC GTGACTATTT TCGTGATATT GGTCGTATGG TTCTTGCTGC CGAGGGTCGC
4570 4580 4590 4600 4610 4620 4630 4640 4650 4660 4670 4680

Fig. 1 *(Continued)*

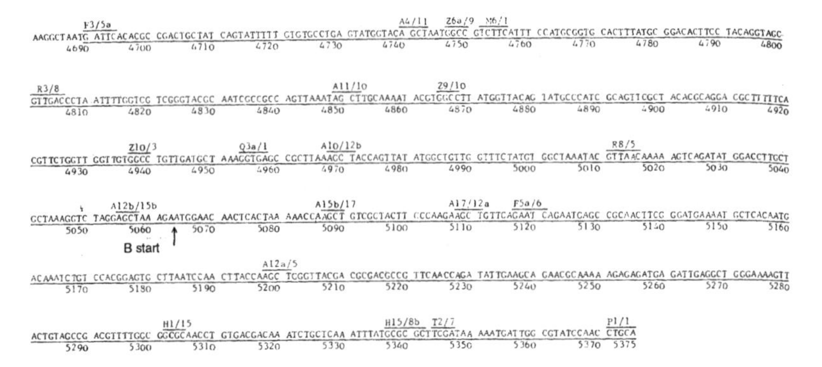

Fig. 1 A provisional nucleotide sequence for the DNA of bacteriophage ΦX174 *am*3 *cs*70. Solid underlining indicates sequences that are fully confirmed; sequences with no underlining probably do not contain more than one mistake per 50 residues. Broken underlining indicates more uncertain sequences. Restriction enzyme recognition sites are indicated (for key to single letter enzyme code see legend to Fig. 2), as are mRNA starts and protein initiation and termination sites. Nucleotides 4,127 to 4,201 have been independently sequenced by van Mansfield *et al.*[58]. The *am*3 codon is at position 587.

Transcription of *Hin*dII fragment 10, amino acid sequence data in the *G* gene, and the plus and minus method using *Hin*dII fragments 2 and 10 as primers, gave a sequence of 195 nucleotides (positions 2,387–2,582, Fig. 1) at the N terminus of gene *G* (ref. 24).

The 'plus and minus' method

Further work on the ΦX sequence has been done using chiefly the plus and minus method primed with restriction fragments. Figure 2 shows the various restriction enzymes used and the fragment maps for each (refs 25–30 and C.A.H., submitted for publication, and N.L.B., C.A.H. and M.S., submitted for publication).

Figure 1 shows the combined results of the sequence work to date. The sequence is numbered from the single cleavage site of the restriction enzyme *Pst*I. As with other methods of sequencing nucleic acids, the plus and minus technique used by itself cannot be regarded as a completely reliable system and occasional errors may occur. Such errors and uncertainties can only be eliminated by more laborious experiments and, although much of the sequence has been so confirmed, it would probably be a long time before the complete sequence could be established. We are not certain that there is any scientific justification for establishing

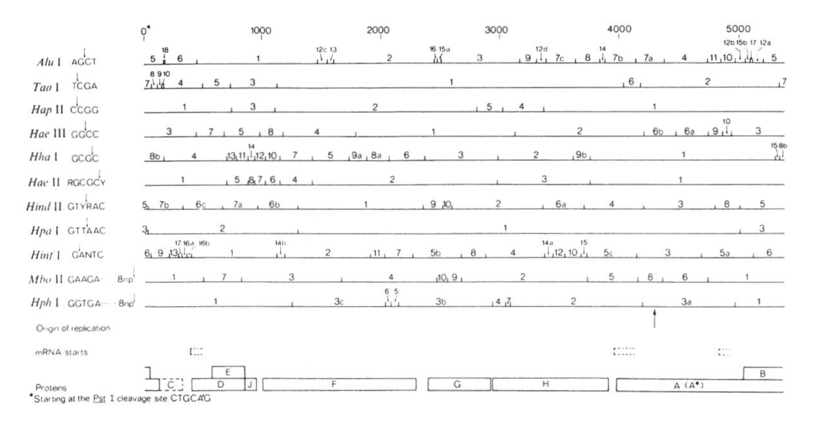

Fig. 2 Fragment maps of restriction enzymes used in the sequence analysis of ΦX174 *am*3 RFI DNA. Fragment maps of ΦX174 have been prepared for *Hind*II (R), *Hae*III (Z) and *Hpa*I + II by Lee and Sinsheimer[25], *Hin*HI and *Hap*II (Y) by Hayashi and Hayashi[26] ,and for *Alu*I (A) by Vereijken *et al.*[27] and for *Pst*I (P) by Brown and Smith[30]. B.G.B., G.M.A., C.A.H. and D. Jaffe prepared the *Hinf*I (F) map, C.A.H. the *Hph*I (Q) map, and Jeppesen *et al.*[28] the *Hha*I (H), *Alu*I, *Hae*II and *Hap*II maps by using a rapid method depending on priming with DNA polymerase. A rapid two-dimensional hybridisation technique has been developed by C.A.H. (submitted for publication) and recently used for mapping *Mbo*II (M) (N.L.B., C.A.H., and M.S., submitted for publication) and *Taq*I (T)[29]. *Hha*I and *Hinf*I maps have also been prepared by Baas *et al.*[52].

every detail and, as it is felt that the results may be useful to other workers, it has been decided to publish the sequence in its present form.

As template we have used both the viral (plus) and complementary (minus) strands of ΦX. Usually it is possible to determine a sequence with a single primer starting at about 15–100 nucleotides from the appropriate restriction enzyme site. In a particularly good experiment the sequence can be read out to 150–200 nucleotides but the results may become less reliable. Most sequences have been derived by priming on both strands; this allows more confidence than when only one strand could be used.

A useful method for confirming runs of the same nucleotide is depurination of ^{32}P-labelled small restriction enzyme fragments or of products of the DNA polymerase priming experiments (ref. 31 and N.L.B. and M.S., in preparation). The most satisfactory way of confirming the DNA sequences is through amino acid sequence data. As the methods used are entirely unrelated, the results of the two approaches complement each other very well and therefore complete sequences can usually be deduced from incomplete data obtained by each method. The complete sequence of genes *G* (ref. 32), *D* (ref. 33), *J* (ref. 33 and Freymeyer, unpublished) and most of *F* have been obtained in this way.

Many of the sequences in Fig. 1 have been amply confirmed and are regarded as established: these are indicated in the figure by underlining. Some sequences are considered to be reasonably accurate and probably contain no more than one mistake in every 50 nucleotides. Sequences that are particularly uncertain—either because of lack of data or conflicting results—are also indicated in Fig. 1.

In considering the sequence of ΦX174 as a functional unit it is convenient to begin in the region between the *H* and *A* genes and to continue around the DNA in the direction of transcription and translation.

A *promoter and terminator*

Sinsheimer *et al.*[34,35] and Axelrod[36] have determined the sequences of the 5′ end of three ΦX *in vitro* mRNA species and have located them on the restriction map. These sequences have been identified on the DNA sequence and one of them (AAATCTTGG) is found only at position 3,954 at which an *in vivo* unstable mRNA start has been located[37]. The sequence to the left of this has some characteristics of typical *E. coli* promoters[38] in that five out of the 'ideal' TATPuATPu residues are present. Nearby, to the right of this mRNA initiation, however, is the sequence TTTTTTA which is similar to sequences found at the 3′ ends of a number of mRNAs (see ref. 39) and seems a likely signal for mRNA termination. The presence of a rho-independent termination site in this approximate position has been suggested[36,37], but the relative positions of the initiating and putative termination signals is rather surprising since the terminator for one mRNA would be expected to precede the initiator for the next. One possibility is that the T_6A might be acting as an 'attenuator' involved in the control of mRNA production in a similar manner to that suggested for the tryptophan operon by Bertrand *et al.*[40]. If indeed it were acting as a transcription terminator one would expect a small RNA of 20 nucleotides to be produced, but no such product has yet been detected. Recent work, however (Rosenberg, unpublished and ref. 41), indicates that termination may require the presence of a base-paired loop structure before the termination site. From the DNA sequence such a loop is probably present before the T_6A sequence, but in mRNA starting from the initiation site at position 3,954 this loop is not formed (Fig. 3). Therefore mRNA that had started at an earlier promoter and extended through

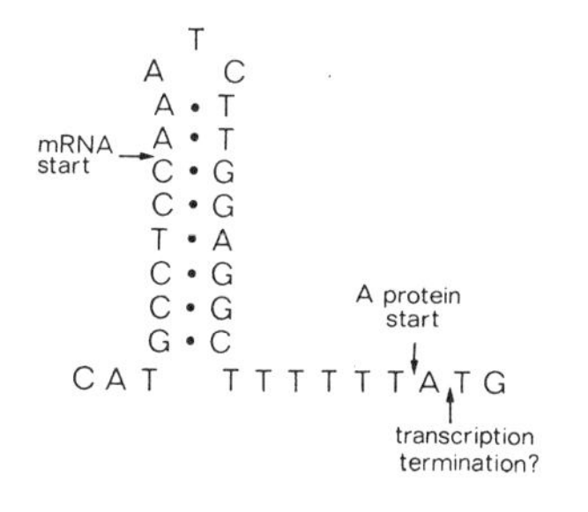

Fig. 3 Potential secondary structure at the *A* mRNA start.

the *H* gene would be expected to terminate here, whereas mRNA newly initiated at position 3,954 would not. This could be a way in which the phage has economised on the use of DNA—by having the ends of the two mRNAs overlapping.

The A protein

Where the amino acid sequence is available there is no problem in relating the DNA sequence to its coding properties, but it is more difficult to do so in the absence of such data, as is the case for the A protein. One way of identifying the reading phase of the DNA is from the distribution of nonsense codons. Over a sufficiently long sequence that is known to be coding for a single protein there is usually one phase that contains no nonsense codons, and this is identified as the reading phase. This requires completely accurate determination of the DNA sequences however: omission of a single nucleotide may give completely erroneous results. Another approach is possible in the case of ΦX. The results with the *F* and *G* genes[23,24,32] showed an unexpectedly high frequency of codons ending in T. Therefore in a coding region there is a tendency for every third nucleotide to be a T and it is then possible to define the reading phase. Figure 4 illustrates how this characteristic was used to help determine the reading phase for the A protein and to identify its initiation codon at position 3,973. In a similar way the distribution of Ts may be used to identify errors in the DNA sequence, provided that such errors occur only infrequently.

A different approach to identifying the initiation site and reading phase in a coding sequence is by looking for a characteristic 'initiation sequence'. Shine and Dalgarno have shown that a common feature of ribosome binding sites is a number of nucleotides (at least three) preceding the ATG that are capable of forming base pairs with a sequence at the 3′ end of 16S rRNA[42,43]. All of the known initiation sites in ΦX174 that have been identified by direct amino acid sequencing (for the F, G, H, J and D proteins) satisfy this criterion (see Table 2) and the fact that the

	Ts in codon position (1)	(2)	(3)
3,910 CCG . TCA . GGA . TTG . ACA . CCC . TCC . CAA . TTG . TAT .	5	2	1
	3	4	3*
3,940 GTT . TTC . ATG . CCT . CCA . AAT . CTT . GGA . GGC . TTT .	2	5	5
	3	5	7*
3,970 TTT . ATG . GTT . CGT . TCT . TAT . TAC . CCT . TCT . GAA .	5	3	7
	5	0	7*
4,000 TGT . CAC . GCT . GAT . TAT . TTT . GAC . TTT . GAG . CGT .	4	2	7

*These figures refer to the last five codons of the previous line and the first five of the next line.

Fig. 4 Identification of the initiation codon for the A protein. Sequences of 30 nucleotides in the region in which the initiation was expected were written down and arbitrarily marked off in triplets. The number of T residues in the first position in each triplet was then counted and listed, and similarly the number of Ts in the second and third positions. The marked preference for Ts in position 3 in the last two lines, as compared with the first two lines, suggests that they are coding for protein and that the triplets are correctly marked off. The most likely initiation codon for the A protein is the ATG in position 3,973.

sequence preceding the ATG in position 3,973 also has this characteristic supports its identification as the initiation site for the A protein.

If, as has been suggested[37], some mRNA from the previous promoter does extend beyond the hairpin structure, initiation of A protein synthesis may be controlled by the inclusion of the region complementary to the 16S rRNA in the hairpin loop. This could explain the presence of two types of mRNA covering the *A* cistron, as suggested by Hayashi *et al.*[37]—one unstable and active and the other stable but inactive. The former would be initiated at the A promoter, the latter at an earlier promoter and result from 'read-through' at the terminator. The postulated reading frame for the A protein was confirmed by sequencing amber mutants mapping in the N-terminal region of gene *A*. *am*86 proved to be a C→T change at position 4,108 and *am*33 a C→T change at position 4,372. These both result in formation of an amber codon (TAG) in the same reading frame as the proposed initiating ATG and the sequence continues to the termination codon at position 133. The A protein, which is the largest coded by ΦX174, is thus 512 amino acids long with a molecular weight of 56,000, in good agreement with SDS gel estimations (see refs 4 and 44). The A* protein, with a molecular weight of about 35,000, is believed to result from an internal translational start in the *A* gene, in the same reading phase[45]. From consideration of possible ribosome binding sequences[42,43] the ATG in position 4,657 seems to be the most likely initiation site for the A* protein.

The origin of replication

The origin of ΦX viral strand DNA synthesis has been located in gene *A*, in restriction fragment Z6b (ref. 46). This origin, while coding for part of the A protein, probably corresponds to the position of the plus strand nick made by the same protein[44]. Gaps in this region that are found in replicating double-stranded (RF) DNAs are probably related to the position of the nick. Eisenberg *et al.*[47] have investigated such gaps by depurination analysis and identified, in particular, the product C_6T. The sequence CTC_5 is found in position 4,285 (Fig. 1) and the location of the origin in this region agrees precisely with the results of Baas *et al.*[46]. It is not possible at present to identify the actual position of the origin nick. The region shows no apparent secondary structure or symmetrical sequences, although there is an AT-rich region (4,298–4,307) between two GC-rich regions which might be of significance. Such a region is found near the origin of replication of SV40 DNA (ref. 48).

B *promoter*

The second of the mRNA 5′ sequences (AUCGC)[34] has been mapped in restriction fragment R8 (Fig. 2), which starts about 300 nucleotides on from the proposed A* initiation. The sequence ATCGC is found at positions 4,832 and 4,888 in Fig. 1. The only way we can choose between them at the moment is that the second is preceded by the sequence TACAGTA (position 4,877), which is more akin to sequences found in known promoters[38] than are sequences preceding the other possible mRNA start. Irrespective of which of these sequences is used, the mRNA has a long 'leader' sequence (232 or 176 nucleotides) before the next proposed initiation codon (gene *B*).

The B protein

From a study of the ribonuclease T_1 digestion products of the ribosome binding sites of ΦX mRNAs[49], it was possible to identify an initiating ATG in position 5,064. From the genetic map[2,3], this would be expected to be gene *B* but, as discussed above, the A protein coding sequence extends right through this region, past the *Pst* site at residue 1 in Fig. 1, and terminates at residue 133. The initiating codon contained in the ribosome-protected sequence is, however, out of phase with the A protein reading frame. The proposed B protein coding sequence is one nucleotide to the left of the A protein phase, and continues until a termination codon occurs at position 49. Therefore the B protein coding sequence is totally contained within the *A* gene. These reading frames have been confirmed by sequencing mutants in genes *A* and *B* (*am*16, N.L.B. and M.S., in preparation; *am*18, *am*35, *ts*116 (ref. 50)). Since the B protein has not been purified no protein sequence data is available. The complete amino acid sequence can be predicted from the DNA sequence however. The protein is 120 amino acids long with a molecular weight of 13,845 (including the N-terminal Met). The molecular weight estimates of the B protein obtained by SDS–gel electrophoresis are mostly greater than this (see review, ref. 4), but the electrophoretic mobility varied with gel concentration and cross linker. Such anomalous behaviour suggests that there may be, for instance, carbohydrate attached to the B protein.

The C protein

The next known gene product, protein C, maps between genes *B* and *D*. Examination of the DNA sequence in this region indicates that the most probable initiating ATG overlaps the termination codon, TGA, of gene A in the sequence ATGA at position 134. A possible termination codon for gene *C* could then be at position 391, although the sequence and phasing is not yet confirmed through this region. There is another possible protein initiation codon (position 51, overlapping the B protein termination codon) which would result in a slightly shorter gene product terminating at nucleotide 219. For the C protein, however, we favour the 'A terminator' start, since only this reading frame contains a CAA sequence, which by a C→T alteration could give the ochre 6 mutant. Ochre 6 is a gene *C* mutant produced by the decay of ^{3}H-cytosine[51] and has been mapped in fragments A6 and F9 (ref. 52); that is, between nucleotides 170 and 205 (Fig. 1).

Sequence following the D promoter

The mRNA 5′ sequence which maps before the *D* gene (GAUGC)[34] is found at position 358 in Fig. 1. The sequence preceding the messenger start has only four of the TATPuATPu nucleotides[38]. Thirty-two nucleotides after the mRNA initiation is the ATG (position 390) that initiates D protein synthesis. The amino acid sequence of the D protein has been determined almost completely, and nucleotide and amino acid sequences can be correlated to the termination codon at position 846 (ref. 33). The D protein, which is involved in capsid assembly, is 151 amino acids in length, with a molecular weight of 16,811. The *D* termination codon overlaps the initiation codon for gene *J* in the sequence TAATG. A similar structure has also been found by Platt and Yanofsky[53] in the tryptophan operon. The DNA sequence following this initiation codon matches the amino acid sequence of the small basic protein (37 amino acids) of the virion determined by D. Freymeyer, P. R. Shank, T. Vanaman, C.A.H. and M. H. Edgell (personal communication). Benbow *et al.*[2,3] suggested that the mutation *am*6 was located in a gene *J*, coding for the small protein of the virion, and mapping immediately before gene *F*. Although marker rescue experiments indicate that *am*6 is not in this region[54], the DNA sequence shows that there is a gene coding for the virion protein and we have defined this as gene *J* (ref. 33). Since the *J* initiation codon overlaps the *D* termination codon we had to look elsewhere for gene *E*, which genetic mapping[2,3] had placed between them. Amber mutants in gene *E* (*am*3, *am*27, *am*34 and *am*N11) were located by the marker rescue technique and sequenced. All were found to be within the *D* coding sequence, with the mutant amber codons one nucleotide to the right of the *D* reading frame[33]. Thus the *E* coding sequence is completely contained within the *D* coding region but in a different reading frame. The proposed initiation and termination codons for the E protein are at nucleotides 568 and 840, respectively[33], giving a protein 91 amino acids in length with a molecular weight of about 9,900 (including the N-terminal methionine).

The F protein

Following the *J* gene is an intercistronic region of 39 nucleotides before initiation of the F protein. There is no known function of this apparently untranslated sequence, although the presence of a hairpin structure (positions 969–984) suggests that it could be the site of the *in vivo* messenger termination signal[37] mapped in this region. The F protein is initiated by the ATG at position 1,001. This is the capsid component of the virion, and almost all the amino acid sequence is known[22,24]. There are regions in this gene where the DNA sequence is not completely established, but the protein is about 424 amino acids in length, giving a molecular weight of $\simeq$46,300.

The G protein region

The termination signal for the F protein (position 2,276) is followed by an unusually long untranslated sequence of 111 nucleotides until the G protein initiation codon[31]. This region contains a looped structure which was postulated to have some functional role, as yet unknown, in the single-stranded DNA or the mRNA.

Initiation of the G protein at position 2,387 is followed by a sequence of 425 nucleotides until termination at position 2,912, giving a spike protein of molecular weight 19,053. The nucleotide and amino acid sequences of this gene and product are known[24,32].

The H protein

The initiation codon for the H protein (position 2,923) was identified first on the basis of the distribution of T nucleotides between the three reading phases, and later confirmed by amino acid sequence analysis. Amino acid sequence data on the H protein is minimal but the five peptide sequences known do correspond to the amino acid sequence, deduced from the DNA sequence by using the high frequency of third position T to help in assigning a reading frame to any given region. The DNA sequence is not entirely confirmed but it is possible to write a reasonably accurate amino acid sequence for the H protein. The protein terminates at nucleotide 3,907, in agreement with carboxypeptidase results, giving a spike protein of molecular weight ≃ 35,600 (326 amino acids). The amino acid sequence at the N terminus seems to be particularly rich in hydrophobic residues, which is consistent with its suggested function as the 'pilot' protein that reacts with the bacterial membrane[55,56]. After H protein termination there are 66 nucleotides before initiation of the A protein at position 3,973.

Coding capacity of the ΦX174 genome

The most striking feature of the ΦX DNA sequence is the way in which the various functions of the genome are compressed within the 5,375 nucleotides. Since the identification of ΦX gene products[2,4] it has been clear that proteins of the accepted molecular weights could not be separately coded on the available length of DNA. However, with the presence of two pairs of overlapping genes (*B* within *A* (ref. 50), *E* within *D* (ref.

Table 1 ΦX174 coding capacity

Gene	Protein molecular weight from SDS gels*	Number of nucleotides (Fig. 1)	Protein molecular weight from sequence information
A	55,000–67,000	1,536	56,000
(*A**)	35,000		
B	19,000–25,000	(360)†	13,845‡
C	7,000		
D	14,500	456	16,811‡
E	10,000–17,500	(273)†	9,940
J	5,000	114	4,097‡
F	48,000	1,275	46,400
G	19,000	525	19,053‡
H	37,000	984	35,800
Non-coding and *C*		485	
Total		5,375	

* See ref. 4.

† Values in parentheses are overlapping sequences and therefore not included in the addition to obtain the total length of DNA.

‡ These values are calculated from the amino acid sequence (in the case of B deduced from the nucleotide sequence). The others are derived using the formula

$$\text{Protein molecular weight} = \frac{\text{No. of nucleotides}}{3 \times 0.00915}$$

33)) the genome has more coding capacity than had been originally supposed on the assumption that each gene was physically separate. Table 1 summarises the molecular weights of the known ΦX-coded proteins. There are other potential initiation sites for polypeptide synthesis (for example, in genes *A, F, G* and *H*) and further genetic work may clarify whether there are in fact other ΦX genes as yet unidentified.

Initiation of protein synthesis

Table 2 lists the protein initiation sequences for genes *A, B, D, E, J, F, G* and *H*. It can be noted that there are no extra precursor sequences in proteins D, J, F, G or H at either the N or C terminus. There seems to be no relationship between the degree of complementarity to the 16S rRNA and the amount of protein synthesised, and we see no other features in the sequence that could explain different efficiencies of translation except where genes overlap.

Table 2 Initiation sequences of ΦX174 coded proteins

D	C-C-A-C-T-A-A-T-A-G-G-T-A-A-G-A-A-A-T-C-A-T-G-A-G-T-C-A-A-G-T-T-A-C-T Ser Gln Val Thr
E	C-T-G-C-G-T-T-G-A-G-G-C-T-T-G-C-G-T-T-T-A-T-G-G-T-A-C-G-C-T-G-G-A-C-T
J	C-G-T-G-C-G-G-A-A-G-G-A-G-T-G-A-T-G-T-A-A-T-G-T-C-T-A-A-A-G-G-T-A-A-A Ser Lys Gly Lys
F	C-C-C-T-T-A-C-T-T-G-A-G-G-A-T-A-A-A-T-T-A-T-G-T-C-T-A-A-T-A-T-T-C-A-A Ser Asn Ile Gln
G	T-T-C-T-G-C-T-T-A-G-G-A-G-T-T-T-A-A-T-C-A-T-G-T-T-T-C-A-G-A-C-T-T-T-T Met Phe Gln Thr Phe
H	C-C-A-C-T-T-A-A-G-T-G-A-G-G-T-G-A-T-T-T-A-T-G-T-T-T-G-G-T-G-C-T-A-T-T Met Phe Gly Ala Ile
A	C-A-A-A-T-C-T-T-G-G-A-G-G-C-T-T-T-T-T-T-A-T-G-G-T-T-C-G-T-T-C-T-T-A-T
B	A-A-A-G-G-T-C-T-A-G-G-A-G-C-T-A-A-A-G-A-A-T-G-G-A-A-C-A-A-C-T-C-A-C-T
16S RNA 3′ end	HO A-U-U-C-C-U-C-C-A-C-U-A-G

Where the protein start has been independently confirmed by protein sequencing data the amino acid sequences are indicated. The other initiation regions were identified as described in the text. Sequences complementary to the 3′ end of 16S rRNA (refs 42, 43) are boxed; broken lines indicate further complementarity if some nucleotides are looped out or not matched. Ribosome binding to mRNA has been demonstrated in these regions for genes *J*, *F*, *G* and *B* (ref. 49).

Transcription of ΦX174

The sequences preceding known mRNA starts[34–36] are shown in Table 3. Other studies on promoter sequences[38] have suggested certain features that they may have in common. Although some of these features are present in the sequences preceding the ΦX transcription initiations others are not, and at present it is difficult to suggest what signal on the DNA determines a promoter site or the efficiency with which it initiates RNA synthesis. It is interesting to note that a polymerase binding site found by Chen *et al.*[57], but not associated with any *in vitro* or *in vivo* mRNA starts, mapped near the region where there is the sequence TATGATG characteristic of promoters[38] (positions 2,705–2,711).

Table 3 Promoter sequences in ΦX174

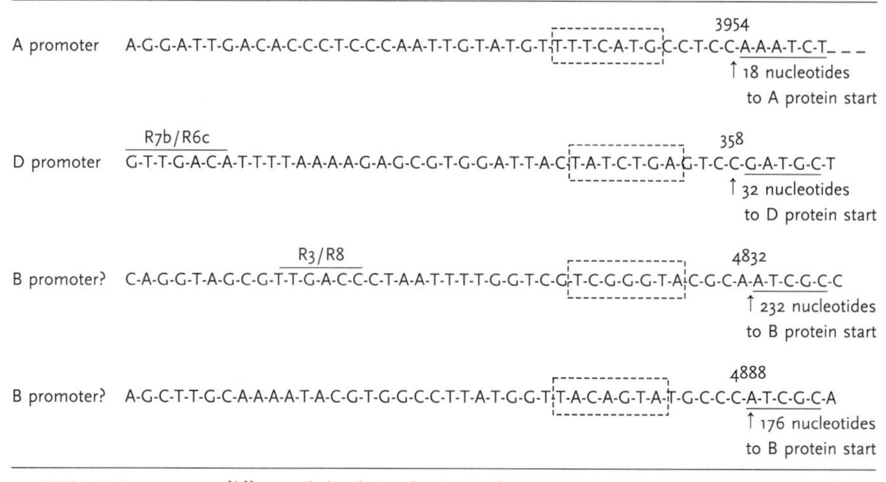

Promoter	Sequence	Position / note
A promoter	A-G-G-A-T-T-G-A-C-A-C-C-C-T-C-C-C-A-A-T-T-G-T-A-T-G-T-T-T-T-C-A-T-G-C-C-T-C-C-A-A-A-T-C-T_ _ _	3954; ↑ 18 nucleotides to A protein start
D promoter	R7b/R6c: G-T-T-G-A-C-A-T-T-T-T-A-A-A-A-G-A-G-C-G-T-G-G-A-T-T-A-C-T-A-T-C-T-G-A-G-T-C-C-G-A-T-G-C-T	358; ↑ 32 nucleotides to D protein start
B promoter?	R3/R8: C-A-G-G-T-A-G-C-G-T-T-G-A-C-C-C-T-A-A-T-T-T-T-G-G-T-C-G-T-C-G-G-G-T-A-C-G-C-A-A-T-C-G-C-C	4832; ↑ 232 nucleotides to B protein start
B promoter?	A-G-C-T-T-G-C-A-A-A-A-T-A-C-G-T-G-G-C-C-T-T-A-T-G-G-T-T-A-C-A-G-T-A-T-G-C-C-C-A-T-C-G-C-A	4888; ↑ 176 nucleotides to B protein start

mRNA initiation sequences[34–36] are underlined. Boxed regions indicate sequences that may correspond to the TATPu-ATPu sequence found in other promoters[38], taking into account the distance from the mRNA starts.

The use of codons in ΦX174

Table 4 shows the codons used in regions where the nucleotide sequence is fully confirmed. It is clear that the pattern established by early observations on non-random use of codons[23,24] is continued now that more information is available. In particular, the preference for T at the third position of the codon is marked throughout the genome, as shown in Table 4. In regions of overlapping genes, one of the pair tends to continue the 'third T' trend (*D* and *B*), thus excluding the other (*E* and *A*). This may give some indication of the order in which overlapping genes evolved[33,50]. Another interesting feature is the very low occurrence of codons starting AG, particularly in non-overlapping regions. The base composition of the sequence of ΦX174 DNA shown in Fig. 1 is: *A*, 23.9%; *C*, 21.5%; *G*, 23.3% and *T*, 31.2%. This is in good agreement with previously determined values (see ref. 4).

We thank D. McCallum and R. Staden for carrying out the computer data storage and analysis of the sequence.

Note added in proof: J. E. Sims and D. Dressler (personal communication) have independently determined the sequence in positions 263–375 and 4,801–4,940. Their results agree with those given in Fig. 1. They have also identified the 'B' mRNA start as being at position 4,888.

MRC Laboratory of Molecular Biology, Hills Road, Cambridge CB2 2QH, UK

Received November 30; accepted December 24, 1976 [Published 24 February 1977]

Table 4 Codons used in ΦX174

Phe	TTT	39	Ser	TCT	35	Tyr	TAT	36	Cys	TGT	12
	TTC	26		TCC	9		TAC	15		TGC	10
Leu	TTA	19		TCA	16	Ter	TAA	3	Ter	TGA	5
	TTG	26		TCG	14		TAG	0	Trp	TGG	16
Leu	CTT	36	Pro	CCT	34	His	CAT	16	Arg	CGT	40
	CTC	15		CCC	6		CAC	7		CGC	29
	CTA	3		CCA	6	Gln	CAA	27		CGA	4
	CTG	24		CCG	21		CAG	34		CGG	8
Ile	ATT	45	Thr	ACT	40	Asn	AAT	37	Ser	AGT	9
	ATC	12		ACC	18		AAC	25		AGC	5
	ATA	2		ACA	13	Lys	AAA	47	Arg	AGA	6
Met	ATG	42		ACG	19		AAG	31		AGG	1
Val	GTT	53	Ala	GCT	64	Asp	GAT	44	Gly	GGT	38
	GTC	14		GCC	17		GAC	35		GGC	28
	GTA	10		GCA	12	Glu	GAA	27		GGA	13
	GTG	11		GCG	12		GAG	34		GGG	3

The totals are derived from sequences in Fig. 1 which are fully confirmed, that is, 377 codons in gene *A*, 120 in gene *B*, 152 in gene *D*, 91 in gene *E*, 38 in gene *J*, 344 in gene *F*, 175 in gene *G* and 49 in gene *H*. Out of a total of 1,346 codons 42.9% terminate in T. The percentages in the different genes are: *A*, 37.1 (non-overlapping region 47.1; overlapping region 15.8); *B*, 34.2; *D*, 42.1; *E*, 14.3; *J*, 47.4; *F*, 52.0; *G*, 54.3; *H*, 49.0. The initiating ATG is included in all cases.

References

1. Sanger, F. & Coulson, A. R. *J. molec. Biol.* **94**, 441–448 (1975).
2. Benbow, R. M., Hutchison, C. A. III, Fabricant, J. D. & Sinsheimer, R. L. *J. Virol.* **7**, 549–558 (1971).
3. Benbow, R. M., Zuccarelli, A. J., Davis, G. C. & Sinsheimer, R. L. *J. Virol.* **13**, 898–907 (1974).
4. Denhardt, D. T. *CRC Crit. Rev. Microbiol.* **4**, 161–222 (1975).
5. Hall, J. B. & Sinsheimer, R. L. *J. molec. Biol.* **6**, 115–127 (1963).
6. Ling, V. *Proc. natn. Acad. Sci. U.S.A.* **69**, 742–746 (1972).
7. Harbers, B., Delaney, A. D., Harbers, K. & Spencer, J. H. *Biochemistry* **15**, 407–414 (1976).
8. Burton, K. & Petersen, G. B. *Biochem. J.* **75**, 17–27 (1960).
9. Chadwell, H. A. Thesis, University of Cambridge (1974).
10. Sadowski, P. D. & Bakyta, I. *J. biol. Chem.* **247**, 405–412 (1972).
11. Ling, V. *FEBS Lett.* **19**, 50–54 (1971).
12. Ziff, E. B., Sedat, J. W. & Galibert, F. *Nature new Biol.* **241**, 34–37 (1973).
13. Galibert, F., Sedat, J. W. & Ziff, E. B. *J. molec. Biol.* **87**, 377–407 (1974).
14. Robertson, H. D., Barrell, B. G., Weith, H. L. & Donelson, J. E. *Nature new Biol.* **241**, 38–40 (1973).
15. Air, G. M. & Bridgen, J. *Nature new Biol.* **241**, 40–41 (1973).
16. Sanger, F., Donelson, J. E., Coulson, A. R., Kössel, H. & Fischer, D. *Proc. natn. Acad. Sci. U.S.A.* **70**, 1209–1213 (1973).

17. Schott, H. *Makromolek. Chem.* **175**, 1683–1693 (1974).
18. Donelson, J. E., Barrell, B. G., Weith, H. L., Kössel, H. & Schott, H. *Eur. J. Biochem.* **58**, 383–395 (1975).
19. Blackburn, E. H. *J. molec. Biol.* **93**, 367–374 (1975).
20. Blackburn, E. H. *J. molec. Biol.* **107**, 417–432 (1976).
21. Sedat, J. W., Ziff, E. B. & Galibert, F. *J. molec. Biol.* **107**, 391–416 (1976).
22. Air, G. M. *J. molec. Biol.* **107**, 433–444 (1976).
23. Air, G. M. *et al. J. molec. Biol.* **107**, 445–458 (1976).
24. Air, G. M., Blackburn, E. H., Sanger, F. & Coulson, A. R. *J. molec. Biol.* **96**, 703–719 (1975).
25. Lee, A. S. & Sinsheimer, R. L. *Proc. natn. Acad. Sci. U.S.A.* **71**, 2882–2886 (1974).
26. Hayashi, M. N. & Hayashi, M. *J. Virol.* **14**, 1142–1152 (1974).
27. Vereijken, J. M., van Mansfeld, A. D. M., Baas, P. D. & Jansz, H. S. *Virology* **68**, 221–233 (1975).
28. Jeppesen, P. G. N., Sanders, L. & Slocombe, P. M. *Nucl. Acids Res.* **3**, 1323–1339 (1976).
29. Sato, S., Hutchison, C. A. III & Harris, J. I. *Proc. natn. Acad. Sci. U.S.A.* (in the press).
30. Brown, N. L. & Smith, M. *FEBS Lett.* **65**, 284–287 (1976).
31. Fiddes, J. C. *J. molec. Biol.* **107**, 1–24 (1976).
32. Air, G. M., Sanger, F. & Coulson, A. R. *J. molec. Biol.* **108**, 519–533 (1976).
33. Barrell, B. G., Air, G. M. & Hutchison, C. A. III *Nature* **264**, 34–41 (1976).
34. Smith, L. H. & Sinsheimer, R. L. *J. molec. Biol.* **103**, 699–735 (1976).
35. Grohmann, K., Smith, L. H. & Sinsheimer, R. L. *Biochemistry* **14**, 1951–1955 (1975).
36. Axelrod, N. *J. molec. Biol.* **108**, 753–779 (1976).
37. Hayashi, M., Fujimura, F. K. & Hayashi, M. *Proc. natn. Acad. Sci. U.S.A.* **73**, 3519–3523 (1976).
38. Pribnow, D. *Proc. natn. Acad. Sci. U.S.A.* **72**, 784–788 (1975).
39. Rosenberg, M., de Crombrugghe, B. & Musso, R. *Proc. natn. Acad. Sci. U.S.A.* **73**, 717–721 (1976).
40. Bertrand, K. *et al. Science* **189**, 22–26 (1975).
41. Sugimoto, K., Sugisaki, H., Okamoto, T. & Takanami, M. *J. molec. Biol.* (in the press).
42. Shine, J. & Dalgarno, L. *Proc. natn. Acad. Sci. U.S.A.* **71**, 1342–1346 (1974).
43. Steitz, J. A. & Jakes, K. *Proc. natn. Acad. Sci. U.S.A.* **72**, 4734–4738 (1975).
44. Henry, T. J. & Knippers, R. *Proc. natn. Acad. Sci. U.S.A.* **71**, 1549–1553 (1974).
45. Linney, E. & Hayashi, M. *Nature* **249**, 345–348 (1974).
46. Baas, P. D., Jansz, H. S. & Sinsheimer, R. L. *J. molec. Biol.* **102**, 633–656 (1976).
47. Eisenberg, S., Harbers, B., Hours, C. & Denhardt, D. T. *J. molec. Biol.* **99**, 107–123 (1975).
48. Subramanian, K. N., Dhar, R. & Weissman, S. M. *J. biol. Chem.* (in the press).
49. Ravetch, J. V., Model, P. & Robertson, H. D. *Nature* **265**, 698–702 (1977).
50. Smith, M. *et al.* (submitted to Nature).
51. Funk, F. & Sinsheimer, R. L. *J. Virol.* **6**, 12–19 (1970).
52. Baas, P. D., van Heusden, G. P. H., Vereijken, J. M., Weisbeek, P. J. & Jansz, H. S. *Nucl. Acids Res.* **3**, 1947–1960 (1976).
53. Platt, T. & Yanofsky, C. *Proc. natn. Acad. Sci. U.S.A.* **72**, 2399–2403 (1975).
54. Weisbeek, P. J., Vereijken, J. M., Baas, P. D., Jansz, H. S. & Van Arkel, G. A. *Virology* **72**, 61–71 (1976).
55. Jazwinski, S. M., Lindberg, A. A. & Kornberg, A. *Virology* **66**, 283–293 (1975).
56. Kornberg, A. *DNA Synthesis* (W. H. Freeman, San Francisco, 1974).

57. Chen, C. Y., Hutchison, C. A. III & Edgell, M. H. *Nature new Biol.* **243**, 233–236 (1973).
58. van Mansfeld, A. D. M., Vereijken, J. M. & Jansz, H. S. *Nucl. Acids Res.* **3**, 2827–2843 (1976).

1980

en meteorite. There are only three meteorites for
ture data assignments differ systematically from
ment; the Mezö–Madaras subtype could be
1, the Parnallee value increased by 0.1 and the
ue increased by 0.2. Clearly there would be no
ning a petrological subtype in the absence of TL
there were other good quality data available.

We thank the donors of meteorites listed in Tab
Wasson for many discussions and encouragement;
Wood, R. T. Dodd, and W. R. Van Schmus fo
Joanna Pai and Clare Marshall for technical ass
research is supported by NASA grant NGR 05
UCLA) and NSF grant EAR78-22440 (at Was
versity).

epted 19 September 1980.

im. cosmochim. Acta 26, 739–7
n Schmus, R. J. geophys. Res. 70, 38
arth Sci. Rev. 5, 145–184 (1969).
im. cosmochim. Acta 33, 161–203 (
Geophys. Space Phys. 10, 711–759 (
. & Wood, J. A. Geochim. cosmochi 31, 747–763
asson, J. T. Geochim. cosmochim. Acta 44, 431–446 (19
. & Ribbe, R. H. Geochim. cosmochim. Acta 32, 968).
Schmus, W. R. & Koffman, D. M. Geochim. Acta 31,
, D. W. & Hearsey, R. J. Phys (E): Sci. Instrum. 10, 51–56 (1977).
preparation).
s (in the press).
ritics 13, 628–632 (1978).
K. & Taylor, G. J. Geochim. cosmochim. Acta (in the press); Meteoritics
78).
6, 1–49 (1967).
ck, G. Materials (Clarendon, Oxford, 1970).
ner, J. Ander Geochim. cosmochim. Acta 31, 1239–12
C apathy, ders, E. & Morgan, J. W. Geochim. co
357).
ger, chim. Acta 32, 209–237 (1968).
er, A. & C rmoluminescence of Geological Materials (e
(Academic, London, New York, 1968).
21. Lalou, C., Nordemann, D. & Labyrie, J. C. r. hebd. Séanc. Acad.
2104–2401 (1970).
22. Vaz, J. E. & Sears, D. W. Meteoritics 12, 47–60 (1977).
23. Houtermans, F. G. & Liener, A. J. geophys. Res. 71, 3387–3396 (196
24. Heymann, D. Icarus 6, 189–221 (1967).
25. Binns, R. A. Nature 213, 1111–1112 (1967).
26. Schultz, L. & Kruse, H. Nucl. Track Det. 2, 65–103 (1976).
27. Moore, C. B. & Lewis, C. J. geophys. Res. 72, 6289–6292.
28. Schultz, L. & Signer, P. Earth planet. Sci. Lett. 36, 363–371 (1977).
29. Hutchison, R. et al. Nature 287, 787–790 (1980).

utations affecting segment number an polarity in *Drosophila*

Christiane Nüsslein-Volhard & Eric Wieschaus
European Molecular Biology Laboratory, PO Box 10.2209, 69 Heidelberg, FRG

earches for embryonic lethal mutants of Drosophila melanogaster *we have identified 15 loci*
he segmental pattern of the larva. These loci probably represent the majority of such genes in Dros
the mutant embryos indicate that the process of segmentation involves at least three level
he entire egg as developmental unit, a repeat unit with the length of two segments, and the individu

ion of complex form from similar repeating
feature of spatial organisation in all higher
ttle is known for any organism about the genes
process. In *Drosophila*, the metameric nature of
most obvious in the thoracic and abdominal
e larval epidermis and we are attempting to
required for the establishment of this pattern.
on of these genes and the description of their
uld lead to a better understanding of the general
sponsible for the formation of metameric

, the anlagen for the individual segments arise as
ubdivisions of the blastoderm, each segment
a transverse strip of about three or four cell
cell lineage restriction between neighbouring
ablished at or soon after this stage[2]. Two basic
on have been described which change the seg-
of the *Drosophila* larva. Maternal effect muta-
dal lead to a global alteration of the embryonic
udal embryos develop two posterior ends
ror-image symmetry, and lack head, thorax and

abdominal segments into mesothoracic segments. H
homeotic loci do not affect the total number, size c
the segments, nor do they point to any other step
intervene between the maternal gradient and the fir
segments.

We have undertaken a systematic search for mu
affect the segmental pattern depending on the zygo
We describe here mutations at 15 loci which show
novel types of pattern alteration: pattern duplica
segment (segment polarity mutants; six loci), patter
alternating segments (pair-rule mutants; six loci) an
a group of adjacent segments (gap mutants; three lo
Fig. 1).

The segmental pattern of the normal *Drosophila* larva

Figure 2 shows the cuticular pattern of a normal
larva shortly after hatching. The larval body is c
three thoracic and eight abdominal segments
differences are observed in different body regions,
have certain morphological features in common. Th

SEVENTEEN

The blueprint of animals revealed

Ginés Morata

In 1980, Christiane Nüsslein-Volhard and Eric Wieschaus published a highly influential paper on the effects of mutations on development of the fruit fly *Drosophila melanogaster.* This work paved the way for tackling the question of how genes establish the patterns of multicellular organisms and so why, for example, a leg is different from an eye. Subsequent work has shown that many of the principles (and the genes) evident in the fly are common to all animals, including humans.

For most of the twentieth century, *Drosophila* has been the traditional organism for investigating the genetics of multicellular creatures. Over the years a large number of inheritable variations (mutations) have accumulated in these little flies, consistently revealing themselves as anomalous traits—for example, a curved wing or an unusual eye color. This is the mutant phenotype.

Many such mutants are "viable," reaching the adult stage, and most studies were based on them. Flies have an external skeleton, the cuticle, which bears a diversity of organs (wings, legs, eyes, and so on) and is largely covered with sensory bristles arranged in characteristic patterns. All of these external structures could be easily inspected with a magnifying glass or a low-power microscope. Inevitably, geneticists were fascinated by the richness in detail of the adult cuticle and the ease of detecting mutations affecting it and the various organs. Classical genetic studies, then, centered on the identification and study of genes involved in the development of adult cuticular structures.

When, during the 1920s, methods were devised to induce mutations artificially, it became clear that although many mutants are viable, many more are not. The great majority of mutations lead to death of a fly in its embryonic, larval, or pupal stages, so that it never becomes an adult. These mutations were simply labeled as "lethals," and for a long time they were largely neglected.

From these studies an important class of developmental control agents emerged. These were the homeotic genes, nowadays referred to as *Hox* genes, which specify the diversity of the body along the axis running from the front of an organism to the back. For example, the adult fly consists of a head, a thorax of three segments, and an abdomen of six (male) or seven (female) segments (fig. 17.1). Each of these segments has its own typical morphology, often referred to as its "identity." The function of the *Hox* genes is to determine the developmental programs followed by the groups of cells that make the different segments, so that the appropriate segment identity is formed in each position. Typically, a mutation in a *Hox* gene transforms the identity of one segment into that of another. For example, in *Ultrabithorax* mutants the third thoracic segment (which has no wings) develops like the second thoracic segment (which has wings). The mutant fly therefore has two pairs of wings, one in the second and another in the third thoracic segment.

The work of Ed Lewis, at the California Institute of Technology in Pasadena, is especially relevant to our story. Lewis identified and characterized several of the *Hox* genes that control the manner in which individual segments develop. However, even though the focus of his work was on the adult fly, in 1978 he published a key paper[1] describing for the first time a transformation in larval patterns that resulted when an entire group of adjacent homeotic genes was missing. This genotype was lethal, but Lewis realized that the mutant larvae reached the stage at which the segments had become evident, so he could describe the homeotic transformations based on the larval segment patterns. These larvae showed transformations affect-

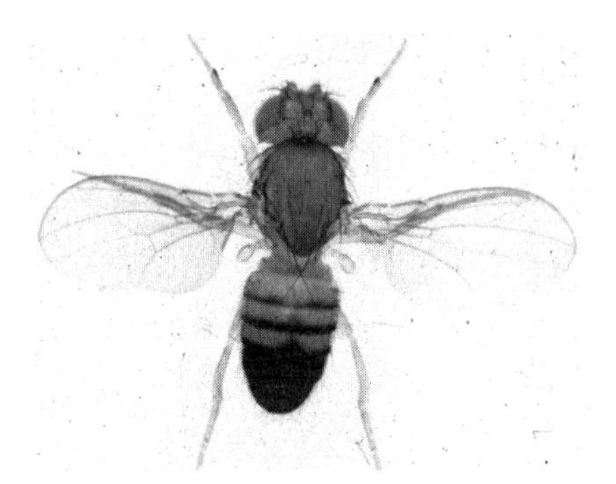

Fig. 17.1. A normal *Drosophila* adult male, showing the head, thorax, and abdomen. Secondary modifications mean that the original segmentation is clearly visible only in the abdomen. Each pair of legs corresponds to a different thoracic segment.

ing part of the thorax and the entire abdomen. One reason why this paper was so striking was that it predicted the existence of further *Hox* genes involved in abdominal development, and these genes were indeed identified some years later.

But the greater significance of Lewis's paper was twofold. It showed that larval patterns were technically easy to study and could serve as well as adult patterns for studying the genetic control of development. But more importantly, it indicated the existence of a process for generating the segments of the body that was independent of the homeotic mechanism. This conclusion stemmed from examination of the mutant larvae: despite the absence of the genes necessary to specify thoracic and abdominal identities, the larvae still had the normal number of segments, although they all developed as thoracic-like segments.

This was the setting for Nüsslein-Volhard and Wieschaus's *Nature* paper[2] of 1980. They were working at the European Molecular Biology Laboratory in Heidelberg and set themselves the aim of understanding the genetic basis of how the fly's body is initially subdivided into individual segments, and understanding which are the fundamental features of the organization of segments that are common to all of them. To achieve this goal, Nüsslein-Volhard and Wieschaus designed experiments to isolate mutations in all of the genes involved in the formation and disposition of *Drosophila* segments. Realizing that their aims could not be achieved using adult flies, their assay was based on larval patterns.

The methods they employed were not novel—the approach was a conventional one, with ethyl methane sulfonate being used as the mutagenic agent. But their painstaking strategy was new. Several years of work were involved, as the analysis required thousands of individual matings of flies, and the inspection of the progeny of each mating under the compound microscope. Moreover, as the intention was to identify all of the genes involved in segmentation, it was necessary to complete the experiment to "saturation"—that is, until all or nearly all susceptible genes had been mutated. Saturation usually involves showing that the experiment has produced several mutations per gene, making it unlikely that there are many genes still to be mutated. The task was huge, and meant recording the consequences of some forty thousand matings.

Analysis of the mutations showed that only fifteen genes, an unexpectedly low number, are involved in segmentation. Some in which mutations affected the adult fly were already known, but most had not been identified before. Nüsslein-Volhard and Wieschaus named them for the larval phenotypes that resulted, and for the first time now-famous names such as *hedge-*

hog, patched, and *even-skipped* appeared in print. Moreover, the genes could be subdivided into three well-defined classes known as "gap," "pair-rule," and "polarity." This subdivision was itself of interest, as it suggested steps in the process of segmentation that result in the finer-grained specification of each segment: first, subdivision of large body areas; then the formation of two-segment units; and finally formation of the individual segment units. These have to have a polarity, in which the anterior region that meets the preceding segment is different from a posterior region that lies next to the following segment.

The end of this process is the formation of a chain of segments that are all alike and have a similar organization. The only difference between them is their position along the antero-posterior axis of the body. It is on this scaffold that the *Hox* genes operate; each *Hox* gene becomes active in a characteristic position where it establishes the appropriate segment identity.

The impact of the paper,[2] first on those working with *Drosophila* and later on the whole of biology, cannot be overemphasized. A first consequence was that it made possible the genetic analysis of developmental patterns, which had previously been impossible because nine out of ten mutations are lethal. It turned out that nearly of all the lethal mutants live long enough to make larval patterns. At a stroke, then, 90 percent of the *Drosophila* genome was opened up for investigation. In regard to this, as biology becomes ever more competitive, it is worth mentioning Nüsslein-Volhard and Wieschaus's willingness (very much in the tradition of *Drosophila* genetics) to share their research material: the last sentence of the paper reads "All mutants are available on request."

The obvious importance of these genes, the straightforwardness of the analysis, and the availability of the mutations triggered general interest in the study of segmentation in the *Drosophila* community. There was a shift from the study of adults to the study of larvae, whose simplicity of development and well-characterized patterns facilitated further genetic analyses of double- and triple-mutant combinations, and of interactions between genes.

Moreover, the timing of the paper was just right, for it came at a crucial juncture in biology. The availability of all these mutant flies coincided precisely with the advent of molecular techniques for cloning DNA, meaning that the genes themselves—the specific stretches of DNA—could be identified. *Drosophila* workers were not renowned for their molecular abilities, but suddenly numbers of well-trained molecular biologists from other fields arrived in search of a problem against which to pit their wits. The result was that in a short time all of the segmentation genes concerned were cloned, and the molecular analysis of developmental patterns began. As Peter Lawrence has pointed out in *The Making of a Fly*, in 1990 half of the presentations at the main meeting on *Drosophila* development were based on the genes discovered by Nüsslein-Volhard and Wieschaus.

The fusion of developmental genetics with molecular biology has produced one of the most exciting findings in biology: that the basic developmental mechanisms in all animals are much the same. Many of the genes identified by Nüsslein-Volhard and Wieschaus have counterparts in all species, including humans, and also have similar functions.[3] This revelation has dramatically changed the way in which questions in biology are approached and understood; developmental biology has become a general discipline, the principles of which apply to all animals. It has also provided new concepts for understanding the history of life on Earth. Knowledge of how the diversity of body parts is generated during development is a great help in interpreting the differences, generated by evolution, that we see in the ten million or so living species of animals.

Finally, there is also a direct connection with human biology and medicine. Many of the genes described by Nüsslein-Volhard and Wieschaus, or by others who followed their approach, can be directly involved in human ill-health. For instance *hedgehog, cubitus interruptus, patched,* and *wingless* are all altered in cancer and in other degenerative diseases.[4–7] The investigations carried out with *Drosophila* will, without doubt, help in the fight against disease.

All in all, Nüsslein-Volhard and Wieschaus's contribution of 1980 was arguably the most influential paper in developmental biology during the second half of the twentieth century. They, together with Lewis, received due acknowledgment of their achievement with the award of the Nobel Prize in Physiology or Medicine in 1995.

References

1. Lewis, E. B. A gene complex controlling segmentation in *Drosophila. Nature* **276,** 565–560 (1978).
2. Nüsslein-Volhard, C. & Wieschaus, E. Mutations affecting segment number and polarity in *Drosophila. Nature* **287,** 795–801 (1980).
3. Marigo, V., Scott, M. P., Johnson, R. L., Goodrich, L. V. & Tabin, C. Conservation of hedgehog signaling: induction of a chicken patched homolog by Sonic Hedgehog in the developing limb. *Development* **122,** 1225–1233 (1996).
4. Potter, C. J., Turenchalk, G. S. & Xu, T. *Drosophila* in cancer research: An expanding role. *Trends Genet.* **16,** 33–39 (2000).
5. Fernandez-Funez, P. *et al.* Identification of genes that modify ataxin-1-induced neurodegeneration. *Nature* **408,** 101–106 (2000).

6. Taipale, J. & Beachy, P. A. The Hedgehog and Wnt signalling pathways in cancer. *Nature* **411**, 349–354 (2001).
7. McMahon, A. P., Ingham, P. W. & Tabin, C. J. Developmental roles and clinical significance of hedgehog signaling. *Curr. Top. Dev. Biol.* **53**, 1–114 (2003).

Further reading

Lawrence, P. A. *The Making of a Fly: The Genetics of Animal Design* (Blackwell Scientific, Oxford, 1992).

Nobel e-Museum. *The Nobel Prize in Physiology or Medicine 1995* (http://www.nobel.se/medicine/laureates/1995/).

1980

Mutations affecting segment number and polarity in *Drosophila*

Christiane Nüsslein-Volhard and Eric Wieschaus

In systematic searches for embryonic lethal mutants of *Drosophila melanogaster* we have identified 15 loci which when mutated alter the segmental pattern of the larva. These loci probably represent the majority of such genes in *Drosophila*. The phenotypes of the mutant embryos indicate that the process of segmentation involves at least three levels of spatial organization: the entire egg as developmental unit, a repeat unit with the length of two segments, and the individual segment.

The construction of complex form from similar repeating units is a basic feature of spatial organisation in all higher animals. Very little is known for any organism about the genes involved in this process. In *Drosophila,* the metameric nature of the pattern is most obvious in the thoracic and abdominal segments of the larval epidermis and we are attempting to identify all loci required for the establishment of this pattern. The identification of these genes and the description of their phenotypes should lead to a better understanding of the general mechanisms responsible for the formation of metameric patterns.

In *Drosophila,* the anlagen for the individual segments arise as equally sized subdivisions of the blastoderm, each segment represented by a transverse strip of about three or four cell diameters[1]. A cell lineage restriction between neighbouring segments is established at or soon after this stage[2]. Two basic types of mutation have been described which change the segmental pattern of the *Drosophila* larva. Maternal effect mutations like *bicaudal* lead to a global alteration of the embryonic pattern[3]. Bicaudal embryos develop two posterior ends arranged in mirror-image symmetry, and lack head, thorax and anterior abdomen. The *bicaudal* phenotype suggests that the initial spatial organisation of the egg established during oogenesis involved a morphogen gradient that defines anteroposterior coordinates in early embryonic pattern formation[3,4]. The subdivision of the embryo into segments is thought to occur by a differential response of the zygotic genome to the maternal gradient. Homeotic mutations (for example, *bithorax*[5,6]) seem to be involved in a final step of this response process. These mutations change the identity of individual segments; for example, *Ultrabithorax* transforms the metathoracic and first abdominal segments into mesothoracic segments. However, the homeotic loci do not affect the total number, size or polarity of the segments, nor do they point to any other step which might intervene between the maternal gradient and the final pattern of segments.

We have undertaken a systematic search for mutations that affect the segmental pattern depending on the zygotic genome. We describe here mutations at 15 loci which show one of three novel types of pattern alteration: pattern duplication in each segment (segment polarity mutants; six loci), pattern deletion in alternating segments (pair-rule mutants; six loci) and deletion of a group of adjacent segments (gap mutants; three loci) (Table 1, Fig. 1).

The segmental pattern of the normal Drosophila *larva*

Figure 2 shows the cuticular pattern of a normal *Drosophila* larva shortly after hatching. The larval body is comprised of three thoracic and eight abdominal segments. Although differences are observed in different body regions, all segments have certain morphological features in common. The anterior of each segment is marked with a band of denticles, most of which point posteriorly. The posterior part of each segment is naked. The segment borders run along the anterior margins of the denticle bands[7], they have no special morphological features. The polarity of the pattern is indicated by the orientation of the denticles and, in the abdomen, by the shape of the bands (Fig. 3). In the thoracic segments the bands are narrow with fine denticles whereas those in the abdominal segments are broader and comprised of thick pigmented denticles (for a detailed description of the cuticular pattern see ref. 1).

Segment polarity mutants: deletions in each segment

Mutants in this class have the normal number of segments. However, in each segment a defined fraction of the normal pattern is deleted and

Table 1 Loci affecting segmentation in *Drosophila*

Class	Locus	Map position*	No. of alleles†	Ref.
Segment-polarity	*cubitus interruptus*^{D} (*ci*^{D})	4–0	(2)	20
	wingless (*wg*)	2–30	6	9
	gooseberry (*gsb*)	2–104	1	This work
	hedgehog (*hh*)	3–90	2	This work
	fused (*fu*)‡	1–59.5	(9)	8, 20
	patch (*pat*)	2–55	8	This work
Pair-rule	*paired* (*prd*)	2–45	3	This work
	even-skipped (*eve*)	2–55	2	This work
	odd-skipped (*odd*)	2–8	2	This work
	barrel (*brr*)	3–27	2	This work
	runt (*run*)	1–65	1	This work
	engrailed (*en*)	2–62	6	11, 20
Gap	*Krüppel* (*Kr*)	2–107.6	6	12, 20
	knirps (*kni*)	3–47	5	This work
	hunchback (*hb*)	3–48	1	This work

* For the new loci (see last column) the map positions are based on recombination between the markers *S, Sp, Bl, cn, bw* for the second chromosome, and *ru, h, th, st, cu, sr,* e^5, *ca* for the third chromosome. For description of markers see ref. 20. The loci *runt, Krüppel* and *knirps* were further mapped using the breakpoints of deficiencies and duplications for the respective regions. All mutants were mapped by scoring the embryonic progeny of single recombinant males backcrossed to heterozygous females from the original mutant stocks.

† The numbers in parentheses refer to the alleles listed in Lindsley and Grell[20]. All other alleles, except the *runt* allele, three *Kr* alleles and one *knirps* allele, were isolated in a screen for embryonic lethal mutants on the second chromosome. 5,800 balanced stocks were established from individual males heterozygous for an ethyl methane sulphonate-treated *cn bw sp* chromosome using the DTS-procedure suggested by Wright[21]. 4,500 of the stocks had one or more new lethal mutations. Unhatched embryos from 2,600 putative embryonic lethal stocks were inspected for cuticular abnormalities[22]. Third chromosomal mutants discovered in the second chromosomal balanced lines were recovered after selection through individual females by balancing individual third chromosomes over TM3. Complementation tests were carried out between mutants with similar phenotypes whereby the occurrence of mutant embryos among the progeny of the crosses served as the criterion for allelism. Three new *Kr* alleles were isolated in a screen for lethals over the original *Kr* of Gloor[12], and one *knirps* allele of presumably spontaneous origin was discovered on a TM1 chromosome. The *runt* allele was isolated in a screen for X-linked lethals.

‡ *fused* is a male-rescuable maternal-effect locus[8]. Thus, the segment polarity reversal is observed in *fu*/*fu* embryos from *fu*/*fu* mothers. The progeny of *fu*/ + females show a normal embryonic pattern regardless of embryonic genotype.

the remainder is present as a mirror-image duplication. The duplicated part is posterior to the 'normal' part and has reversed polarity (Figs 1–3).

Six such loci have been identified. Three loci, *fused*[8], *wingless* (ref. 9 and G. Struhl, personal communication) and *cubitus interruptus*^{D}, were previously known whereas *gooseberry, hedgehog* and *patch* are new (Ta-

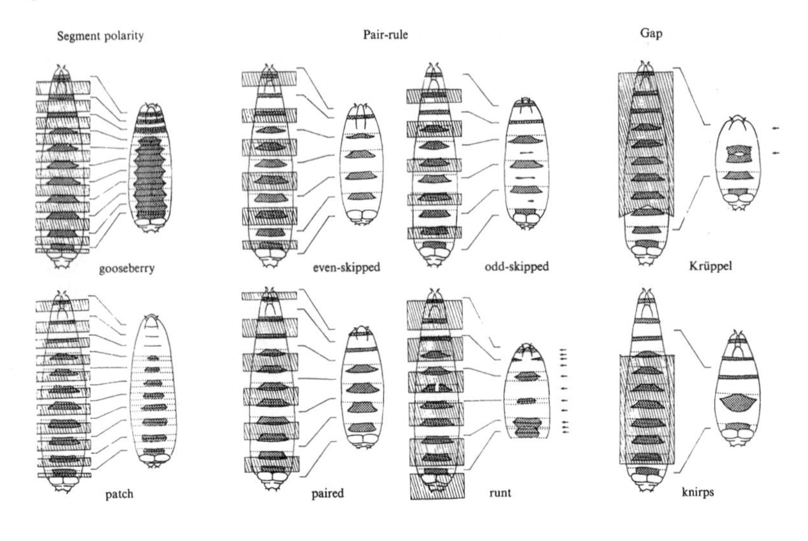

Fig. 1 Semi-schematic drawings indicating the regions deleted from the normal pattern in mutant larvae. Dotted regions indicate denticle bands, dotted lines the segmental boundaries. The regions missing in mutant larvae are indicated by the hatched bars. The transverse lines connect corresponding regions in mutant and normal larvae. Planes of polarity reversal in *runt* and *Krüppel* are indicated by the arrows. The two segment polarity loci *patch* and *gooseberry* are represented at the left. For indication of the polarity of the patterns, see Fig. 3. The patterns of *fused* and *ci*^{D} (not shown) look similar to the *gooseberry* pattern, whereas in *hedgehog* and *wingless* the deleted regions are somewhat larger, cutting into the denticle bands at either side. Four pair-rule mutants are shown in the centre. The interpretation of their phenotypes is based on the study of weak as well as strong alleles, combinations with *Ubx* (see text) and, in the case of *runt*, on gynandromorphs (unpublished). They probably represent the extreme mutant condition at the respective loci. The phenotypes of all known *barrel* and *engrailed* alleles (not shown) are somewhat variable and further studies are needed to deduce the typical phenotype. At the right, the two gap loci *Krüppel* and *knirps* are shown. Both patterns represent the amorphic phenotype as observed in embryos homozygous for deficiencies of the respective loci. The only known *hunchback* allele (not shown) deletes the meso- and metathorax.

ble 1). All the mutations in this class are zygotic lethals and the phenotypes are only produced in homozygous embryos. One of the loci, *fused,* also shows a maternal effect in that a wild-type allele in the mother is sufficient to rescue the mutant phenotype of the homozygous embryos.

In all mutants except *patch* the region deleted includes the naked posterior part of the pattern and the duplication involves a substantial fraction of the anterior denticle band. In mutant larvae, the ventral side of each segment is almost entirely covered with denticles, the posterior fraction of which point anteriorly. The segment identity seems to be normal, the denticles of the abdominal segments being large and pigmented whereas those of the thorax are short and pale (Figs 1, 2). The anterior

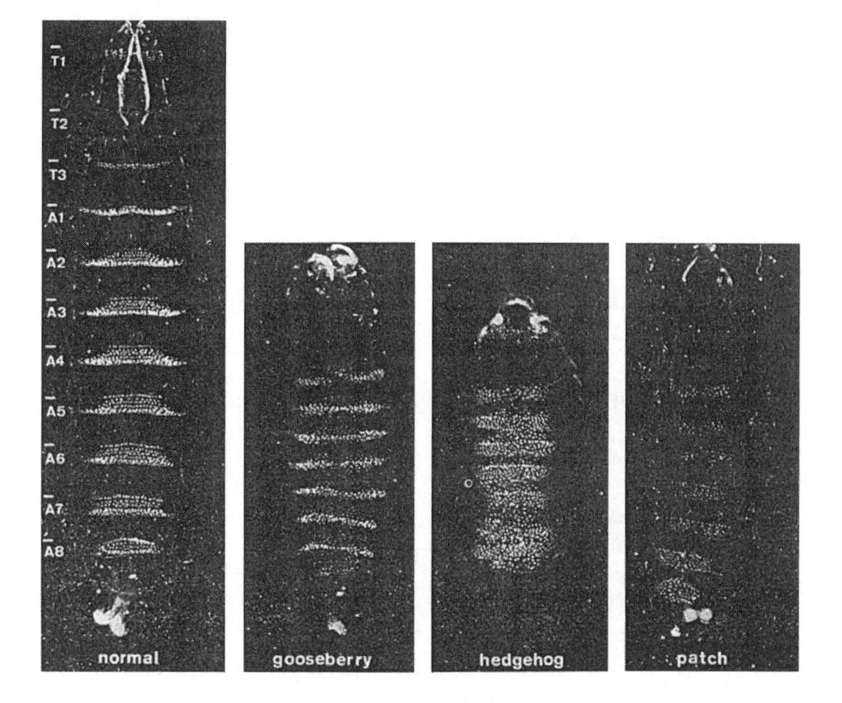

Fig. 2 Ventral cuticular pattern of (from left to right) a normal *Drosophila* larva shortly after hatching, and larvae homozygous for *gooseberry*, *hedgehog* and *patch*. The mutant larvae were taken out of the egg case before fixation. All larvae were fixed, cleared and mounted as described in ref. 22. A, abdominal segment; T, thoracic segment. For further description see text and Fig. 3. ×140.

margin of the region duplicated in these mutants coincides with the segment boundary in only two cases, *fused* and *gooseberry* (Fig. 3). In *wingless* and *hedgehog* it lies posterior to the boundary, such that these larvae apparently lack all segment boundaries.

The phenotype of embryos homozygous for *patch* contrasts with that produced by the five loci described above in that the duplicated region includes some naked cuticle anterior to each denticle band. The duplicated unit thus involves structures of two adjacent segments. *Patch* larvae, despite the normal number of denticle bands, have twice the normal number of segment boundaries (Figs 1, 3).

Despite these differences, the common feature of all mutants in this class is that a defined fraction of the pattern in each segment is deleted. This deletion is associated with a mirror image duplication of the remaining part of the pattern. We suggest that these loci are involved in the specification of the basic pattern of the segmental units.

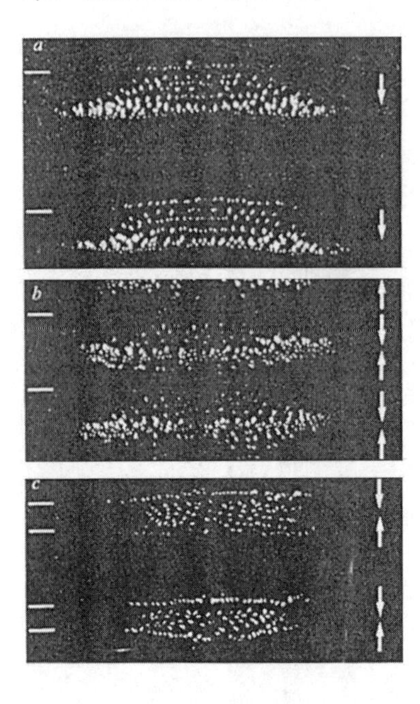

Fig. 3 Details from the ventral abdomen of a normal (*a*), a *gooseberry* (*b*), and a *patch* (*c*) larva. The positions of the segment boundaries are indicated at the left by the transverse lines. The arrows at the right indicate the polarity of the pattern as judged by the orientation of the denticles as well as the shape of the denticle bands.

Pair-rule mutants: deletions in alternating segments

In mutants of this class homologous parts of the pattern are deleted in every other segment. Each of the six loci is characterized by its own specific pattern of deletions (Table 1, Figs 1, 4). For example, in *even-skipped* larvae, the denticle bands and adjacent naked cuticle of the pro- and metathoracic, and the 2nd, 4th, 6th and 8th abdominal segments are lacking. This results in larvae with half the normal number of denticle bands separated by enlarged regions of naked cuticle (Figs 1, 4). In *paired* larvae the apparent reduction in segment number results essentially from the deletion of the naked posterior part of the odd-numbered and the anterior denticle bands of the even-numbered segments (Figs 1, 4). The double segments thus formed are composites of anterior mesothorax and posterior metathorax, anterior first abdominal and posterior second abdominal segment, etc. The identification of the regions present or deleted in mutant larvae is based on the phenotypes produced by alleles with lower expressivity. In such embryos the deletions are in general smaller and variable in size. In a 'leaky' allele of *even-skipped,* the denticle bands of the even-numbered segments are frequently incomplete, whereas in *paired*^{2} (*prd*^{2}) the pattern deletions often involve only part of the naked cuticle of the odd-numbered segments, leading to a pairwise fusion of denticle bands (Fig. 4).

Further support for the composite nature of the segments in *prd* larvae was obtained in combinations with *Ultrabithorax* (*Ubx*), a homeotic mutation which when homozygous causes the transformation of the first abdominal segment into mesothorax[5]. In such *prd*;*Ubx* larvae, only the *denticle band* of the first double segment in the abdomen is transformed. The posterior margin of this band and the naked cuticle which follows remain abdominal in character and lack, for example, the typical meso-

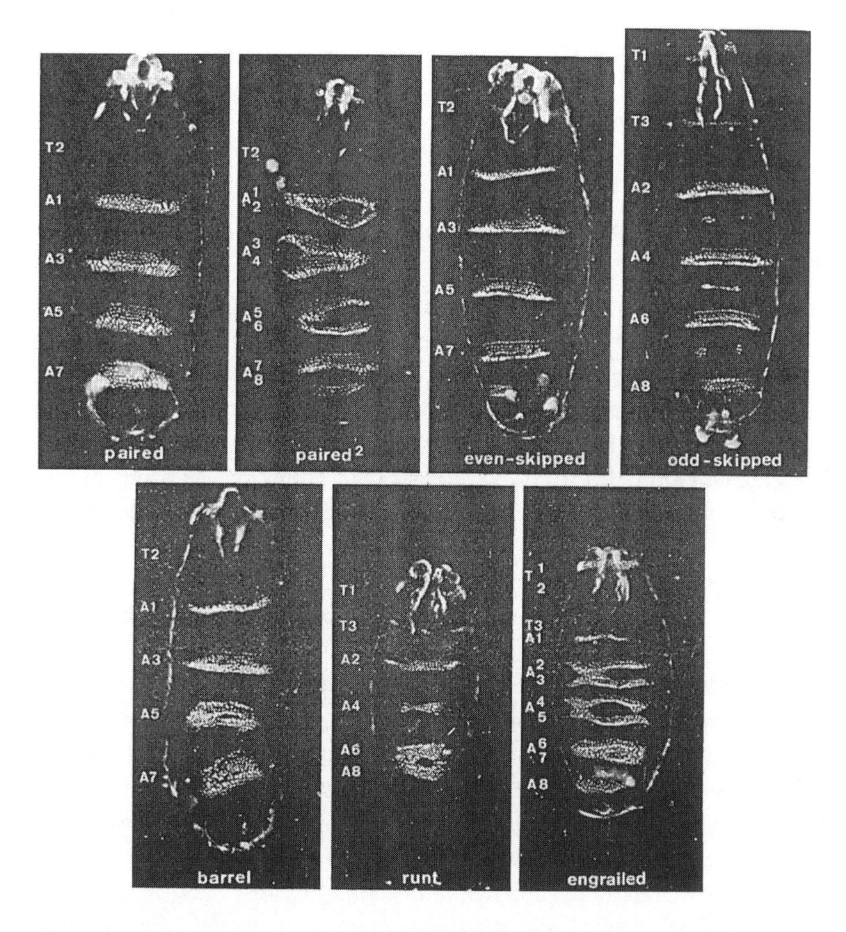

Fig. 4 Larvae homozygous for mutations at the six pair-rule loci. The segmental identity of the denticle bands is indicated at the left of each picture. A, abdominal band; T, thoracic band. For comparison with the normal pattern see Figs 1 and 2.

thoracic sense organs. The composite nature of the segments in *paired* shows that the establishment of segmental identity does not require the establishment of individual segments as such.

The segmentation pattern in *prd;Ubx* larvae is typical of that of *paired* alone and is not affected by the *Ubx* homeotic transformation. Similarly, in *prd;Polycomb* larvae the naked cuticle of alternating segments is deleted just as it is in *paired* alone, even though in *Polycomb* embryos all the thoracic and abdominal segments have an 8th abdominal character[5]. These combinations, and similar ones with *even-skipped,* indicate that the observed grouping of segments in pairs depends on the position of segments within the segmental array rather than the segmental identity. These combinations thus provide evidence that mutations such as *paired* and *even-skipped* affect different processes from those altered in homeotic mutants.

Other mutants in this class show different deletion patterns. The phenotype of *odd-skipped* is similar to that of *even-skipped.* However, in this case it is the odd-numbered denticle bands that are affected. In *odd-skipped,* the deleted region is smaller and is restricted to a more posterior part of the segment than in *even-skipped* (Figs 1, 4). *Barrel* has a phenotype similar to *paired* although the pattern is often less regular (Fig. 4). *Runt,* on the X chromosome, is the only pair-rule mutant showing mirror-image duplications. *Runt* embryos have half the normal number of denticle bands, each a mirror-image duplication of the anterior part of a normal band (similar to the duplications found in *patch*). The bands in *runt* embryos, as well as the region of naked cuticle separating them, are of unequal sizes (Figs 1, 4).

To the list of pair-rule loci we have added the *engrailed*[10] locus. Lethal *engrailed* alleles[11] lead to a substantial deletion of the posterior region of even-numbered segments. In addition, the anterior margin and adjacent cuticle of each segment are affected. Thus, the defect pattern in *engrailed* shows repeats which are spaced at both one- and two-segment intervals.

Each of the six different pair-rule loci affects a different region within a double segmental repeat. In no case does the margin of the deleted region coincide with a segment boundary. When the deleted region corresponds in size up to one entire segment (*paired, even-skipped*) it includes parts of two adjacent segments.

The phenotypes of the pair-rule mutants suggest that at some stage during normal development the embryo is organized in repeating units, the length of which corresponds to two segmental anlagen.

Gap mutants: one continuous stretch of segments deleted

One of the striking features of the mutations of the first two classes is that the alteration in the pattern is repeated at specific intervals along the antero-posterior axis of the embryo. No such repeated pattern is found in mutants of the third class and instead a single group of up to eight adjacent segments is deleted from the final pattern. Three loci have been identified which cause such gaps in the pattern (Table 1, Figs 1, 5). *Krüppel* (*Kr*) was originally described by Gloor[12]. Embryos homozygous for *Kr* lack thorax and anterior abdomen. The posterior-terminal region with the abdominal segments 8, 7 and 6 is normal, although probably somewhat enlarged. Anterior to the 6th abdominal segment is a plane of mirror-image symmetry followed by one further segment band with the character of a 6th segment oriented in reversed polarity. The exact position of the plane of symmetry varies and does not usually coincide with a segmental boundary. A large part of the *Krüppel* pattern is reminiscent of the pattern observed in embryos produced by the maternal-effect mutant *bicaudal,* although no maternal component is involved in the production of the *Krüppel* phenotype (in preparation).

In the other two loci of this class, the gap in the pattern occurs in specific morphologically defined subregions of the larval pattern, in the

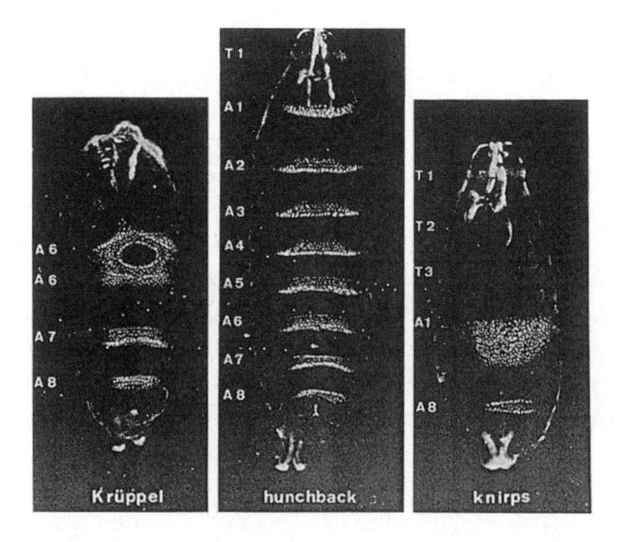

Fig. 5 Larvae homozygous for mutations at the three gap loci. A, abdominal segment; T, thoracic segment.

thorax and in the abdomen, respectively. In *hunchback,* the meso- and metathoracic segments are deleted. In embryos homozygous for *knirps,* only two rather than eight denticle bands are formed in the abdomen. The posterior-terminal region including the 8th abdominal segment seems normal whereas the anterior abdominal denticle band is considerably enlarged. The anterior margin of the denticle band is morphologically similar to the first abdominal segment but combinations with *Ubx* show that the band is a composite with more than one segmental identity.

All three loci are required for a normal segmental subdivision of one continuous body region. The lack of a repeated pattern of defects suggests that the loci are involved in processes in which position along the antero-posterior axis of the embryo is defined by unique values.

When are genes affecting segment number active?

The phenotypes described above are observed only in homozygous embryos, indicating that the loci identified by these mutations are active after fertilization and are crucial for the normal segmental organisation of the embryo. We have described the mutations in terms of their effect on the differentiated pattern. However, in many instances their effect can be observed much earlier in development. In normal embryos segmentation is first visible 1 h after the onset of gastrulation as a repeated pattern of bulges in the ventral ectoderm (unpublished observations). In *paired* and *even-skipped* embryos the number of bulges is reduced and corresponds to the number of segments observed in the differentiated mutant embryos. *Krüppel, runt* and *knirps* embryos can be identified 15 min after the onset of gastrulation. All three mutations cause shorter germ bands, a phenomenon clearly related to the strong reduction in segment number observed in the differentiated larvae. Further evidence for an early activity at the *paired* locus was obtained using a temperature-sensitive allele. The extreme phenotype is only obtained when the embryo is kept at the restrictive temperature during the blastoderm stage. All these results indicate that the wildtype genes defined by the mutations are active before the end of gastrulation during normal development.

Discussion

Segmentation in *Drosophila* proceeds by the transition from a single field into a repeated pattern of homologous smaller subfields. Mutant alleles at the 15 loci we have described interfere with this process at various points. Although each locus has its own distinct phenotype, we were able to distribute the mutations in three classes. In one class only a single large subregion of the embryo is affected, whereas in mutants of the other two classes a reiteration of defects is produced with a repeat length of one or two segments, respectively. This suggests that the process of segmentation involves three different units of spatial organization.

The organization of the egg is thought to be controlled by a monotonic gradient set up during oogenesis under the control of the maternal genome[3,4]. All the mutants we have described depend on the embryonic rather than the maternal genome. None of them alters the overall polarity of the embryo, the head always being at the anterior and the telson at the posterior end of the egg. The most dramatic alterations of the pattern are produced in the gap mutants and involve one large subregion of the egg. *Hunchback* and *knirps* affect the development of thorax and abdomen, respectively, and might be involved in the establishment of these large morphologically unique subregions of the embryo. The large mirror-image duplications found in the posterior pattern of *Krüppel* embryos are similar to those of *bicaudal*. The *bicaudal* phenotype has been interpreted as resulting from an instability in the maternal gradient[3,13], and thus *Krüppel* might be involved in the maintenance of elaboration of this gradient in the posterior egg region after fertilization.

The smallest repeat unit, the individual segment, is affected in mutants at the six segment-polarity loci. The pattern alteration consists of a mirror-image duplication of part of the normal pattern with the remainder deleted. In these mutants the deleted region corresponds in size to more than half a segment. Mutations causing smaller deletions in the segmental pattern do not lead to polarity reversals (unpublished observations). The mirror-image duplications produced by the pair-rule mutation *runt* are also associated with a deletion of more than half a repeat unit, that is, more than one segment. The tendency of partial fields containing less than half the positional values to duplicate is well described for imaginal disks in *Drosophila*[14], as well as for the larval epidermis in other insects[15,16]. On the other hand, the same types of pattern duplication are also produced in conditions which do not involve cell death and regeneration, but rather a reorganization of the positional information in the entire field[3,17]. More detailed studies of early development in mutant embryos may reveal which mechanism is responsible for the different mutant phenotypes.

Given the evidence for a homology between segments, the existence of a class of mutants affecting corresponding regions in each individual segment is perhaps not surprising. The discovery of a mutant class affecting corresponding regions in every other segment was not expected and suggests the existence at some time during development of homologous units with the size of two segmental anlagen. It is possible that the double segmental unit corresponds to transitory double segmental fields which are established early during embryogenesis and are later subdivided into individual segments. At the blastoderm stage the epidermal primordium giving rise to thorax and abdomen is only about 40 cells long[1]. An initial subdivision into double segments would avoid problems of accuracy encountered in a simultaneous establishment of segment boundaries every three to four cells. A stepwise establishment of segments implies that the borders defining double segments be made before the intervening ones. The mutant phenotypes do not definitely show which, if any, of the segment borders define a primary double segmental division in normal development. The mutant phenotypes which come closest to a pattern one would expect if the transition from the double segment to the individual segment stage were blocked are the patterns of *paired* and *even-skipped*. Both suggest the frame meso- and metathorax, 1st and 2nd, 3rd and 4th abdominal segment, etc.

It is also possible that the double segmental units are never defined by distinct borders in normal development. The existence of a double segmental homology unit may merely reflect a continuous property such as a wave with a double segmental period responsible for correct spacing of segmental boundaries (see, for example, refs 18, 19). We have not found any mutations showing a repeat unit larger than two segments. This may indicate that the subdivision of the blastoderm proceeds directly by the double segmental repeat with no larger intervening homology units. However, the failure to identify such larger units may reflect the incompleteness of our data.

Drosophila has been estimated to have about 5,000 genes and only a very small fraction of these when mutated result in a change of the segmental pattern of the larva. Some of the loci described here were known previously but only in the case of *Krüppel* has the embryonic phenotype been recognized as affecting segmentation.[12] The majority of the mutants described here have been isolated in systematic searches for mutations affecting the segmentation pattern of the *Drosophila* larva. These experiments are still incomplete. Most of the alleles on the second chromosome were isolated in one experiment which yielded an average allele frequency of four or five alleles per locus (based on 42 embryonic lethal loci). From this yield and similar calculations for the third and first chromosomes, we estimate that we have identified almost all segmentation loci on the second chromosome and about 50% each of those on the third and first chromosome. Our sample of 15 loci should therefore represent the majority of the loci affecting segmentation in the *Drosophila* genome. Thus, in *Drosophila* it would seem feasible to identify all genetic components involved in the complex process of embryonic pattern formation.

We thank Hildegard Kluding for excellent technical assistance, Adelheid Schneider, Maria Weber and Gary Struhl for help during various parts of the mutant screens, Gerd Jürgens for stimulating discussion and our colleagues from the *Drosophila* laboratories in Cambridge, Freiburg, Heidelberg and Zürich for critical comments on the manuscript, and Claus Christensen for the photographic prints. Thomas Kornberg and Gary Struhl provided us with lethal alleles of *engrailed* and *wingless* respectively which facilitated the identification of our alleles. All mutants are available on request.

Note added in proof: All known *barrel* alleles fail to complement *hairy*[20], suggesting that the *barrel* mutations are alleles at the *hairy* locus.

European Molecular Biology Laboratory, PO Box 10.2209, 69 Heidelberg, FRG
Received 26 June; accepted 29 August 1980 [Published 30 October]

References

1. Lohs-Schardin, M., Cremer, C. & Nüsslein-Volhard, C. *Devl. Biol.* **73**, 239–255 (1979).
2. Wieschaus, E. & Gehring, W. *Devl. Biol.* **50**, 249–263 (1976).
3. Nüsslein-Volhard, C. in *Determinants of Spatial Organisation* (eds Subtelney, S. & Konigsberg, I. R.) 185–211 (Academic, New York, 1979).
4. Sander, K. *Adv. Insect Physiol.* **12**, 125–238 (1976).
5. Lewis, E. B. *Nature* **276**, 565–570 (1978).
6. Garcia-Bellido, A. *Am. Zool.* **17**, 613–629 (1977).
7. Szabad, J., Schüpbach, T. & Wieschaus, E. *Devl. Biol.* **73**, 256–271 (1979).
8. Counce, S. *Z. Induktive Abstammungs-Vererbungslehre* **87**, 462–81 (1958).
9. Sharma, R. P. & Chopra, V. L. *Devl. Biol.* **48**, 461–465 (1976).
10. Lawrence, P. A. & Morata, G. *Devl. Biol.* **50**, 321–337 (1976).
11. Kornberg, T., in preparation.
12. Gloor, H. *Arch. Julius-Klaus-Stift. VererbForsch* **25**, 38–44 (1950).
13. Meinhardt, H. *J. Cell Sci.* **23**, 117–139 (1977).
14. Bryant, P. J. *Ciba Fdn Symp.* **29**, 71–93 (1975).
15. Wright, D. & Lawrence, P. A., in preparation.
16. Lawrence, P. A. in *Developmental Systems: Insects* (eds Counce, S. & Waddington, C. H.) 157–209 (Academic, London, 1973).
17. Jürgens, G. & Gateff, E. *Wilhelm Roux Arch.* **186**, 1–25 (1979).
18. Meinhardt, H. & Gierer, A. *J. Cell. Sci.* **15**, 321–346 (1974).

19. Kaufmann, S. A., Shymko, R. M. & Trabert, K. *Science* **199**, 259–270 (1978).
20. Lindsley, D. & Grell, E. H. *Genetic Variations of Drosophila melanogaster* (Carnegie, Washington, 1968).
21. Wright, T. R. F. *Drosoph. Inf. Serv.* **45**, 140 (1970).
22. Vander Meer, J. *Drosoph. Inf. Serv.* **52**, 160 (1977).

massive binary system comprising an O7III–V [21,26] with a mass of ~17 $M_\odot$ together with a n star[27] in orbit with a 1.408-day period. The as occasional flaring episodes during which X rays are pulsed[27] with a period of 13.5 s, pin period of the neutron star. The inclination[27] ~66° and the X-ray source is eclipsed between 0.08. If UHE protons are accelerated in the mpact object UHE γ rays could be produced actions in the atmosph the e, we would expect γ-ray nd 0.08 when our line-of- the star's surface. The ob of 0.90–0.95 is consistent with this p sence of UHE emission at phases is lar situation exists for Cyg X-3 and Vela X-1 burst of UHE γ rays is observed per orbit). The s high and low states associated with a 30.5-day ted to precession of an accretion disk. This picture is supported by ultraviolet observations[28] an almost constant X-ray heating of the stellar

not appear to fill its Roche lobe[21] and seems low stellar wind[21,28]. Mass transfer may occur ing accretion stream[21] which may feed the Evidence for this comes from variable obscur- npanion star[27] between orbital phases 0.6 and g to matter trailing behind the neutron star by 0.3 of an orbit. If this matter is of considerable orbital plane it could be the target material for UHE protons produced near the neutron star. would produce γ rays only at a phase of ~0.9

that we have found evidence for UHE γ-ray LMC X-4 system modulated with the 1.408-day The emission occurs when the neutron star is eclipse by the companion star and could result eractions in the atmosphere of the companion s produced near the neutron star. Alternatively, ion stream could provide target material and ence of UHE γ rays on leaving eclipse. Further progess above 10^{15} and 10^{16} eV coupled with m a more southerly site at about 10^{14} eV could the universality of the microwave background, direct measurement of the magnetic field between l the Large Magellanic Cloud, or provide an asurement of the distance to the Large Magel-

dge the efforts of P. R. Gerhardy in obtaining nalysed in the present work. Others particularly the development of the Buckland Park array Prescott, J. R. Patterson, A. G. Gregory, P. C. F. Liebing. We thank D. F. Liebing and A. A. ful comments and K. J. Orford and K. E. Turver ssions. R.J.P. thanks the Australian Government abeth II Fellowship. This work was supported ustralian Research Grants Committee.

1984; accepted 6 March 1985.

nm, W. *Astrophys. J. Lett.* **268**, L17–L22 (1983).
l. *Nature* **305**, 784–787 (1983).
y, R. W. & Gerhardy, P. R. *Astrophys. J. Lett.* **280**, L47–L50 (1984).
nd, W. T. *Nature* **307**, 613–614 (1984).

19. Gould, R. J. *Astrophys. J. Lett.* **274**, L23–L25 (1983).
20. van der Klis, M. *et al. Astr. Astrophys.* **106**, 339–344 (1982).
21. Huchings, J. B., Crampton, D. & Cowley, A. P. *Astrophys. J.* **225**, 548–
22. Lang, F. L. *et al. Astrophys. J. Lett.* **246**, L21–L25 (1981).
23. Bradt, H. V. & McClintock, J. E. *A. Rev. Astr. Astrophys.* **21**, 13–66 (1
24. Prescott, J. R. *et al. Proc. 18th int. Cosmic Ray Conf.* Bangalore **6**, 257
25. Cawley, M. F., Gibbs, K. & Weekes, T. C. *Proc. 18th int. Cosmic Ray* 69–72 (1983).
26. Chevalier, C. & Ilovaisky, S. A. *Astr. Astrophys.* **59**, L9–L12 (1977).
27. R. L. *et al. Astrophys. J.* **264**, 568–574 (1983).
lis, M. *et al. Astr. Astrophys.* **106**, 339–344 (1982).
K. *Space Sci. Rev.* **30**, 441–446 (1981).
& van Paradijs, J. *Astr. Astrophys.* **126**,
D. & Cowley, A. P. *Astrophys. J. Lett.*, **275**
A. P. pton, D. & Huchings, J. B. *Astrophys. J.* **256**, 605
T. *strophys. J.* **264**, 563–567 (1983).
ll, J. *Astrophys. J.* **273**, 709–715 (1983).
M. *et al. R. astr. Soc.* **185**, 29P–32P (1978).
P. J. N. *et al. Not. R. astr. Soc.* **181**, 73P–79P (1977).
37. Gottlieb, E. W., Wri & Liller, W. *Astrophys. J. Lett.* **195**, L33
38. Middleditch, *s. J.* **244**, 1001–1021 (1981).
39. van Paradijs, *Astrophys. Suppl.* **55**, 1–14 (1984).

Large losses of total ozone in An reveal seasonal ClO_x/NO_x intera

J. C. Farman, B. G. Gardiner & J. D. Shankli

British Antarctic Survey, Natural Environment Research High Cross, Madingley Road, Cambridge CB3 0ET, UK

Recent attempts[1,2] to consolidate assessments of human activities on stratospheric ozone (O_3) dimensional models for 30° N have suggested that of total O_3 will remain small for at least the next de from such models are often accepted by default as mates[3]. The inadequacy of this approach is here ma observations that the spring values of total O_3 in An now fallen considerably. The circulation in the lower is apparently unchanged, and possible chemical ca considered. We suggest that the very low tempera prevail from midwinter until several weeks after the sp make the Antarctic stratosphere uniquely sensitive inorganic chlorine, ClX, primarily by the effect of th the NO_2/NO ratio. This, with the height distribu irradiation peculiar to the polar stratosphere, could the O_3 losses observed.

Total O_3 has been measured at the British Anta stations, Argentine Islands 65° S 64° W and 76° S 27° W, since 1957. Figure 1*a* shows data from The mean and extreme daily values from October 19 1973 and the supporting calibrations have been dis where[4,5]. The mean daily value for the four late observing seasons (October 1980–March 1984 individual daily values for the current observing detailed in Fig. 1. The more recent data are provisi Very generous bounds for possible corrections ±30 matm cm. There was a changeover of spectro at the station in January 1982; the replacement inst been calibrated against the UK Meteorological Offi in June 1981. Thus, two spectrophotometers have sho values of total O_3 to be much lower than March valu entirely lacking in the 1957–73 data set. To interpre ence as a seasonal instrumental effect would be

EIGHTEEN

A hole in Earth's shield

Richard S. Stolarski

Ozone in the stratosphere shields the Earth from ultraviolet radiation. Worries about the possible depletion of this protective layer led, in the early 1980s, to international attempts to control chlorofluorocarbons, the manmade chemicals thought to be responsible. But it took the dramatic discovery of the Antarctic "ozone hole," in 1985, to show that humans really were modifying the global environment. The resulting international agreements have put the atmosphere on the road to recovery and provide a model for environmental stewardship.

Chlorofluorocarbons (CFCs) were invented in the 1930s, and at one time seemed a miracle of modern chemistry. They are carbon-based compounds like methane and ethane, but with the hydrogen atoms replaced by chlorine or fluorine. The simplest CFCs, and the most commonly produced in the 1970s, when the ozone depletion story begins, are CFC-11 ($CFCl_3$) and CFC-12 (CF_2Cl_2). These compounds are inert, insoluble, and transparent. They were perfect for use in refrigeration and as spray-can propellants. Too perfect! Their inertness meant that they were long-lived, and they began to accumulate in the atmosphere. They would be around for a long time.

In the late 1960s and early 1970s, atmospheric chemists began to think about the role of chemical catalysis in the destruction of ozone. Paul Crutzen highlighted the importance of reactions involving nitrogen oxides,[1] and Harold Johnston argued that nitrogen oxide emissions from the proposed fleet of supersonic aircraft would seriously deplete the ozone layer.[2] Around the same time, Ralph Cicerone and I considered ozone destruction reactions involving chlorine,[3] but the known sources of chlorine in the stratosphere—

volcanoes and the Space Shuttle—were not significant enough to warrant concern.

Mario Molina and F. Sherwood Rowland were the first to identify industrially produced CFCs as a culprit in ozone destruction. They proposed that chlorine derived from CFCs was destroying stratospheric ozone at a steadily increasing rate[4]—a prescient contribution for which they were to receive, with Crutzen, the 1995 Nobel Prize in Chemistry.

With the chemical reactions highlighted by Molina and Rowland, chemists were able to calculate the expected effects of CFCs on the ozone layer using mathematical models of the atmosphere. The calculations showed that, at the rates at which CFCs were being introduced into the atmosphere, the amount of ozone in the upper stratosphere would have decreased by about one percent by the mid-1980s. This was too small a change to be detectable.

What was to be done? If our understanding was correct, by the time chlorine concentrations in the stratosphere would be large enough to have a clearly detectable effect, we would have a serious global problem that would be with us for a century. Action was taken: the United Nations Environment Programme (UNEP) convened talks that led to the adoption, in March 1985, of the Vienna Convention for the Protection of the Ozone Layer. The convention provided for cooperation in research and monitoring of the atmosphere, but did not require countries to reduce their production of ozone-depleting substances.

Then came the paper that changed everything. In May 1985, Joe Farman and colleagues at the British Antarctic Survey[5] reported in *Nature* that, in recent years, the springtime (October) atmospheric ozone concentration over their station at Halley Bay, Antarctica, had declined by 40 percent from its average value throughout the previous few decades. This decline had begun in the late 1970s and grew through the early 1980s. Its growth mirrored the increase in chlorine in the stratosphere from CFCs.

Farman *et al.*'s paper was based on data that went back to late 1956, when an ozone-measuring instrument known as a Dobson spectrophotometer had been installed at Halley Bay. The power of their measurements came not just from the rapid decrease observed, but from the long, continuous dataset, which allowed them to conclude that a truly new phenomenon was being seen.

The British Antarctic Survey wasn't the only organization monitoring atmospheric ozone. In November 1978, the U.S. space agency NASA had launched two ozone-measuring devices on the Nimbus 7 satellite: the Total

Ozone Mapping Spectrometer (TOMS) and the Solar Backscatter Ultraviolet (SBUV) instrument. These instruments produced maps of ozone over the sunlit portions of the globe. So why did Farman *et al.* discover the ozone hole while NASA scientists did not?

Part of the answer lies in the sheer volume of data generated by the TOMS/SBUV instruments. About two hundred thousand measurements were made each day, placing a great strain on the computer capacity of the day; it took a year or two to process the data and get them into the hands of scientists. Then there was a long list of interesting patterns to investigate, and the researchers knew that not all of them would turn out to be real. The low ozone amounts seen in the Antarctic spring were below what was considered to be the lowest possible amount of ozone and were flagged as potential "bad" data.

Before the discovery of the ozone hole, the scientific community showed little interest in the TOMS data. My office was on the same hallway as those of the main scientists on the TOMS/SBUV team, but I didn't look at their data until after the Farman *et al.* paper had been published. Perhaps if more scientists had examined the data, the significance of the springtime Antarctic ozone measurements would have been recognized sooner.

As the British team watched the measured ozone concentrations decline each spring, they too worried that their instrument was not working correctly. But they had the advantage of data from another instrument—at the Argentine Islands station on the Antarctic peninsula—to provide a check on the Halley Bay data. When they began to see a similar effect in the Argentine Islands data, they knew they were seeing something real.

Farman *et al.* theorized that the decline in ozone was due to the increase in chlorine from CFCs. They turned out to be right, even though the chemical mechanism they proposed was not. Most importantly, they pointed out that this unexpectedly large effect had not been predicted by the mathematical models of the atmosphere that were then being used to forecast the effect of human activities on stratospheric ozone. This was a wake-up call for both scientists and governments.

The low amounts of ozone observed in the TOMS data were reported in an abstract for a 1985 meeting in Prague. The abstract, by P. K. Bhartia and Don Heath, was submitted in February 1985; the meeting took place in August. I had first seen the Farman *et al.* paper when it was passed around at a UNEP ozone assessment meeting in July of that year. When I heard Bhartia give the talk in Prague, the significance of the findings began to sink in. This was a real effect over a significant area, and we had the data that

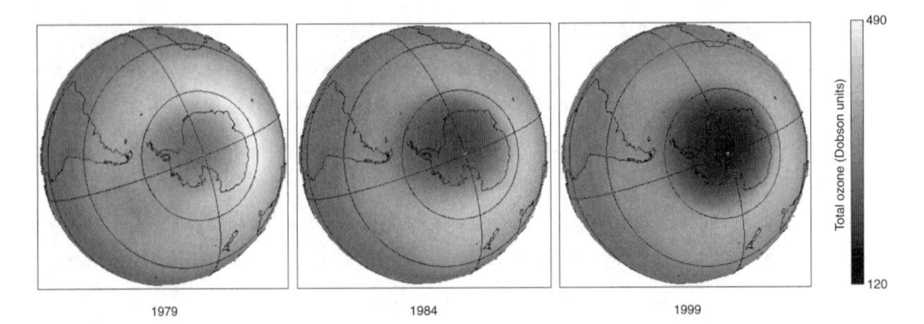

Fig. 18.1. The evolution of the Antarctic ozone hole, as mapped by the TOMS instrument on NASA's Nimbus 7 and Earth Probe satellites. Each image shows ozone levels in the atmosphere of the Southern Hemisphere, averaged over the month of October. The shade of gray corresponds to the total amount of ozone above a given point on the globe, with the largest amounts shown in light gray to white, and the smallest amounts in dark gray to black. In 1979 (left), the first year for which October TOMS data exist, Antarctic ozone depletion was not yet noticeable. The ring of relatively high ozone amounts surrounding a shallow minimum near the pole reflects the dynamics of the atmosphere, not the chemical destruction of ozone. By 1984 (center), when the ozone hole was discovered, the amount of ozone over the pole had decreased by 40 percent, from above 300 to about 190 Dobson units (DU). By 1999 (right), the hole had deepened and expanded further, with ozone levels down to below 130 DU at the center of the hole.

would confirm this remarkable finding and map its areal extent. Our paper[6] reporting the satellite data (from both SBUV and TOMS) was published in *Nature* in August 1986.

When we first submitted the paper, we included the term "Antarctic ozone hole" in the title, as it was beginning to be used as a shorthand to describe the phenomenon. One of the peer-reviewers objected, so we replaced it with "springtime Antarctic ozone decrease." But the continent-wide extent of the depletion, combined with its extreme magnitude (fig. 18.1) makes the term "hole" an apt description.

The discovery of the ozone hole gave a huge jolt to the field. We had been looking for barely visible trends; now we had an enormous effect that was obvious from casual inspection of the data. Theorists were quick to put forward a variety of ideas to explain the effect. Chemists thought it might be caused by chemical reactions, driven by sunlight. Dynamicists thought it might be the result of ozone-poor air flowing upward from the lower atmosphere. Others thought the cause might be an unusually active phase of the eleven-year solar cycle, which would increase the amount of ozone-destroying nitrogen oxides in the atmosphere. This last hypothesis was ruled out by the discovery that nitrogen oxides were in fact depleted in the Antarctic stratosphere—evidence, as it turns out, that the real culprits were chlorofluorocarbons.

The chemical explanations came thick and fast,[7–9] all based on a key realization: that chemical reactions on the surfaces of stratospheric clouds in the cold Antarctic atmosphere could convert inert forms of chlorine to active forms that would begin to destroy ozone as soon as the Sun rose in the spring. But none of the mechanisms was capable of explaining the size of the observed decline until Luisa and Mario Molina[10] pointed out in 1987 that, at parts-per-billion concentrations, chlorine monoxide (ClO) would react with itself, forming a dimer (Cl_2O_2). The dimer would then decompose in sunlight to give back chlorine atoms, allowing a catalytic, ozone-destroying cycle to proceed (fig. 18.2).

Meanwhile, expeditions were mounted to obtain more data. The most dramatic finding came from repeated ozone and chlorine monoxide measurements made in the Antarctic stratosphere by a NASA ER-2 aircraft flying from Punta Arenas, Chile. As the plane flew south toward Antarctica in early August at about twenty kilometers altitude, the amount of ClO suddenly jumped from a few tens of parts per trillion to about one part per billion—a thirtyfold increase. No variation was seen in ozone. Six weeks later, in late September, the same jump was seen in ClO, but now it was associated with a large decrease in ozone. Between the two flights, sunlight had been driving the chlorine reactions that destroy ozone. This smoking gun was the clearest demonstration that chlorine was the primary cause of what was by this time widely known as the Antarctic ozone hole.

The Farman *et al.* paper[5] has had a significant and lasting effect on global environmental issues. When the paper was published, the Vienna Convention had just been adopted. In the following year negotiations resumed, leading to the adoption, in September 1987, of the Montreal Protocol on

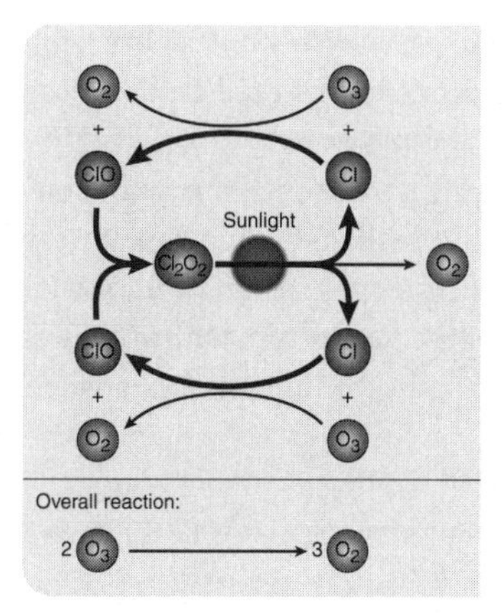

Fig. 18.2. Chemical reactions in the Antarctic stratosphere leading to ozone loss. Chlorine atoms catalyze the destruction of ozone; they are not used up in the ozone-depletion reaction, but can participate over and over again. The cycle starts with a reaction between chlorine (Cl) and ozone (O_3), to produce oxygen molecules (O_2) and chlorine monoxide (ClO). At first, atmospheric chemists could not figure out how chlorine atoms could be regenerated fast enough to destroy ozone at the observed rate. The solution came when Luisa and Mario Molina[10] pointed out that, at the high concentrations found in the Antarctic atmosphere, ClO molecules can react with each other to form Cl_2O_2. Then, when the Sun rises in the spring, the Cl_2O_2 splits, regenerating chlorine atoms for further ozone destruction.

Substances that Deplete the Ozone Layer, which bound its signatories to reducing the production and consumption of CFCs. The protocol was based, not on the spectacular ozone hole, but on the precautionary view that unacceptable damage could result from waiting for clear evidence of ozone loss. Nevertheless, we can be sure that the observation of the rapid decline of Antarctic ozone was a strong influence on the parties to the protocol to take this issue seriously. It helped give impetus to nations to view global environmental impacts as worthy of their attention. By the time the parties to the protocol met to amend it, in June 1990, the ozone hole loomed large in the public consciousness. The result was a much stricter agreement, according to which developed countries agreed to end the use of CFCs (and related compounds known as halons) by the year 2000, and a fund was set up to help developing countries take similar steps.

The paper's impact on stratospheric science was just as great. Following the recognition of the crucial role played by chemical reactions occurring on cloud particles in the Antarctic atmosphere, we now routinely consider the contribution of such reactions at other latitudes. Laboratories that had been devoted to studying reactions that take place between gas molecules began to attack the problem of characterizing reactions on particles. The formation of particles also has to be considered: for example, what exactly are the conditions under which a particle will form? And the Arctic atmosphere, where the more variable conditions make the interpretation of ozone loss more difficult, has become a center of attention. We wonder what will happen when climate change brings about colder temperatures in the Arctic—might the seasonal depletion there someday escalate into a full-blown Arctic ozone hole?

Fortunately, the answer is that it probably will not. As I write, the levels of ozone-depleting chemicals in the atmosphere have been level or slightly decreasing for several years. The Antarctic ozone hole is predicted to disappear for good by 2050, if the provisions of the Montreal protocol continue to be followed. The ozone hole was the first clear evidence that man could damage the global environment, but it also provided the first case of concerted international action to counteract such an effect. Its story provides both a cautionary tale and a guiding light for the future, when other environmental threats to the planet will have to be faced by humanity as a whole.

References

1. Crutzen, P. J. Influence of nitrogen oxides on atmospheric ozone content. *Q. J. R. Meteorol. Soc.* **96,** 320–325 (1970).

2. Johnston, H. S. Reductions of stratospheric ozone by nitrogen oxide catalysts from supersonic transport exhaust. *Science* **173,** 517–522 (1971).
3. Stolarski, R. S. & Cicerone, R. J. Stratospheric chlorine: possible sink for ozone. *Can. J. Chem.* **52,** 1610–1615 (1974).
4. Molina, M. J. & Rowland, F. S. Stratospheric sink for chlorofluoromethanes: chlorine atom-catalysed destruction of ozone. *Nature* **249,** 810–812 (1974).
5. Farman, J. C., Gardiner, B. G. & Shanklin, J. D. Large losses of total ozone in Antarctica reveal seasonal ClO_x/NO_x interaction. *Nature* **315,** 207–210 (1985).
6. Stolarski, R. S. *et al.* Nimbus 7 satellite measurements of the springtime Antarctic ozone decrease. *Nature* **322,** 808–811 (1986).
7. Solomon, S., Garcia, R. R., Rowland, F. S. & Wuebbles, D. J. On the depletion of Antarctic ozone. *Nature* **321,** 755–758 (1986).
8. McElroy, M. B., Salawitch, R. J., Wofsy, S. C. & Logan, J. A. Reductions of Antarctic ozone due to synergistic interactions of chlorine and bromine. *Nature* **321,** 759–762 (1986).
9. Crutzen, P. J. & Arnold, F. Nitric acid cloud formation in the cold Antarctic stratosphere: a major cause for the springtime 'ozone hole'. *Nature* **324,** 651–655 (1986).
10. Molina, L. T. & Molina, M. J. Production of Cl_2O_2 from the self-reaction of the ClO radical. *J. Phys. Chem.* **91,** 433–436 (1987).

Further reading

Benedick, R. E. *Ozone Diplomacy: New Directions in Safeguarding the Planet* (Harvard University Press, 1998).

Middlebrook, A. M. & Tolbert, M. A. *Stratospheric Ozone Depletion* (Palgrave, Basingstoke, 2001).

Nobel e-Museum. *The Nobel Prize in Chemistry 1995* (http://www.nobel.se/chemistry/laureates/1995/).

1985

Large losses of total ozone in Antarctica reveal seasonal ClO_x/NO_x interaction

J. C. Farman, B. G. Gardiner, and J. D. Shanklin

Recent attempts[1,2] to consolidate assessments of the effect of human activities on stratospheric ozone (O_3) using one-dimensional models for 30° N have suggested that perturbations of total O_3 will remain small for at least the next decade. Results from such models are often accepted by default as global estimates[3]. The inadequacy of this approach is here made evident by observations that the spring values of total O_3 in Antarctica have now fallen considerably. The circulation in the lower stratosphere is apparently unchanged, and possible chemical causes must be considered. We suggest that the very low temperatures which prevail from midwinter until several weeks after the spring equinox make the Antarctic stratosphere uniquely sensitive to growth of inorganic chlorine, ClX, primarily by the effect of this growth on the NO_2/NO ratio. This, with the height distribution of UV irradiation peculiar to the polar stratosphere, could account for the O_3 losses observed.

Total O_3 has been measured at the British Antarctic Survey stations, Argentine Islands 65° S 64° W and Halley Bay 76° S 27° W, since 1957. Figure 1*a* shows data from Halley Bay. The mean and extreme daily values from October 1957 to March 1973 and the supporting calibrations have been discussed elsewhere[4,5]. The mean daily value for the four latest complete observing seasons (October 1980–March 1984) and the individual daily values for the current observing season are detailed in Fig. 1. The more recent data are provisional values. Very generous bounds for possible corrections would be ±30 matm cm. There was a changeover of spectrophotometers at the station in January 1982; the replacement instrument had been calibrated against the UK Meteorological Office standard in June 1981. Thus, two spectrophotometers have shown Octo-

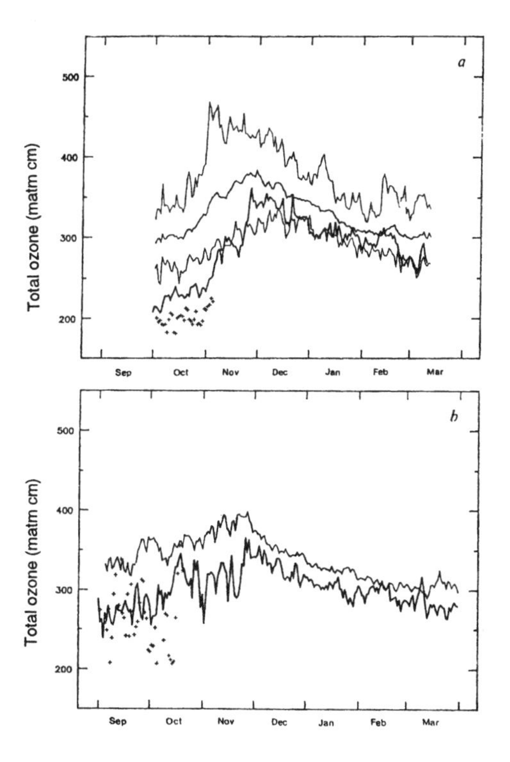

Fig. 1 Daily values of total O_3. *a*, Halley Bay: thin lines, mean and extreme values for 16 seasons, 1957–73; thick line, mean values for four seasons, 1980–84; + [points at lower left in each panel], values for October 1984. Observing season: 1 October to 13 March. *b*, Argentine Islands: as for Halley Bay, but extreme values for 1957–73 omitted. Observing season: 1 September to 31 March.

ber values of total O_3 to be much lower than March values, a feature entirely lacking in the 1957–73 data set. To interpret this difference as a seasonal instrumental effect would be inconsistent with the results of routine checks using standard lamps. Instrument temperatures (recorded for each observation) show that the March and October operating conditions were practically identical. Whatever the absolute error of the recent values may be, within the bounds quoted, the annual variation of total O_3 at Halley Bay has undergone a dramatic change.

Figure 1*b* shows data from Argentine Islands in a similar form, except that for clarity the extreme values for 1957–73 have been omitted. The values for 1980 to the present are provisional, the extreme error bounds again being ±30 matm cm. The changes are similar to those seen at Halley Bay, but are much smaller in magnitude.

Upper-air temperatures and winds are available for these stations from 1956. There are no indications of recent departures from established mean values sufficient to attribute the changes in total O_3 to changes in the circulation. The present-day atmosphere differs most prominently from that of previous decades in the higher concentrations

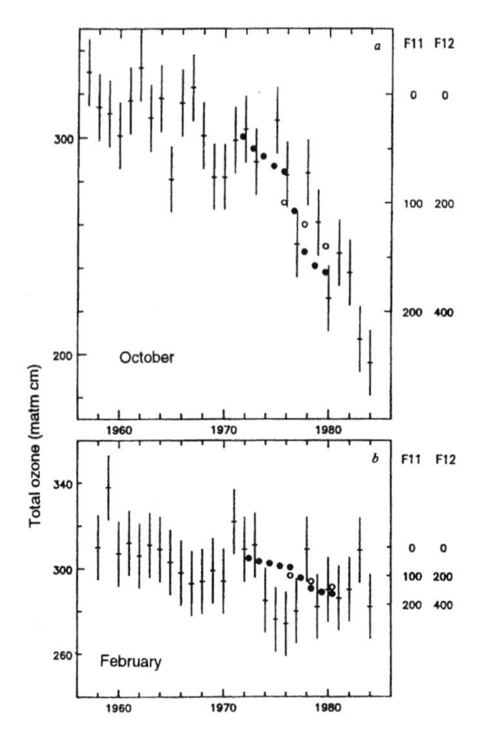

Fig. 2 Monthly means of total O_3 at Halley Bay, and Southern Hemisphere measurements of F-11 (●, p.p.t.v. (parts per thousand by volume) $CFCl_3$) and F-12 (○, p.p.t.v. CF_2Cl_2). *a*, October, 1957–84. *b*, February, 1958–84. Note that F-11 and F-12 amounts increase down the figure.

of halocarbons. Figure 2*a* shows the monthly mean total O_3 in October at Halley Bay, for 1957–84, and Fig. 2*b* that in February, 1958–84. Tropospheric concentrations of the halocarbons F-11 ($CFCl_3$) and F-12 (CF_2Cl_2) in the Southern Hemisphere[3] are also shown, plotted to give greatest emphasis to a possible relationship. Their growth, from which increase of stratospheric ClX is inferred, is not evidently dependent on season. The contrast between spring and autumn O_3 losses and the striking enhancement of spring loss at Halley Bay need to be explained. In Antarctica, the lower stratosphere is ~40 K colder in October than in February. The stratosphere over Halley Bay experiences a polar night and a polar day (many weeks of darkness, and of continuous photolysis, respectively); that over Argentine Islands does not. Figure 3 shows calculated amounts of NO_x in the polar night and the partitioning between the species[6]. Of these, only NO_3 and NO_2 are dissociated rapidly by visible light. The major reservoir, N_2O_5, which only absorbs strongly below 280 nm, should be relatively long-lived. Daytime levels of NO and NO_2 should be much less in early spring, following the polar night, than in autumn, following the polar day. Recent measurements[7] support these inferences. The effect of

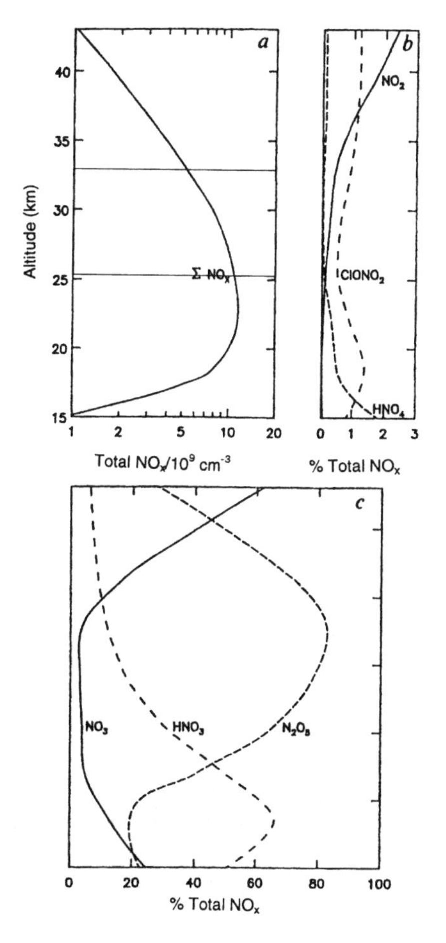

Fig. 3 NO_x during the polar night. *a*, Total NO_x cm^{-3}, from 15 to 43 km. *b*, NO_2, $ClONO_2$ and HNO_4 as percentages of total NO_x. *c*, NO_3, HNO_3 and N_2O_5 as percentages of total NO_x.

these seasonal variations on the strongly interdependent ClO_x and NO_x cycles is examined below.

The O_3 loss rate resulting from NO_x and ClO_x may be written[8]

$$L = N + C = 2\,k_2[O][NO_2] + 2\,k_6[O][ClO] \tag{1}$$

L accounts for over 85% of O_3 destruction in the altitude range 20–40 km. At 40 km, N and C are roughly equal. Lower down, C decreases rapidly to 10% of L at 30 km, 3% at 20 km (refs 6, 8). Equation (1) is based on two steady-state approximations (see Table 1*a* for the reactions involved):

$$\psi = \frac{[NO_2]}{[NO]} \sim \frac{k_1[O_3] + k_4[ClO]}{k_2[O] + j_3} \tag{2}$$

Table 1 Reaction list

a Governing ψ and χ (see text)		
	$NO + O_3 \rightarrow NO_2 + O_2$	(1)
	$NO_2 + O \rightarrow NO + O_2$	(2)
	$NO_2 + h\nu \rightarrow NO + O$	(3)
	$NO + ClO \rightarrow NO_2 + Cl$	(4)
	$Cl + O_3 \rightarrow ClO + O_2$	(5)
	$ClO + O \rightarrow Cl + O_2$	(6)
b Governing [Cl + ClO]		
	$HCl + OH \rightarrow Cl + H_2O$	(7)
	$ClONO_2 + h\nu \rightarrow ClO + NO_2$	(8)
	$HOCl + h\nu \rightarrow Cl + OH$	(9)
	$ClO + NO_2 + M \rightarrow ClONO_2 + M$	(10)
	$ClO + HO_2 \rightarrow HOCl + O_2$	(11)
	$Cl + CH_4 \rightarrow HCl + CH_3$	(12)
	$Cl + HO_2 \rightarrow HCl + O_2$	(13)
c	$HCl + ClONO_2 \rightarrow Cl_2 + HNO_3$	(14)

and

$$\chi = \frac{[Cl]}{[ClO]} \sim \frac{k_6[O] + k_4[NO]}{k_5[O_3]} \tag{3}$$

valid in daytime, with [O] in steady state with $[O_3]$. Reaction (4) has a negative temperature coefficient, whereas reaction (1) has large positive activation energy[9], with the result that ψ is strongly dependent on [ClO] at low temperature, as shown in Fig. 4. [ClO] is not simply proportional to total ClX, because $ClONO_2$ formation (reaction (10)) intervenes. Throughout the stratosphere, $\chi \ll 1$, so that [ClO] ~ [Cl + ClO]. From a steady-state analysis of the reactions given in Table 1*b*,

$$[Cl + ClO] \sim \frac{k_7[HCl][OH] + j_8[ClONO_2] + j_9[HOCl]}{k_{10}[NO_2] + k_{11}[HO_2] + \chi(k_{12}[CH_4] + k_{13}[HO_2])} \tag{4}$$

Values of ψ, χ and [Cl + ClO] obtained from equations (2), (3) and (4) are in good accord with full one-dimensional model results for late summer in Antarctica[6]. Neglecting seasonal effects other than those resulting from temperature and from variation of $[NO + NO_2]$, it is possible to solve simultaneously for $[NO_2]$ and [ClO], and to derive *L*. Results are shown in Table 2 as relaxation times[8], $[O_3]/L$, for various conditions. The spring values (lines 2, 3 and 4) are highly dependent on ClX amount (compare columns *a* and *b*), the autumn values (line 1) much less so. At

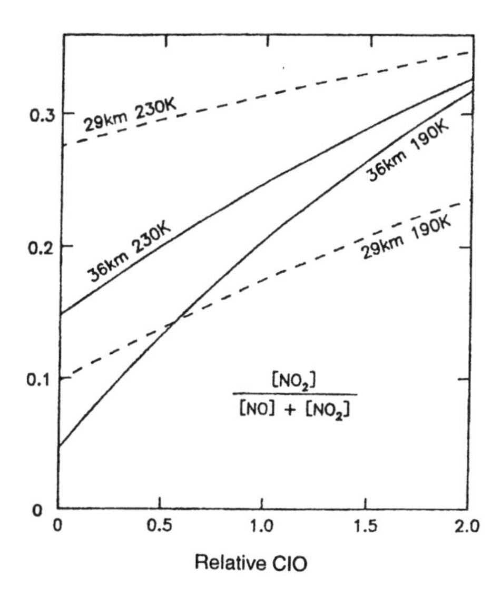

Fig. 4 $[NO_2]/[NO + NO_2]$ has the status of an efficiency factor for O_3 destruction by the NO_x cycle. In terms of the ratio ψ in the text, it is $\psi/(1 + \psi)$. The figure shows how this factor varies with [ClO] at 190 K and at 230 K, at altitudes of 29 and 36 km. Values of $[O_3]$, [O], [ClO] and j_3 were taken from one-dimensional model results[6] (maximum chlorine 2.7 p.p.b.v.) for noon, 25 March at 75.5° S, solar elevation 12.5°. The abscissa is [ClO] relative to the model value. The rate-limiting reaction, (2) in Table 1, for the NO_x cycle has zero activation energy. Note how, nevertheless, O_3 was protected against destruction by NO_x at low temperatures in a stratosphere with small amounts of ClX, but is losing this protection as ClX grows.

Argentine Islands, the sensitivity to ClX growth should resemble that seen in line 2, attributable solely to low temperature. Lines 3 and 4 show the enhanced sensitivity possible at stations within the Antarctic Circle, such as Halley Bay, arising from slow release of $[NO + NO_2]$ following the polar night. It remains to be shown how stable O_3 budgets were achieved with the relaxation times for the lower chlorine level (Table 2, *a*).

Much O_3 destruction is driven by visible light, but production requires radiation below 242 nm. On the dates shown (Table 2), destruction persists for some 11 h, while, because of the long UV paths, production is weak (except around noon) at 29 km, and is virtually absent below that altitude. Line 1 of Table 2 then demands O_3 transport in autumn from the upper to the lower stratosphere, which is consistent with inferred

Table 2 Relaxation times in days, $[O_3]/L$, for maximum chlorine levels 1.5 p.p.b.v. (*a*) and 2.7 p.p.b.v. (*b*) (1980)

	Altitude (km)	22		25.5		29		32.5		36		39.5		43	
Date	Relative NO + NO_2	*a*	*b*	*a*	*b*	*a*	*b*	*a*	*b*	*a*	*b*	*a*	*b*	*a*	*b*
25 March	1	105	103	38	37	16	15	6.9	5.7	2.6	2.0	1.2	0.9	0.83	0.57
17 September	1	224	210	86	77	32	27	12.2	9.2	4.9	3.4	2.6	1.7	1.90	1.22
17 September	0.75	288	265	107	93	39	31	14.2	10.4	5.7	3.9	3.0	1.9	2.13	1.33
17 September	0.5	398	353	141	118	47	36	17.0	12.0	6.9	4.5	3.5	2.2	2.41	1.46

Noon values at 75.5° S, solar elevation 12.5°. Lower stratosphere at 230 K on 25 March; 190 K on 17 September. The altitudes shown apply to the summer temperature profile used in the model[6].

thermally-driven lagrangian-mean circulations[10]. A mean vertical velocity of 45 m per day is in good accord with calculations of net diabatic cooling[11] and gives a realistic total O_3 decay rate in an otherwise conventional one-dimensional model[6]. The short relaxation times in the lower stratosphere in autumn are tolerable, with adequate transport compensating for lack of O_3 production.

In early spring, on the other hand, wave activity scarcely penetrates the cold dense core of the Antarctic polar vortex and with very low temperatures the net diabatic cooling is very weak[11]. Lagrangian transport in the vortex should then be almost negligible. (The virtual exclusion of Agung dust from the vortex supports this view[5].) The final warming signals the end of this period of inactivity and is accompanied by large dynamically induced changes in O_3 distribution. However, before the warming, with low chlorine, total O_3 was in a state of near-neutral equilibrium, sustained primarily by the long relaxation times. With higher chlorine, relaxation times of the order seen in line 4, Table 2, entail more rapid O_3 losses. With negligible production below 29 km and only weak transport, large total O_3 perturbation is possible. The extreme effects could be highly localized, restricted to the period with diurnal photolysis between polar night and the earlier of either the onset of polar day or the final spring warming. At the pole [NO + NO_2] rises continuously after the polar night, with the Sun. The final warming always begins over east Antarctica and spreads westwards across the pole. At Halley Bay the warming is typically some 14 days later than at the pole. Maximum O_3 depletion could be confined to the Atlantic half of the zone bordered roughly by latitudes 70 and 80°S.

Comparable effects should not be expected in the Northern Hemisphere, where the winter polar stratospheric vortex is less cold and less stable than its southern counterpart. The vortex is broken down, usually well before the end of the polar night, by major warmings. These are accompanied by large-scale subsidence and strong mixing, in the course of which peak O_3 values for the year are attained. Hence, sensitivity to ClX growth should be minimal if, as suggested above, this primarily results from O_3 destruction at low temperatures in regions where O_3 transport is weak.

We have shown how additional chlorine might enhance O_3 destruction in the cold spring Antarctic stratosphere. At this time of the year, the long slant paths for sunlight make reservoir species absorbing strongly only below 280 nm, such as N_2O_5, $ClONO_2$ and HO_2NO_2, relatively long-lived. The role of these reservoir species should be more readily demonstrated in Antarctica, particularly the way in which they hold the balance between the NO_x and ClO_x cycles. An intriguing feature could be the homogeneous reaction (Table 1*c*) between HCl and $ClONO_2$. If this process has a rate constant as large as 10^{-16} cm^3 s^{-1} (ref. 2) and a negligible temperature coefficient, the reaction would go almost to completion in the polar night, leaving inorganic chlorine partitioned between HCl and Cl_2, almost equally at 22 km for example. Photolysis of Cl_2 at near-visible wavelengths would provide a rapid source of [Cl + ClO] at sunrise, not treated in equation (4). The polar-night boundary is, therefore, the natural testing ground for the theory of nonlinear response to chlorine[1,2]. It might be asked whether a nonlinear response is already evident (Fig. 2*a*). An intensive programme of trace-species measurements on the polar-night boundary could add greatly to our understanding of stratospheric chemistry, and thereby improve considerably the prediction of effects on the ozone layer of future halocarbon releases.

We thank B. A. Thrush and R. J. Murgatroyd for helpful suggestions.

British Antarctic Survey, Natural Environment Research Council, High Cross, Madingley Road, Cambridge CB3 0ET, UK

Received 24 December 1984; accepted 28 March 1985 [Published 16 May]

References

1. Cicerone, R. J., Walters, S. & Liu, S. C. *J. geophys. Res.* **88**, 3647–3661 (1983).
2. Prather, M. J., McElroy, M. B. & Wofsy, S. C. *Nature* **312**, 227–231 (1984).
3. *The Stratosphere 1981, Theory and Measurements* (WMO Global Ozone Research and Monitoring Project Rep. No. 11, 1981).
4. Farman, J. C. & Hamilton, R. A. *Br. Antarct. Surv. Sci. Rep.* No. 90 (1975).
5. Farman, J. C. *Phil. Trans. R. Soc.* **B279**, 261–271 (1977).
6. Farman, J. C., Murgatroyd, R. J., Silnickas, A. M. & Thrush, B. A. *Q. Jl R. met. Soc.* (submitted).
7. McKenzie, R. L. & Johnston, P. V. *Geophys. Res. Lett.* 11, 73–75 (1984).
8. Johnston, H. S. & Podolske, J. *Rev. Geophys. Space Phys.* **16**, 491–519 (1978).
9. *Chemical Kinetics and Photochemical Data for Use in Stratospheric Modelling, Evaluation No. 6* (JPL Publ. 83–62, 1983).
10. Dunkerton, T. *J. atmos. Sci.* **35**, 2325–2333 (1978).
11. Dopplick, T. G. *J. atmos. Sci.* **29**, 1278–1294 (1972).

insterfullerene

J. R. Heath, S. C. O'Brien, R. F. Curl
y

itute and Departments of Chemistry and Electrical
University, Houston, Texas 77251, USA

Fig. 1 A football (in the United States, a soccerball) on Texas grass. The C_{60} molecule featured in this letter is suggested to have the truncated icosahedral structure formed by ... each vertex on the ... such a ball by a ... carbo...

nts aimed at understandin... carbon molecules are forme... shells[1], graphite has bee... cing a remarkably stable cluster consist... . Concerning the question of what ... cture might give rise to a superstable species, we ed icosahedron, a polygon with 60 vertices and ich are pentagonal and 20 hexagonal. This object untered as the football shown in Fig. 1. The C_{60} sults when a carbon atom is placed at each vertex as all valences satisfied by two single bonds and has many resonance structures, and appears to

used to produce and detect this unusual s the vaporization of carbon species from the d disk of graphite into a high-density helium used pulsed laser. The vaporization laser was onic of Q-switched Nd:YAG producing pulse nJ. The resulting carbon clusters were expanded olecular beam, photoionized using an excimer ted by time-of-flight mass spectrometry. The mber is shown in Fig. 2. In the experiment the opened first and then the vaporization laser recisely controlled delay. Carbon species were e helium stream, cooled and partially equili- ansion, and travelled in the resulting molecular ization region. The clusters were ionized by n excitation with a carefully synchronized lse. The apparatus has been fully described

on of carbon has been studied previously in a ratus[6]. In that work clusters of up to 190 carbon rved and it was noted that for clusters of more nly those containing an even number of atoms n the mass spectra displayed in ref. 6, the C_{60} st for cluster sizes of >40 atoms, but it is not nant. We have recently re-examined this system nder certain clustering conditions the C_{60} peak ut 40 times larger than neighbouring clusters. s a series of cluster distributions resulting from vaporization conditions evolving from a cluster lar to that observed in ref. 3, to one in which inant. In Fig. 3c, where the firing of the vaporiz- elayed until most of the He pulse had passed, n distribution of large, even-numbered clusters ns resulted. The C_{60} peak was largest but not 3b, the vaporization laser was fired at the time ium density; the C_{60} peak grew into a feature s stronger than its neighbours, with the excep- ig. 3a, the conditions were similar to those in ddition the integrating cup depicted in Fig. 2

graphite fus... mbered ring structure. We b distribution in Fig. 3c is fairly representative o distribution of larger ring fragments. When these ho are left in contact with high-density helium, the c brate by two- and three-body collisions towards th species, which appears to be a unique cluster o atoms.

When one thinks in terms of the many fused with unsatisfied valences at the edges that would n from a graphite fragmentation, this result seems there is not much to choose between such isomer stability. If one tries to shift to a tetrahedral diamo the entire surface of the cluster will be covered wi valences. Thus a search was made for some ot structure which would satisfy all sp^2 valences. Only structure appears likely to satisfy this criterion, an minster Fuller's studies were consulted (see, for 7). An unusually beautiful (and probably unique) truncated icosahedron depicted in Fig. 1. As ment all valences are satisfied with this structure, and appears to be aromatic. The structure has the sym icosahedral group. The inner and outer surfaces with a sea of π electrons. The diameter of this C_6 ~7 Å, providing an inner cavity which appears t of holding a variety of atoms[8].

Assuming that our somewhat speculative structu there are a number of important ramifications aris existence of such a species. Because of its stability under the most violent conditions, it may be widel in the Universe. For example, it may be a major c circumstellar shells with high carbon content. It constituent of interstellar dust and a possible m

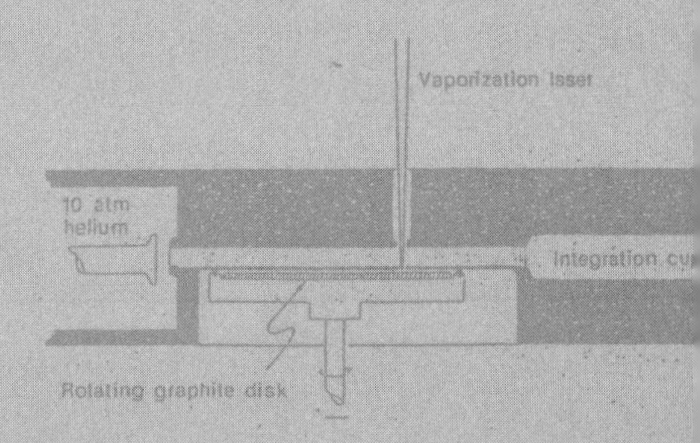

Fig. 2 Schematic diagram of the pulsed supersonic

NINETEEN

Carbon cages and carbon tubes

Philip Ball

Pure carbon was long thought to exist in only two crystalline forms: diamond and graphite. But in 1985, a team at Rice University in the United States discovered "buckminsterfullerene"—sixty-atom carbon clusters reminiscent of the geodesic domes of Buckminster Fuller. Since then, the family of fullerenes and related carbon molecules has grown. Particular technological interest now centers on the long, hollow fibers known as carbon nanotubes, discovered in the early 1990s in Japan.

How, before 1985, could we not even have suspected that fullerenes exist? How obvious, how familiar they now seem—as do so many great discoveries. There is an aura of classical antiquity about the geometry of carbon.

But wasn't it the Greek philosopher Plato who put the hex on the pentagon? His earth, air, fire, and water were made up of geometrical atoms with triangular and square faces, four of the five regular solids. Yet the fifth, the twelve-sided dodecahedron assembled from pentagons, was banished to become (was it promotion or relegation?) an ethereal symbol of the cosmos.

Pentagons are, however, the key to the hollow carbon cages of the fullerene molecules. Conventional thinking had long since fixed on two other structural motifs: the fourfold tetrahedral coordination found in diamond, and the six-atom hexagonal rings that Friedrich August Kekulé dreamed up to explain the structure of benzene (if you believe that story). Side by side, these rings make up the patchwork quilt of graphite's carbon sheets. That was how carbon atoms arranged themselves: either diamondlike or flat.

Of course, carbon was long known to form five-membered rings—even regular pentagons, for example in the pentacene ions of the ferrocene molecule. But when Harry Kroto and his collaborators[1] made fullerenes in

1985, they had to rediscover for themselves that pentagons put a curve into graphite-like carbon.

Leonhard Euler knew this already in the eighteenth century—but that was geometry, not chemistry. Ernst Haeckel recognized the same when he drew pentagons in the domed shells of the radiolarian *Aulonia hexagona,* a single-celled marine organism. Most significantly for the fullerene discoverers, the U.S. architect Richard Buckminster Fuller knew that he needed pentagons among the hexagons of his geodesic domes of the 1950s and 1960s (fig. 19.1). Fuller's structures showed Kroto and colleagues how to make a sixty-atom carbon cluster, the most abundant in their fullerene mixture, that had no edges or unused ("dangling") chemical bonds. This was C_{60}, which made its debut on *Nature*'s cover two months after its discovery.

C_{60} is the loveliest fullerene because it is virtually spherical, with the same symmetry properties as the twenty-sided regular icosahedron. It is the smallest cage in which all twelve of the pentagons needed for closure are isolated among hexagons (fig. 19.2)—a stabilizing situation. But it is formed amidst a whole family of carbon-cage molecules with fewer or more hexagons. At the smallest extreme is a dodecahedral cage composed only of pentagons, a twenty-atom cluster that, until its recent synthesis,[2] was thought to be too strained to be stable. With all of its carbon atoms hydrogenated this becomes dodecahedrane, $C_{20}H_{20}$, made by organic chemical synthesis three

Fig. 19.1. An example of the geodesic domes designed by Richard Buckminster Fuller. This dome served as the U.S. pavilion at Expo '67 in Montreal. (Figure courtesy of Jimmy Fox/Science Photo Library.)

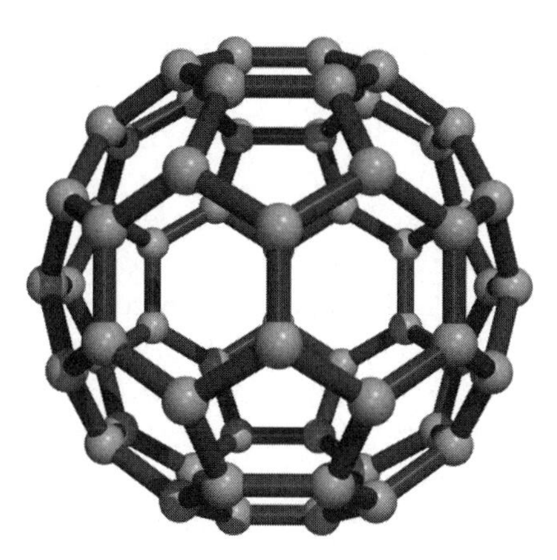

Fig. 19.2. C_{60}—the loveliest fullerene. In this computer representation of the molecule, the spheres represent carbon atoms and the rods are the bonds between them. (Image courtesy of P. R. C. Kent and University of Cambridge.)

years before C_{60}'s discovery. At the other extreme of size there are new surprises, such as multi-shelled "carbon onions" and new nonspherical shapes (described below).

Creating fullerenes was an accidental pursuit that began among the stars. When Kroto, of the University of Sussex, U.K., went to work with Richard Smalley at Rice University in Texas, he was intent on making linear carbon molecules thought to be responsible for the "diffuse interstellar bands" seen in absorption spectra of the tenuous substance between the stars. Smalley's team had the experimental means of making small atomic clusters, which they used to study metals and semiconductors. Spectroscopist Bob Curl at Rice was the go-between; Jim Heath, Sean O'Brien, and, initially, Yuan Liu were the graduate students who knew best of all how to make the equipment work.

The technique they used was laser vaporization coupled to mass spectrometry, a technique that indicates only the atomic mass of the products. When graphite was vaporized, the C_{60} peak stood out from a forest of others periodically separated by the mass of two carbon atoms (see p. 310, figure 3). Careful tinkering produced almost pure C_{60}, but in such small amounts that mass spectrometry was the sole means of characterization. For five years after that, the putative hollow-shelled C_{60} molecule remained only a curious hypothesis.

I first laid eyes on the reddish powder of solid, unrefined C_{60} in 1992, in northern China. By that time, thanks to work by Wolfgang Krätschmer, Don Huffman, and their coworkers reported in 1990,[3] any laboratory in the world could set up its own fullerene generator at minimal cost: a couple of graphite rods, separated by a small gap across which an electric arc discharged. The carbon gas that evaporates from the graphite rods condenses into C_{60}-rich soot. The fullerene component can be separated from the junk because it is soluble in organic solvents such as benzene. From the rich red solution crystals will form, within which the positions of the atoms can be deduced by X-ray diffraction.

It was this "arc-discharge" method that Krätschmer and colleagues described in 1990. Much less sophisticated than the technique originally used to make and detect C_{60}, its feasibility was perhaps previously obscured by its very simplicity. As it happened, several other groups had converged on the same track when Krätschmer and Huffman's paper appeared. Within a few years other low-tech preparative methods were identified, such as the high-temperature breakdown (pyrolysis) of hydrocarbons—suggesting that combustion scientists had probably been making small quantities of fullerenes inadvertently for decades, if not centuries.

The relative ease of synthesis raised the stakes. Fullerenes were identified in the unusual carbon-rich mineral shungite, formed by lightning strikes. They turned up in carbonaceous meteorites, as well as in carbonaceous material in sediments found at the geological boundary between the Cretaceous and Tertiary periods; these sediments are thought to contain the fallout from a giant meteorite impact. Pyrolytic formation of fullerenes on impact is a likely explanation here, but formation or deposition on the parent bodies in space is also possible: putative sightings of C_{60} in spectra of the interstellar medium abound. Small wonder that C_{60} was soon being invoked (with rather awkward logic) in discussions about the origin of life.

The mass production of fullerenes changed everything, not least by providing a crystal structure to lay to rest the skepticism that had always dogged the structural hypothesis of the 1985 paper. Here was a new material—and what delightful things one could do with it. Kroto foresaw a "round chemistry" of fullerene balls studded with reactive chemical groups like the appendages of viruses. (This duly materialized, though C_{60} is not the most amenable of molecules to chemical modification.) Atoms can be trapped inside the cages; helium will squeeze through the walls. C_{60} pearls were strung into polymeric necklaces and pendants, or gently coddled in basketlike molecules.

It is not easy to recall, now that the pace of fullerene research has settled into a more leisurely stride, the feeling of almost limitless possibilities that pervaded the work during the early 1990s. Dope it with alkali metals, and C_{60} conducts electricity. A month later and—guess what?—its electrical resistance vanishes and it becomes a superconductor,[4] with the "high" transition temperature of −255 °C, later pushed up to −228 °C. (Most metals superconduct only within a few degrees of absolute zero, −273 °C.) Here is C_{60} providing an organic ferromagnet; there it is inhibiting the AIDS virus (though, sadly, not competitively enough). There can be little question that such a fertile discovery warranted the 1996 Nobel Prize in Chemistry, which quite properly went to Kroto, Smalley, and Curl (although three people, the limit for award of a Nobel Prize, is as ever too few).

Yet C_{60} may prove to be of greatest significance not as a destination but as a signpost. Graphite-like carbon, it tells us, is a material for molecular-scale architecture. Defects open up the flat sheets to the third dimension: pentagons through positive (bowl-like) curvature, heptagons through negative (saddle-like) curvature. A bowl that curves into a sphere is but one option. Numerous ways of folding the graphite sheet using such defects have now been postulated by theorists, from hollow rings to ordered carbon foams. The easiest to make, and potentially the most useful, is the tube.

Sumio Iijima at the NEC Corporation in Japan was engaged in fullerene-inspired arc-discharge experiments in 1991 when he discovered carbon nanotubes.[5] These cylindrical tubes of graphitic carbon came initially in the form of nested, Russian-doll-like structures: multiwalled, concentric tubes just a few nanometers in diameter. Now single-walled versions are preferred, being more amenable to theory. Again, Buckminster Fuller had been there already: for what is the tunnel protruding from his geodesic dome for the Union Tank Car Company, if not a scaled-up nanotube complete with hemispheric fullerene-like end cap?

Carbon nanotubes have now supplanted fullerenes as the focus of research on nanoscale carbon.[6] As superstrong fibers (with a stiffness and strength potentially greater than those of diamond), as molecular wires (some are semiconducting), as nanoscale probes (attached to the atomic force microscope), as field-emission devices for display technology or hydrogen storage vessels for fuel cells, they offer tremendous scope for practical applications. For carbon, the future is perhaps not round but straight and narrow.

References

1. Kroto, H. W., Heath, J. R., O'Brien, S. C., Curl, R. F. & Smalley, R. E. C_{60}: Buckminsterfullerene. *Nature* **318,** 162–163 (1985).
2. Prinzbach, H. *et al.* Gas-phase production and photoelectron spectroscopy of the smallest fullerene, C_{20}. *Nature* **407,** 60–63 (2000).
3. Krätschmer, W., Lamb, L. D., Fostiropoulos, K. & Huffman, D. R. Solid C_{60}: a new form of carbon. *Nature* **347,** 354–358 (1990).

4. Hebard, A. F. *et al.* Superconductivity at 18 K in potassium-doped C_{60}. *Nature* **350**, 600–601 (1991).
5. Iijima, S. Helical microtubules of graphitic carbon. *Nature* **354**, 56–58 (1991).
6. Ebbesen, T. W. (ed.) *Carbon Nanotubes* (CRC Press, Boca Raton, 1997).

Further reading

Baggott, J. *Perfect Symmetry* (Oxford University Press, 1994).
Nobel e-Museum. *The Nobel Prize in Chemistry 1996* (http://www.nobel.se/chemistry/laureates/1996/).

1985

C_{60}: buckminsterfullerene

H. W. Kroto*, J. R. Heath, S. C. O'Brien, R. F. Curl,
and R. E. Smalley

During experiments aimed at understanding the mechanisms by which long-chain carbon molecules are formed in interstellar space and circumstellar shells[1], graphite has been vaporized by laser irradiation, producing a remarkably stable cluster consisting of 60 carbon atoms. Concerning the question of what kind of 60-carbon atom structure might give rise to a superstable species, we suggest a truncated icosahedron, a polygon with 60 vertices and 32 faces, 12 of which are pentagonal and 20 hexagonal. This object is commonly encountered as the football shown in Fig. 1. The C_{60} molecule which results when a carbon atom is placed at each vertex of this structure has all valences satisfied by two single bonds and one double bond, has many resonance structures, and appears to be aromatic.

The technique used to produce and detect this unusual molecule involves the vaporization of carbon species from the surface of a solid disk of graphite into a high-density helium flow, using a focused pulsed laser. The vaporization laser was the second harmonic of Q-switched Nd:YAG producing pulse energies of ~30 mJ. The resulting carbon clusters were expanded in a supersonic molecular beam, photoionized using an excimer laser, and detected by time-of-flight mass spectrometry. The vaporization chamber is shown in Fig. 2. In the experiment the pulsed valve was opened first and then the vaporization laser was fired after a precisely controlled delay. Carbon species were vaporized into the helium stream,

* Permanent address: School of Chemistry and Molecular Sciences, University of Sussex, Brighton BN1 9QJ, UK.

Fig. 1 A football (in the United States, a soccerball) on Texas grass. The C_{60} molecule featured in this letter is suggested to have the truncated icosahedral structure formed by replacing each vertex on the seams of such a ball by a carbon atom.

cooled and partially equilibrated in the expansion, and travelled in the resulting molecular beam to the ionization region. The clusters were ionized by direct one-photon excitation with a carefully synchronized excimer laser pulse. The apparatus has been fully described previously[2–5].

The vaporization of carbon has been studied previously in a very similar apparatus[6]. In that work clusters of up to 190 carbon atoms were observed and it was noted that for clusters of more than 40 atoms, only those containing an even number of atoms were observed. In the mass spectra displayed in ref. 6, the C_{60} peak is the largest for cluster sizes of >40 atoms, but it is not completely dominant. We have recently reexamined this system and found that under certain clustering conditions the C_{60} peak can be made about 40 times larger than neighbouring clusters.

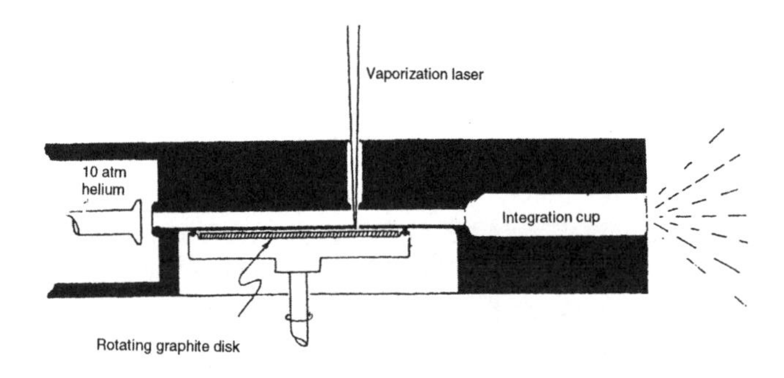

Fig. 2 Schematic diagram of the pulsed supersonic nozzle used to generate carbon cluster beams. The integrating cup can be removed at the indicated line. The vaporization laser beam (30–40 mJ at 532 nm in a 5-ns pulse) is focused through the nozzle, striking a graphite disk which is rotated slowly to produce a smooth vaporization surface. The pulsed nozzle passes high-density helium over this vaporization zone. This helium carrier gas provides the thermalizing collisions necessary to cool, react and cluster the species in the vaporized graphite plasma, and the wind necessary to carry the cluster products through the remainder of the nozzle. Free expansion of this cluster-laden gas at the end of the nozzle forms a supersonic beam which is probed 1.3 m downstream with a time-of-flight mass spectrometer.

Figure 3 shows a series of cluster distributions resulting from variations in the vaporization conditions evolving from a cluster distribution similar to that observed in ref. 3, to one in which C_{60} is totally dominant. In Fig. 3*c,* where the firing of the vaporization laser was delayed until most of the He pulse had passed, a roughly gaussian distribution of large, even-numbered clusters with 38–120 atoms resulted. The C_{60} peak was largest but not dominant. In Fig. 3*b,* the vaporization laser was fired at the time of maximum helium density; the C_{60} peak grew into a feature perhaps five times stronger than its neighbours, with the exception of C_{70}. In Fig. 3*a,* the conditions were similar to those in Fig. 3*b* but in addition the integrating cup depicted in Fig. 2 was added to increase the time between vaporization and expansion. The resulting cluster distribution is completely dominated by C_{60}, in fact more than 50% of the total large cluster abundance is accounted for by C_{60}; the C_{70} peak has diminished in relative intensity compared with C_{60}, but remains rather prominent, accounting for ~5% of the large cluster population.

Our rationalization of these results is that in the laser vaporization, fragments are torn from the surface as pieces of the planar graphite fused six-membered ring structure. We believe that the distribution in Fig. 3*c* is fairly representative of the nascent distribution of larger ring fragments. When these hot ring clusters are left in contact with high-density helium, the clusters equilibrate by two- and three-body collisions towards the most stable species, which appears to be a unique cluster containing 60 atoms.

When one thinks in terms of the many fused-ring isomers with unsatisfied valences at the edges that would naturally arise from a graphite fragmentation, this result seems impossible: there is not much to choose between such isomers in terms of stability. If one tries to shift to a tetrahedral diamond structure, the entire surface of the cluster will be covered with unsatisfied valences. Thus a search was made for some other plausible structure which would satisfy all sp^2 valences. Only a spheroidal structure appears likely to satisfy this criterion, and thus Buckminster Fuller's studies were consulted (see, for example, ref. 7). An unusually beautiful (and probably unique) choice is the truncated icosahedron depicted in Fig. 1. As mentioned above, all valences are satisfied with this structure, and the molecule appears to be aromatic. The structure has the symmetry of the icosahedral group. The inner and outer surfaces are covered with a sea of π electrons. The diameter of this C_{60} molecule is ~7 Å, providing an inner cavity which appears to be capable of holding a variety of atoms[8].

Assuming that our somewhat speculative structure is correct, there

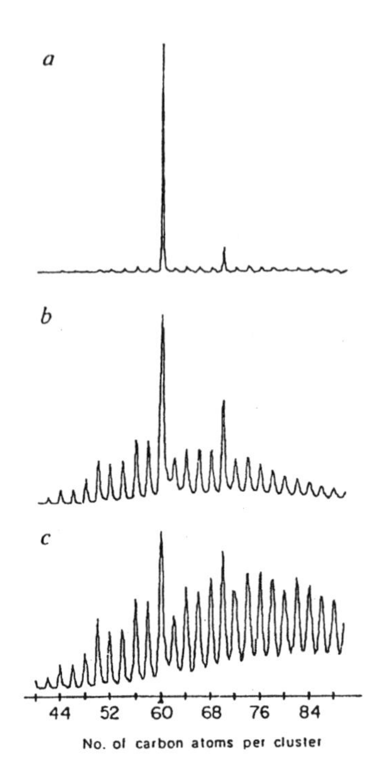

Fig. 3 Time-of-flight mass spectra of carbon clusters prepared by laser vaporization of graphite and cooled in a supersonic beam. Ionization was effected by direct one-photon excitation with an ArF excimer laser (6.4 eV, 1 mJ cm^{-2}). The three spectra shown differ in the extent of helium collisions occurring in the supersonic nozzle. In *c*, the effective helium density over the graphite target was less than 10 torr—the observed cluster distribution here is believed to be due simply to pieces of the graphite sheet ejected in the primary vaporization process. The spectrum in *b* was obtained when roughly 760 torr helium was present over the graphite target at the time of laser vaporization. The enhancement of C_{60} and C_{70} is believed to be due to gas-phase reactions at these higher clustering conditions. The spectrum in *a* was obtained by maximizing these cluster thermalization and cluster-cluster reactions in the 'integration cup' shown in Fig. 2. The concentration of cluster species in the especially stable C_{60} form is the prime experimental observation of this study.

are a number of important ramifications arising from the existence of such a species. Because of its stability when formed under the most violent conditions, it may be widely distributed in the Universe. For example, it may be a major constituent of circumstellar shells with high carbon content. It is a feasible constituent of interstellar dust and a possible major site for surface-catalysed chemical processes which lead to the formation of interstellar molecules. Even more speculatively, C_{60} or a derivative might be the carrier of the diffuse interstellar lines[9].

If a large-scale synthetic route to this C_{60} species can be found, the chemical and practical value of the substance may prove extremely high. One can readily conceive of C_{60} derivatives of many kinds—such as C_{60} transition metal compounds, for example, $C_{60}Fe$ or halogenated species like $C_{60}F_{60}$ which might be a super-lubricant. We also have evidence that an atom (such as lanthanum[8] and oxygen[1]) can be placed in the interior, producing molecules which may exhibit unusual properties. For example, the chemical shift in the NMR of the central atom should be remarkable because of the ring currents. If stable in macroscopic, condensed phases, this C_{60} species would provide a topologically novel aromatic nucleus for new branches of organic and inorganic chemistry. Finally, this especially stable and symmetrical carbon structure provides a possible catalyst and/or intermediate to be considered in modelling prebiotic chemistry.

We are disturbed at the number of letters and syllables in the rather fanciful but highly appropriate name we have chosen in the title to refer to this C_{60} species. For such a unique and centrally important molecular structure, a more concise name would be useful. A number of alternatives come to mind (for example, ballene, spherene, soccerene, carbosoccer), but we prefer to let this issue of nomenclature be settled by consensus.

We thank Frank Tittel, Y. Liu and Q. Zhang for helpful discussions, encouragement and technical support. This research was supported by the Army Research Office and the Robert A. Welch Foundation, and used a laser and molecular beam apparatus supported by the NSF and the US Department of Energy. H.W.K. acknowledges travel support provided by SERC, UK. J.R.H. and S.C.O'B. are Robert A. Welch Predoctoral Fellows.

Rice Quantum Institute and Departments of Chemistry and Electrical Engineering, Rice University, Houston, Texas 77251, USA

Received 13 September; accepted 18 October 1985 [Published 14 November]

References

1. Heath, J. R. *et al. Astrophys. J.* (submitted).
2. Dietz, T. G., Duncan, M. A., Powers, D. E. & Smalley, R. E. *J. chem. Phys.* **74**, 6511–6512 (1981).
3. Powers, D. E. *et al. J. phys. Chem.* **86**, 2556–2560 (1982).
4. Hopkins, J. B., Langridge-Smith, P. R. R., Morse, M. D. & Smalley, R. E. *J. chem. Phys.* **78**, 1627–1637 (1983).
5. O'Brien, S. C. *et al. J. chem. Phys.* (submitted).
6. Rohlfing, E. A., Cox, D. M. & Kaldor, A. *J. chem. Phys.* **81**, 3322–3330 (1984).
7. Marks, R. W. *The Dymaxion World of Buckminster Fuller* (Reinhold, New York, 1960).
8. Heath, J. R. *et al. J. Am. chem. Soc.* (in the press).
9. Herbig, E. *Astrophys. J.* **196**, 129–160 (1975).

1995

r Forschungsinstitut Senckenberg **181,** 55–63 (1995).
aceous Res. **6,** 271–278 (1985).
ogy **12,** 383–399 (1986).
States Geological Exploration of the 40th Parallel 1–201 (Government
ington, 1880).
Ornithol. **1,** 7–50 (1993) (in Russian).
hort Papers, IV Symp. Mes. Terrestr. Ecosyst. (eds Sun, A. & Wang,
Ocean, Beijing, 1995).
th Sci. **11,** 1335–1338 (1974).
on, P. M. J. Vert. Paleont. **11,** 90–107 (1991).
. **31,** 35–56 (1982).
a **207,** 1–20 (1991).
ussi, C. P. Ameghiniana **32,** 57–61 (1995).
eontol. **12,** 122–124 (1992).
l Congr. Int. Ornithol. 55–70 (1963).
D. Smith. Contr. Paleobiol. **63,** 1–22 (1987).
Sin. **27,** 1296–1302 (1984).
, T. Trav. Mus. d'Hist. Nat. Grigore An
, M., Barsbold, R., Clark, J. M. & Norel
ymp. Gond. Dinos. Mem. Queens. Mu
Paleont. **14,** 480–519 (1994).
Mus. Arg. Cien. Nat. "Bernardino Ri

73. Chinsamy, A., Chiappe, L. M. & Dodson, P. Paleobiology **21** (in the pres
74. Chinsamy, A. & Dodson, P. Am. Sci. **83,** 174–180 (1995).
75. Houck, M. A., Gauthier, J. A. & Strauss, R. E. Science **247,** 195–198 (1
76. Campbell, B. & Lack, E. (eds) A Dictionary of Birds (Buteo, Vermillion, 1
77. Elzanowski, A. Acta XVIII Congr. Int. Ornithol. **1,** 178–183 (1985).
78. Norell, M. A. et al. Science **266,** 779–782 (1994).
79. Prosser, C. L. (ed.) Comparative Animal Physiology (Saunders, New York
80. Randolph, S. E. Zool. J. Linn. Soc. **112,** 389–397 (1994).
81. Ruben, J. Evolution **45,** 1–17 (1991).
82. Gatesy, S. M. in Functional Morphology in Vertebrate Paleontology (ed. T
234 (Cambridge Univ. Press, 1995).
83. Holland, S. M. Paleobiology **21,** 92–109 (1995).
84. Norell, M. A. in Extinction and Phylogeny (eds Novacek, M. J. & Wheel
(Columbia Univ. Press, New York, 1992).
85. Gradstein, F. M. et al. Geophys. Res. **99,** 24,051–24,074 (1994).
86. Cracraft, J. in The Phylogeny and Classification of the Tetrapods Volum
Birds J.) 339–361 (Clarendon, Oxford, 1988).
A. & Science **255,** 1690–1693 (1992).

NOW MEN k A. Chinsamy, P. Dandonoli, D. Frost, M.
Witm nts and discussions. P. Conversano and M
Research ed by the Frick Fund of the American M
History, hilip M. M a Foundation, and the National Sci
(DER 9)

AR

ter-mass companion to a solar-type

ayor & Didier Queloz

y, 51 Chemin des Maillettes, CH-1290 Sauverny, Switzerland

e of a Jupiter-mass companion to the star 51 Pegasi is inferred from obse variations in the star's radial velocity. The companion lies only about eigh rom the star, which would be well inside the orbit of Mercury in our Solar might be a gas-giant planet that has migrated to this location throug r from the radiative stripping of a brown dwarf.

ten years, several groups have been examining ies of dozens of stars, in an attempt to identify induced by the presence of heavy planetary The precision of spectrographs optimized for and currently in use is limited to about reflex motion of the Sun due to Jupiter is rrent searches are limited to the detection of east the mass of Jupiter (M_J). So far, all precise have failed to detect any jovian planets or

94 we have monitored the radial velocity of 142 stars with a precision of 13 m s^{-1}. The stars in lected for their apparent constant radial velocity on) from a larger sample of stars monitored for 18 months of measurements, a small number nificant velocity variations. Although most can- dditional measurements, we report here the dis- panion with a minimum mass of 0.5 M_J, orbiting d the solar-type star 51 Peg. Constraints origin- bserved rotational velocity of 51 Peg and from pheric emission give an upper limit of 2 M_J for the mass of the companion. Alternative explanat observed radial velocity variation (pulsation or sp are unlikely.

The very small distance between the companion is certainly not predicted by current models of formation[8]. As the temperature of the companic 1,300 K, this object seems to be dangerously close thermal evaporation limit. Moreover, non-thermal effects are known to be dominant[9] over thermal on ian-mass companion may therefore be the result of of a very-low-mass brown dwarf.

The short-period orbital motion of 51 Peg also dis period perturbation, which may be the signature low-mass companion orbiting at larger distance.

Discovery of Jupiter-mass companion(s)

Our measurements are made with the new fibre-fed trograph ELODIE of the Haute-Provence France[10]. This instrument permits measurements velocity with an accuracy of about 13 m s^{-1} of stars in an exposure time of <30 min. The radial velocity

TWENTY

Seeking other solar systems

Gordon A. H. Walker

It is one of the most fundamental questions in astronomy: do habitable planets exist outside our Solar System? Decades of fruitless searches for planets orbiting stars like our Sun ended in 1995, with the discovery of a Jupiter-like planet orbiting a star in the constellation Pegasus. Dozens of similar discoveries rapidly followed. But detecting planets the size of the Earth remains a tougher challenge.

When, in November 1995, Michel Mayor and Didier Queloz reported their discovery of a Jupiter-like planet orbiting the star 51 Pegasi,[1] it galvanized astronomers and the general public alike. Yet this wasn't the first report of a planet outside our Solar System. Almost four years earlier, Alex Wolszczan and Dale Frail[2] had discovered at least two Earth-like planets orbiting a bizarre type of star called a pulsar (see chapter 11). But the intense, lethal radiation emitted by a pulsar—quite unlike the warm glow of our Sun—makes it highly unlikely that life as we know it could exist on planets orbiting a pulsar. Although Mayor and Queloz's planet—a gas giant like Jupiter, rather than a rocky planet like the Earth—is too close to its hot parent star to harbor life, it provided the first evidence of a planetary companion to a Sun-like star, and thus the first step toward finding a habitable planet outside our Solar System.

If astronomy tells us anything, it is that our Solar System is unlikely to be unique. There are tens of billions of other suns in our Galaxy, which itself resembles billions of other galaxies. Planetary systems are thought to be common, in part because the formation of planets can help stars to form. As stars start to condense from large, rotating gas clouds, they should spin ever faster as they contract, ultimately spinning too fast to collapse further and ignite nuclear reactions. If two or more stars form from the same cloud (as appears to happen about half the time), the excess rotational motion, or "angular momentum," goes into the orbital motion of the stars around each other. For a single star, planets orbiting at a large distance can do the same job of siphoning off angular momentum, allowing a star to form. In our Solar System, Jupiter's motion stores almost all of the angular momentum; perhaps unseen planets play this role for the other single stars that we can see in the sky.

The search for planets outside our Solar System (known as "extrasolar planets") started in the 1940s, but for nearly fifty years raised only false alarms. The problem is that the signature of a planet orbiting another star is extremely subtle. In our Solar System, for example, the Sun contains nearly all the mass and is a billion times brighter than the largest planet, Jupiter. Even for the stars nearest to us, the image of a Jupiter would be lost in the star's glare, while the movement of the star in response to the gravitational tug of the orbiting planet (see fig. 20.1) would be one thousandth the size of the planet's orbit around the star.

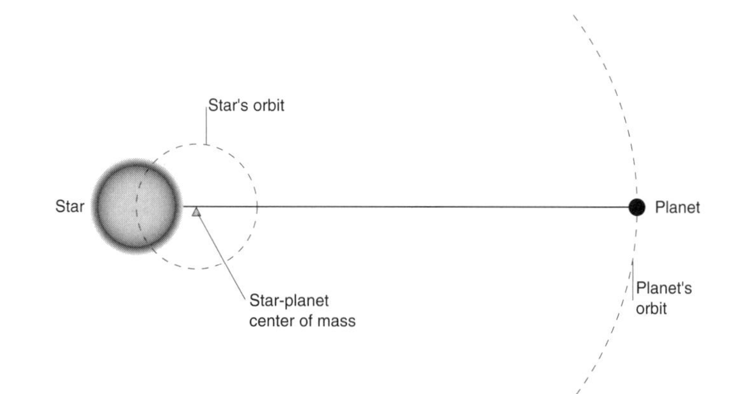

Fig. 20.1. A star and a planet, held in orbit by their mutual gravitational attraction. Although we think of the planet as orbiting a fixed star, in fact, both planet and star are orbiting the same fixed point in space, known as the "center of mass" of the star-planet system. One can think of the star and the planet as being at opposite ends of a rod pivoted at their center of mass. In the case of our Sun and Jupiter, the star is one thousand times as massive as the planet, so the pivot point is just one thousandth of the distance to Jupiter from the center of the Sun—in fact, right on the Sun's surface. Seen from another star, our Sun would appear to pivot slowly about a point on its surface, taking nearly twelve years (the orbital period of Jupiter) to complete an orbit at a speed of less than fifty kilometers an hour. Because extrasolar planets cannot be imaged from the Earth, this "reflex orbit" of the star has been the basis for all of the detections of extrasolar planets so far. The minimum mass of the planet can be derived from the period and size of the star's orbit.

Nevertheless, the best way to find planets has been to look for this slow periodic motion of the star in space. In the 1940s, and then again, with more data, in the 1960s, astronomers at the Sproul Observatory in Swarthmore, Pennsylvania, reported that five nearby stars had tiny periodic motions on the sky of the type expected from systems of giant planets in highly elliptical orbits. But all were later discounted, with the signals explained as the result of periodic cleaning and adjustment of the Sproul telescope primary lens.

Measuring such tiny motions of stars on the sky is extremely challenging. But if the motion has a component toward or away from the observer (a "radial" component), astronomers can detect it using the Doppler effect—the same effect that changes the pitch of a siren as it rushes by. In 1991, Wolszczan and Frail[2] used this technique to detect two (or possibly three) small planets orbiting a pulsar. Pulsars—highly condensed stars as massive as the Sun but only about twenty kilometers in diameter—rotate very rapidly, and keep perfect time. The pulsar PSR1257+12 rotates 161 times a second, with each rotation registering as a radio click. By accurately timing the clicks, Wolszczan and Frail detected the small changes in clock rate (the Doppler shift) caused by the pulsar's motion in response to the orbiting planets.

At a conference on extrasolar planets in December 1993, the pulsar planets were not even mentioned in the program, even though they were the first planets to have been discovered outside our Solar System. It seemed clear, to me at least, that the real preoccupation was the discovery of planets orbiting stars like our own Sun. But such stars do not emit radio clicks; to find planets orbiting them, one has to measure a Doppler shift in the starlight itself. Wolszczan and Frail were able to measure the speed of their pulsar with an error of less than a millimeter per second—little more than a snail's pace! Stellar astronomers, with no cosmic clocks, can at best achieve precisions of a few meters per second—more typical of a bicycle.

In the 1970s, the advent of electronic detectors—now so commonplace in digital cameras—allowed the spectrum of starlight to be measured much more precisely than had been possible using photographic plates. Conventionally, spectral lines in starlight, produced by elements such as iron in the star's atmosphere, are compared in wavelength with calibration lines produced by an iron arc discharge near the detector. But this technique is prone to large errors, which mask the subtle changes in wavelength induced by an orbiting planet. To eliminate these errors, in 1979 Bruce Campbell and I at the University of British Columbia passed the starlight through a tube of gas (highly noxious hydrogen fluoride!) to produce artificial absorption lines, for calibration, directly in the stellar spectrum. From 1980 to 1992 we used the Canada-France-Hawaii telescope, on the summit of Mauna Kea in Hawaii, to search for planetary companions to some twenty-one single Sun-like stars, but we found nothing convincing at the time. (Later observations have shown that two of our stars do indeed have Jupiter-mass planets.)

Meanwhile, Bill Cochran and Artie Hatzes at McDonald Observatory, Texas, and Geoff Marcy and Paul Butler at Lick Observatory, California, had begun (in 1986 and 1987, respectively) independent planet searches on a much longer list of stars, using iodine vapor for the artificial lines. We were all looking for planets like Jupiter—in other words, a giant planet, which because of its large mass would cause a stellar motion large enough for us to detect. By analogy with Jupiter, we expected such a planet to be in a circular orbit with an orbital period of about a decade, requiring observations extending over at least that period to identify a planet with certainty. In consequence, because time at large telescopes is at a premium, we were assigned just a few nights spread over each year. These years of tedious and exacting observations, with endless computation, brought no joy. It was clear that Jupiter analogues were much rarer than had been expected, and perhaps nonexistent.

Then came the break we had been waiting for. In 1994, Mayor and Queloz, of the Geneva Observatory, Switzerland, began to search for low-mass companions (planets or the failed stars known as brown dwarfs) with a telescope at l'Observatoire d'Haute Provence in southern France. Within a year they announced the detection of 51 Pegasi b—a planet with the mass of Jupiter, orbiting a Sun-like star. Mayor and Queloz's discovery,[1] published in *Nature* on 23 November 1995, showed that we had all been looking in the wrong place: 51 Peg b was extremely close to its parent star—one-seventh the distance of Mercury to the Sun—and had an orbital period of four days, not twelve years! Our programs had been ill designed to sample such periods; even the first announcement by Mayor and Queloz was based on snapshots from different cycles (see p. 324, figure 2). By the time their paper was published, they, and several other groups, had fully confirmed the radial velocity curve from single orbits (fig. 20.2).

Ironically, the Geneva discovery was made by the more conventional technique of comparing line positions in a star's spectrum with arc-lamp lines, rather than by imposing artificial absorption lines, which the rest of us had considered essential. Mayor and Queloz used ELODIE, a highly stable spectrograph designed by André Baranne at l'Observatoire de Marseille and built at l'Observatoire d'Haute Provence. It was ideally suited to detect short-period planets; and 51 Peg b's large mass and small orbit gives rise

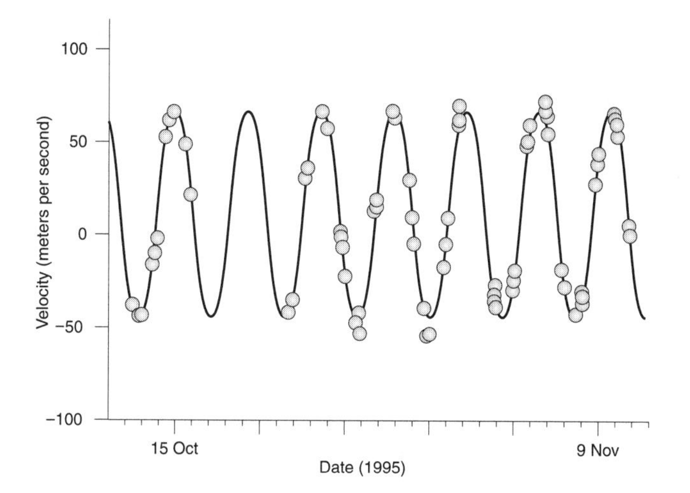

Fig. 20.2. Radial velocity curve for 51 Pegasi, obtained by Paul Butler and Geoff Marcy at Lick Observatory, California, during a one-month interval in late 1995. These data, collected after Mayor and Queloz had made their discovery but before their *Nature* paper was published, helped to confirm the existence of the planet 51 Peg b. The vertical axis shows the velocity in meters per second, and the horizontal axis is the day of the month. The solid line shows how the velocity is expected to change if the star is accompanied by a planet with half the mass of Jupiter in a 4.23-day circular orbit. Mayor and Queloz and others have now accumulated velocity measurements for 51 Pegasi equivalent to several hundred orbits of the planet, and the velocity cycles still synchronize exactly with the 4.23-day period. This is strong proof of the presence of a planetary companion.

to such large stellar motions (radial velocity changes of about fifty meters per second each day) that the signal was quite unambiguous.

But was a planet the only explanation for the observed periodic change in radial velocity? I have to admit to being an early skeptic! How could a Jupiter end up so close to a star, when our own Jupiter is so much farther away from the Sun? Could not a spot, or other perturbation, on the surface of a rotating star generate a similar periodic signal?

In their paper,[1] Mayor and Queloz reviewed all of the other options, including the possibility that they were viewing a binary star system almost pole-on. (If the star's companion were orbiting in a plane almost perpendicular to the observer's line of sight, the star's radial velocity would be a gross underestimate of its real velocity, leading to a similar underestimate of the mass of the companion. In this way, a companion star might conceivably masquerade as a planet.) But they were able to show that a planet with the mass of Jupiter in a circular orbit fit the observations best. When, two years later, David Gray presented data[3] suggesting that the Doppler results might instead be an artifact of stellar rotation or pulsation, there was quite a heated argument! But such salty disputes give science a real flavor, and better data eventually ruled out such an artifact. The radial velocity curve of 51 Pegasi has continued to repeat like clockwork (fig. 20.2)—there really seems to be a planet there. But the second, more distant Jupiter-like companion suggested by Mayor and Queloz (p. 324) has not materialized.

Once the discovery of 51 Peg b had pointed the way toward planetary companions very close to their stars, a harvest of discoveries of Jupiter-like planets followed. As I write, about a hundred planets have been claimed to be orbiting Sun-like stars;[4] some of the systems contain more than one planet. More than half of the planets are in short-period orbits (days or weeks), like that of 51 Peg b, and more than one can be seen (from a periodic dimming of starlight) to be regularly crossing the face of its parent star.[5,6] None of the 1,500 or so stars being monitored has yet revealed a planet with an orbital period nearly as long as that of Jupiter (the maximum period seen so far is seven years, as compared with twelve for Jupiter), although this could be a consequence of the large search programs not having operated for more than a decade. The longer-period extrasolar planets that have been found are mostly in highly elliptical orbits, whereas Jupiter's is nearly circular.

The discovery of these Jupiter-like planets, triggered by Mayor and Queloz's discovery of 51 Peg b, has rekindled the fervor to search for habitable planets. The U.S. space agency NASA has given high priority to the discovery and characterization of extrasolar planets from space—the only environment from which there is any hope of making a direct detection of an Earth-like companion. Ironically, current theory suggests that the presence of a giant planet close to its parent star or in a highly elliptical orbit probably excludes the presence of a habitable planet in a roughly one-year orbit. But we won't know until we look.

References

1. Mayor, M. & Queloz, D. A Jupiter-mass companion to a solar-type star. *Nature* **378**, 355–359 (1995).
2. Wolszczan, A. & Frail, D. A. A planetary system around the millisecond pulsar PSR1257+12. *Nature* **355**, 145–147 (1992).
3. Gray, D. F. Absence of a planetary signature in the spectra of the star 51 Pegasi. *Nature* **385**, 795–796 (1997).
4. *The Extrasolar Planets Encyclopaedia* (http://www.obspm.fr/encycl/encycl.html).

5. Henry, G. W., Marcy, G. W., Butler, R. P. & Vogt, S. S. A transiting "51 Peg–like" planet. *Astrophys. J.* **529,** L41–L44 (2000).
6. Konacki, M., Torres, G., Jha, S. & Sasselov, D. D. An extrasolar planet that transits the disk of its parent star. *Nature* **421,** 507–509 (2003).

Further reading

Boss, A. *Looking for Earths: The Race to Find New Solar Systems* (Wiley, New York, 1998).

Cameron, A. C. Extrasolar planets. *Physics World* **14** (1), 25–31 (2001).

Dorminey, B. *Distant Wanderers: The Search for Planets beyond the Solar System* (Springer, New York, 2002).

Lissauer, J. J. How common are habitable planets? *Nature* **402,** C11–C14 (2000).

Marcy, G. & Butler, P. Hunting planets beyond. *Astronomy* 42–47 (March 2000).

1995

A Jupiter-mass companion to a solar-type star

Michel Mayor and Didier Queloz

The presence of a Jupiter-mass companion to the star 51 Pegasi is inferred from observations of periodic variations in the star's radial velocity. The companion lies only about eight million kilometres from the star, which would be well inside the orbit of Mercury in our Solar System. This object might be a gas-giant planet that has migrated to this location through orbital evolution, or from the radiative stripping of a brown dwarf.

For more than ten years, several groups have been examining the radial velocities of dozens of stars, in an attempt to identify orbital motions induced by the presence of heavy planetary companions[1–5]. The precision of spectrographs optimized for Doppler studies and currently in use is limited to about 15 m s^{-1}. As the reflex motion of the Sun due to Jupiter is 13 m s^{-1}, all current searches are limited to the detection of objects with at least the mass of Jupiter (M_J). So far, all precise Doppler surveys have failed to detect any jovian planets or brown dwarfs.

Since April 1994 we have monitored the radial velocity of 142 G and K dwarf stars with a precision of 13 m s^{-1}. The stars in our survey are selected for their apparent constant radial velocity (at lower precision) from a larger sample of stars monitored for 15 years[6,7]. After 18 months of measurements, a small number of stars show significant velocity variations. Although most candidates require additional measurements, we report here the discovery of a companion with a minimum mass of 0.5 M_J, orbiting at 0.05 AU around the solar-type star 51 Peg. Constraints originating from the observed rotational velocity of 51 Peg and from its low chromospheric emission give an upper limit of 2 M_J for the mass of the companion. Alternative explanations for the observed radial velocity variation (pulsation or spot rotation) are unlikely.

The very small distance between the companion and 51 Peg is certainly not predicted by current models of giant planet formation[8]. As the temperature of the companion is above 1,300 K, this object seems to be dangerously close to the Jeans thermal evaporation limit. Moreover, nonthermal evaporation effects are known to be dominant[9] over thermal ones. This jovian-mass companion may therefore be the result of the stripping of a very-low-mass brown dwarf.

The short-period orbital motion of 51 Peg also displays a long-period perturbation, which may be the signature of a second low-mass companion orbiting at larger distance.

Discovery of Jupiter-mass companion(s)

Our measurements are made with the new fibre-fed echelle spectrograph ELODIE of the Haute-Provence Observatory, France[10]. This instrument permits measurements of radial velocity with an accuracy of about 13 m s^{-1} of stars up to 9 mag in an exposure time of <30 min. The radial velocity is computed with a cross-correlation technique that concentrates the Doppler information of about 5,000 stellar absorption lines. The position of the cross-correlation function (Fig. 1) is used to compute the radial velocity. The width of the cross-correlation function is related to the star's rotational velocity. The very high radial-velocity accuracy achieved is a result of the scrambling effect of the fibres, as well as monitoring by a calibration lamp of instrumental variations during exposure.

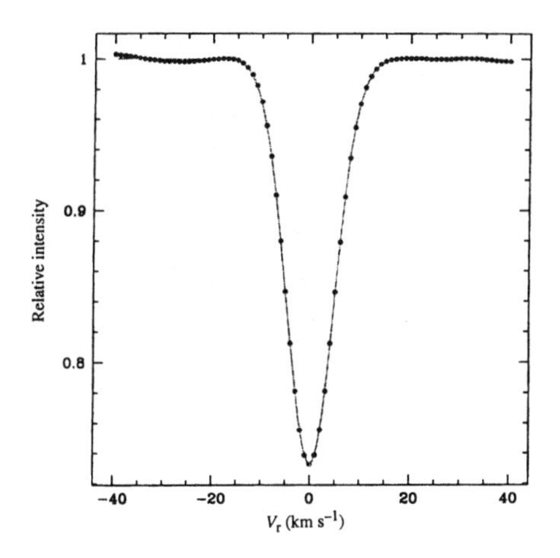

Fig. 1 Typical cross-correlation function used to measure the radial velocity. This function represents a mean of the spectral lines of the star. The location of the gaussian function fitted (solid line) is a precise measurement of the Doppler shift.

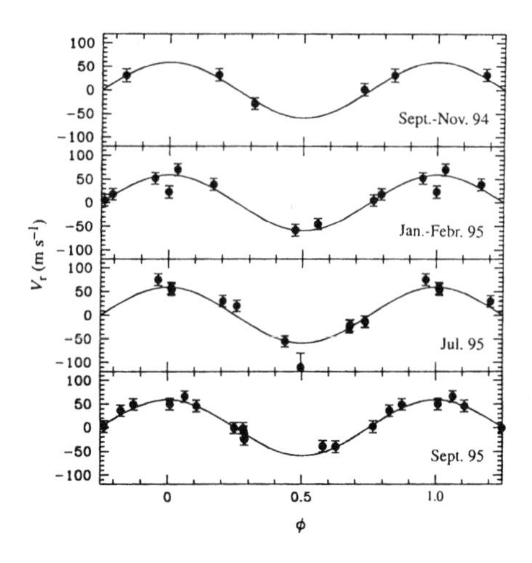

Fig. 2 Orbital motion of 51 Peg at four different epochs corrected from the γ-velocity. The solid line represents the orbital motion fitted on each time span with only the γ-velocity as a free parameter and with the other fixed parameters taken from Table 1.

The first observations of 51 Peg started in September 1994. In January 1995 a first 4.23-day orbit was computed and confirmed by intensive observations during eight consecutive nights in July 1995 and eight in September 1995. Nevertheless, a 24 m s^{-1} scatter of the orbital solution was measured. As this is incompatible with the accuracy of ELODIE measurements, we adjusted an orbit to four sets of measurements carried out at four different epochs with only the γ-velocity as a free parameter (see Fig. 2). The γ-velocity in Fig. 3 shows a significant variation that cannot be the result of instrumental drift in the spectrograph. This slow perturbation of the short-period orbit is probably the signature of a second low-mass companion.

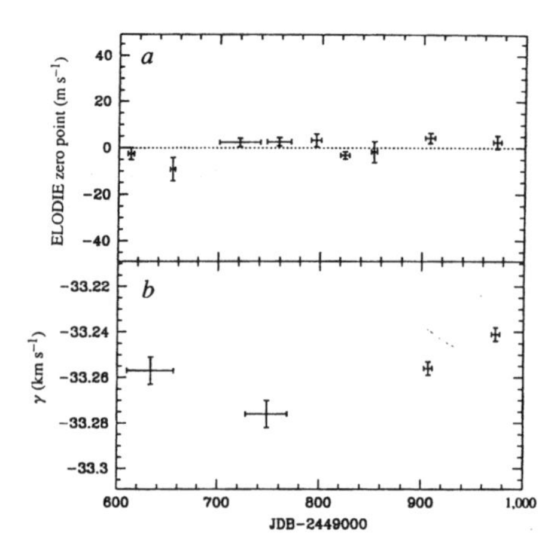

Fig. 3 *a*, ELODIE zero point computed from 87 stars of the sample having more than two measurements and showing no velocity variation. No instrumental zero point drift is detected. *b*, Variation of the γ-velocity of 51 Peg computed from the orbital fits displayed in Fig. 2. Considering the long-term stability of ELODIE this perturbation is probably due to a low-mass companion.

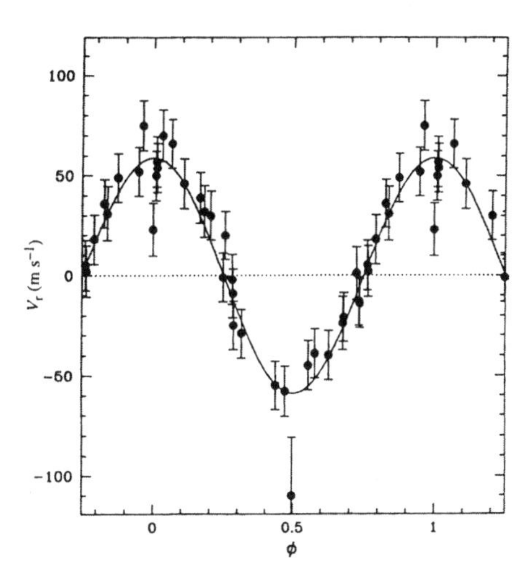

Fig. 4 Orbital motion of 51 Peg corrected from the long-term variation of the γ-velocity. The solid line represents the orbital motion computed from the parameters of Table 1.

The long-period orbit cannot have a large amplitude. The 26 radial velocity measurements made during >12 years with the CORAVEL spectrometer do not reveal any significant variation at a 200 m s^{-1} level. Intensive monitoring of 51 Peg is in progress to confirm this long-period orbit.

In Fig. 4 a short-period circular orbit is fitted to the data after correction of the variation in γ-velocity. Leaving the eccentricity as a free parameter would have given $e = 0.09 \pm 0.06$ with almost the same standard deviation for the r.m.s. residual (13 m s^{-1}). Therefore we consider that a circular orbit cannot be ruled out. At present the eccentricity range is between 0 and about 0.15. Table 1 lists the orbital parameters of the circular-orbit solution.

An orbital period of 4.23 days is rather short, but short-period binaries are not exceptional among solar-type stars. (Five spectroscopic binaries have been found with a period <4 days in a volume-limited sample of

Table 1 Orbital parameters of 51 Peg

P	4.2293 ± 0.0011 d
T	$2{,}449{,}797.773 \pm 0.036$
e	0 (fixed)
K_1	0.059 ± 0.003 km s^{-1}
$a_1 \sin i$	$(34 \pm 2)\ 10^5$ m
$f_1(m)$	$(0.91 \pm 0.15)\ 10^{-10}\ M_\odot$
N	35 measurements
$(O - C)$	13 m s^{-1}

P, period; T, epoch of the maximum velocity; e, eccentricity; K_1, half-amplitude of the velocity variation; $a_1 \sin i$, where a_1 is the orbital radius; $f_1(m)$, mass function; N, number of observations; $(O - C)$, r.m.s. residual.

Table 2 Physical parameters of 51 Peg compared with those of the Sun

		51 Peg		
	Sun	Geneva photometry*	Spectroscopy†	Strömgren photometry and spectroscopy[11]
T_{eff} (K)	5,780	5,773	5,724	5,775
log g	4.45	4.32	4.30	4.18
Fe/H	0		0.19	0.06‡
M/H	0	0.20		
M_v	4.79	4.60		
$R/R_\odot$	1	1.29		

M/H is the logarithmic ratio of the heavy element abundance compared to the Sun (in dex).
* M. Grenon (personal communication).
† J. Valenti (personal communication).
‡ But other elements such as Na I, Mg I, Al I are overabundant, in excess of 0.20.

164 G-type dwarfs in the solar vicinity[6].) Although this orbital period is not surprising in binary stars, it is puzzling when we consider the mass obtained for the companion:

$$M_2 \sin i = 0.47 \pm 0.02\ M_J$$

where i is the (unknown) inclination angle of the orbit.

51 Peg (HR8729, HD217014 or Gliese 882) is a 5.5 mag star, quite similar to the Sun (see Table 2), located 13.7 pc (45 light yr) away. Photometric and spectroscopic analyses indicate a star slightly older than the Sun, with a similar temperature and slight overabundance of heavy elements. The estimated age[11] derived from its luminosity and effective temperature is typical of an old galactic-disk star. The slight overabundance of heavy elements in such an old disk star is noteworthy. But this is certainly not a remarkable peculiarity in view of the observed scatter of stellar metallicities at a given age.

Upper limit for the companion mass

A priori, we could imagine that we are confronted with a normal spectroscopic binary with an orbital plane almost perpendicular to the line of sight. Assuming a random distribution of binary orbital planes, the probability is less than 1% that the companion mass is larger than 4 M_J, and 1/40,000 that it is above the hydrogen-burning limit of 0.08 $M_\odot$. Although these probability estimates already imply a low-mass companion for 51 Peg, an even stronger case can be made from considerations of rotational velocity. If we assume that the rotational axis of 51 Peg is aligned with the orbital plane, we can derive sin i by combining the observed projected rotational velocity ($\upsilon \sin i$) with the equatorial velocity $V_{equ} = 2\pi R/P$ ($\upsilon \sin i = V_{equ} \cdot \sin i$).

Three independent precise $\upsilon \sin i$ determinations of 51 Peg have been made: by line-profile analysis[12], $\upsilon \sin i = 1.7 \pm 0.8$ km s^{-1}; by using the cross-correlation function obtained with the CORAVEL spectrometer[13], $\upsilon \sin i = 2.1 \pm 0.6$ km s^{-1}; and by using the cross-correlation function obtained with ELODIE, $\upsilon \sin i = 2.8 \pm 0.5$ km s^{-1}. The unweighted mean $\upsilon \sin i$ is 2.2 ± 0.3 km s^{-1}. The standard error is probably not significant as the determination of very small $\upsilon \sin i$ is critically dependent on the supposed macroturbulence in the atmosphere. We accordingly prefer to admit a larger uncertainty: $\upsilon \sin i = 2.2 \pm 1$ km s^{-1}.

51 Peg has been actively monitored for variability in its chromospheric activity[14]. Such activity, measured by the re-emission in the core of the Ca II lines, is directly related to stellar rotation via its dynamo-generated magnetic field. A very low level of chromospheric activity is measured for this object. Incidentally, this provides an independent estimate of an age of 10 Gyr (ref. 14), consistent with the other estimates. No rotational modulation has been detected so far from chromospheric emission, but a 30-day period is deduced from the mean chromospheric activity level S-index. A V_{equ} value of 2.2 ± 0.8 km s^{-1} is then computed if a 25% uncertainty in the period determination is assumed.

Using the mean $\upsilon \sin i$ and the rotational velocity computed from chromospheric activity, we finally deduce a lower limit of 0.4 for sin i. This corresponds to an upper limit for the mass of the planet of 1.2 M_J. Even if we consider a misalignment as large as 10°, the mass of the companion must still be less than 2 M_J, well below the mass of brown dwarfs.

The 30-day rotation period of 51 Peg is clearly not synchronized with the 4.23-day orbital period of its low-mass companion, despite its very short period. (Spectroscopic binaries with similar periods are all synchronized.) The lack of synchronism on a timescale of 10^{10} yr is a consequence of the q^{-2} ($q = M_2/M_1$) dependence of the synchronization timescale[15]. In principle this can be used to derive an upper limit to the mass of the companion. It does at least rule out the possibility of the presence of a low-mass stellar companion.

Alternative interpretations?

With such a small amplitude of velocity variation and such a short period, pulsation or spot rotation might explain the observations equally well[16,17].

We review these alternative interpretations below and show that they can probably be excluded.

Spot rotation can be dismissed on the basis of the lack of chromospheric activity and the large period derived from the S chromospheric index, which is clearly incompatible with the observed radial-velocity short period. A solar-type star rotating with a period of 4.2 days would have a much stronger chromospheric activity than the currently observed value[14]. Moreover, a period of rotation of 4.2 days for a solar-type star is typical of a very young object (younger than the Pleiades) and certainly not of an old disk star.

Pulsation could easily yield low-amplitude velocity variations similar to the one observed, but would be accompanied by luminosity and colour variations as well as phase-related absorption line asymmetries. The homogeneous photometric survey made by the Hipparcos satellite provides a comprehensive view of the intrinsic variability of stars of different temperatures and luminosities. The spectral type of 51 Peg corresponds to a region of the Hertzsprung-Russell diagram where the stars are the most stable[18].

Among solar-type stars no mechanisms have been identified for the excitation of pulsation modes with periods as long as 4 days. Only modes with very low amplitude ($\ll 1$ m s^{-1}) and periods from minutes to 1 h are detected for the Sun.

Radial velocity variations of a few days and <100 m s^{-1} amplitude have been reported for a few giant stars[19]. Stars with a similar spectral type and luminosity class are known to be photometric variables[18]. Their observed periods are in agreement with predicted pulsation periods for giant stars with radii $>20\ R_{\odot}$. 51 Peg, with its small radius, can definitely not be compared to these stars. These giant stars also pulsate simultaneously in many short-period modes, a feature certainly not present in the one-year span of 51 Peg observations. It is worth noticing that 51 Peg is too cold to be in the δ Scuti instability strip.

G. Burki *et al.* (personal communication) made 116 photometric measurements of 51 Peg and two comparison stars in the summer of 1995 at ESO (la Silla) during 17 almost-consecutive nights. The observed magnitude dispersions for the three stars are virtually identical, respectively $V = 0.0038$ for 51 Peg, and $V = 0.0036$ and 0.0039 for the comparison stars. The fit of a sine curve with a period of 4.2293 days to the photometric data limits the possible amplitude to 0.0019 for V magnitude and 0.0012 for the $[B_2 - V_1]$ Geneva colour index. Despite the high precision of these photometric measurements we cannot completely rule out, with these photometric data alone, the possibility of a very low-amplitude pulsation. In the coming months, stronger constraints can be expected from the numerous Hipparcos photometric data of this star.

Pulsations are known to affect the symmetry of stellar absorption lines. To search for such features we use the cross-correlation technique, as this technique is a powerful tool for measuring mean spectral line characteristics[20]. The difference in radial velocity of the lower and upper parts of the cross-correlation function is an indicator of the line asymmetry. The amplitude of a 4.2-day sine curve adjusted to this index is less than 2 m s^{-1}. The bisector of the cross-correlation function does not show any significant phase variation.

From all the above arguments, we believe that the only convincing interpretation of the observed velocity variations is that they are due to the orbital motion of a very-low-mass companion.

Jupiter or stripped brown dwarf?

At the moment we certainly do not have an understanding of the formation mechanism of this very-low-mass companion. But we can make some preliminary comments about the importance of evaporation as well as the dynamic evolution of the orbit.

If we compare 51 Peg b with other planets or G-dwarf stellar companions (Fig. 5) it is clear that the mass and the low orbital eccentricity of this object are in the range of heavy planets, but this certainly does not imply that the formation mechanism of this planet was the same as for Jupiter.

Present models for the formation of Jupiter-like planets do not allow the formation of objects with separations as small as 0.05 AU. If ice grains are involved in the formation of giant planets, the minimum semi-major axis for the orbits is about 5 AU (ref. 8), with a minimum period of the order of 10 yr. A Jupiter-type planet probably suffers some orbital decay during its formation by dynamic friction. But it is not clear that this could produce an orbital shrinking from 5 AU to 0.05 AU.

All of the planets in the Solar System heavier than 10^{-6} $M_{\odot}$ have almost circular orbits as a result of their origin from a protoplanetary gaseous disk. Because of its close separation, however, the low eccentricity of 51 Peg b is not a proof of similar origin. Tidal dissipation acting on the convective envelope is known[15] to circularize the orbit and produce

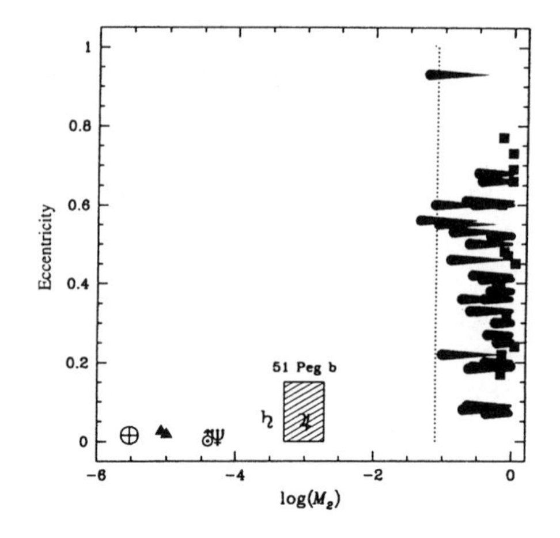

Fig. 5 Orbital eccentricities of planets as well as companions of G-dwarf binaries[8] in the solar vicinity as a function of their mass M_2. The planets of the Solar System are indicated with their usual symbols. The planets orbiting around the pulsar[24,25] PSR B 1257+12 are indicated by filled triangles. The uncertainties on the mass of SB1 (single-spectrum spectroscopic binaries), owing to their unknown orbital inclination, are indicated by an elongated line that thins to a sin *i* probability of 99%. SB2s are indicated by filled squares. (Only the stellar orbits not tidally circularized with periods larger than 11 days are indicated.) Note the discontinuity in the orbital eccentricities when planets and binary stars are compared, and the gap in masses between the giant planets and the lighter secondaries of solar-type stars. The dotted line at 0.08 $M_\odot$ indicates the location of the minimum mass for hydrogen burning. The position of 51 Peg b with its uncertainties is indicated by the hatched rectangle.

a secular shrinking of the semi-major axis of binary systems. The characteristic time is essentially proportional to $q^{-1}P^{16/3}$. For stars of the old open cluster M67, orbital circularization is observed for periods lower than 12.5 days (ref. 21). We derive for 51 Peg a circularization time of a few billion years, shorter than the age of the system. The low orbital eccentricity of 51 Peg b could result from the dynamic evolution of the system and not necessarily from its formation conditions.

A Jupiter-sized planet as close as 0.05 AU to 51 Peg should have a rather high temperature of about 1,300 K. To avoid a significant evaporation of a gaseous atmosphere, the escape velocity V_e has to be larger than the thermal velocity V_{th}: $V_e > \alpha V_{th}$. This imposes a minimum mass for a gaseous planet at a given separation:

$$\frac{M_p}{M_J} > \alpha^2 \left(\frac{kT_*}{m}\right)\left(\frac{GM_J}{R_p}\right)^{-1}(11 - \gamma)^{1/4}\left(\frac{R_*}{2a}\right)^{1/2}$$

where γ denotes the albedo of the planet, R_p and M_p are its radius and mass, m is the mass of atoms in the planet atmosphere, and R_* and T_* are the radius and effective temperature of the star.

Our lack of knowledge of the detailed structure of the atmosphere of the planet prevents us from making an accurate estimate of α. A first-order estimate of $\alpha \approx 5$–6 is nevertheless made by analogy with planets of the Solar System[22]. We find that with a planetary radius probably increased by a factor of 2–3 owing to the high surface temperature (A. Burrows, personal communication), gaseous planets more massive than 0.6–1.0 M_J are at the borderline for suffering thermal evaporation. Moreover, for the Solar-System planets, non-thermal evaporative processes are known to be more efficient than thermal ones[9]. The atmosphere of 51 Peg b has thus probably been affected by evaporation.

Recent work[23] on the fragmentation of molecular clouds shows that binary stars can be formed essentially as close to each other as desired, especially if the effect of orbital decay are considered. We can thus speculate that 51 Peg b results from a strong evaporation of a very close and low-mass brown dwarf. In such a case 51 Peg b should mostly consist of heavy elements. This model is also not free of difficulties, as we expect that a brown dwarf suffers less evaporation owing to its larger escape velocity.

We are eager to confirm the presence of the long-period companion and to find its orbital elements. If its mass is in the range of a few times that of Jupiter and its orbit is also quasi-circular, 51 Peg could be the first example of an extrasolar planetary system associated with a solar-type star.

The search for extrasolar planets can be amazingly rich in surprises. From a complete planetary system detected around a pulsar[24,25], to the rather unexpected orbital parameters of 51 Peg b, searches begin to reveal the extraordinary diversity of possible planetary formation sites.

Note added in revision: After the announcement of this discovery at a meeting held in Florence, independent confirmations of the 4.2-day period radial-velocity variation were obtained in mid-October by a team at Lick Observatory, as well as by a joint team from the High Altitude Observatory and the Harvard–Smithsonian Center for Astrophysics. We are deeply grateful to G. Marcy, P. Butler, R. Noyes, T. Kennelly and T. Brown for having immediately communicated their results to us.

Geneva Observatory, 51 Chemin des Maillettes, CH-1290 Sauverny, Switzerland

Received 29 August; accepted 31 October 1995 [Published 23 November]

References

1. Walker, G. A. H., Walker, A. R. & Irwin, A. W., *Icarus* **116**, 359–375 (1995).
2. Cochran, W. D. & Hatzes, A. P. *Astrophys. Space Sci.* **212**, 281–291 (1994).
3. Marcy, G. W. & Butler, R. P. *Publ. astr. Soc. Pacif.* **104**, 270–277 (1992).
4. McMillan, R. S., Moore, T. L, Perry, M. L & Smith, P. H. *Astrophys. Space Sci.* **212**, 271–280 (1994).
5. Marcy, G. W. & Butler, R. P. in *The Bottom of the Main Sequence and Beyond* (ESO Astrophys. Symp.) (ed. Tinney, C. G.) 98–108 (Springer, Berlin, 1995).
6. Duquennoy, A. & Mayor, M. *Astr. Astrophys.* **248**, 485–524 (1991).
7. Mayor, M., Duquennoy, A., Halbwachs, J. L. & Mermilliod, J. C. in *Complementary Approaches to Double and Multiple Star Research* (eds McAlister, A. A. & Hartkopf, W. I.) (ASP Conf. Ser. **32**, 73–81 (Astr. Soc. Pacific, California, 1992).
8. Boss, A. P. *Science* **267**, 360–362 (1995).
9. Hunten, D. H., Donahue, T. M., Walker, J. C. G. & Kasting, J. F. in *Origin and Evolution of Planetary and Satellite Atmospheres* (eds Atreya, S. K., Pollack, J. B. & Matthews, M. S.) 386–422 (Univ. of Arizona Press, Tucson, 1989).
10. Baranne, A. *Astrophys. J. Suppl.* (submitted).
11. Edvardsson, B. *et al. Astr. Astrophys.* **275**, 101–152 (1993).
12. Soderblom, D. R. *Astrophys. J. Suppl. Ser.* **53**, 1–15 (1983).
13. Baranne, A., Mayor, M. & Poncet, J. L. *Vistas Astr.* **23**, 279–316 (1979).
14. Noyes, R. W., Hartmann, L. W., Baliunas, S. L., Duncan, D. K. & Vaughan, A. H. *Astrophys. J.* **279**, 763–777 (1984).
15. Zhan, J. P. *Astr. Astrophys.* **220**, 112–116 (1989).
16. Walker, G. A. H. *et al. Astrophys. J.* **396**, L91–L94 (1992).
17. Larson, A. M. *et al. Publs astr. Soc. Pacif.* **105**, 825–831 (1993).
18. Eyer, L., Grenon, M., Falin, J. L., Froeschlé, M. & Mignard, F. *Sol. Phys.* **152**, 91–96 (1994).
19. Hatzes, A. P. & Cochran, W. D. *Proc. 9th Cambridge Workshop* (ed. Pallavicini, R.) (Astronomical Soc. of the Pacific) (in the press).
20. Queloz, D. in *New Developments in Array Technology and Applications* (eds Davis Philip, A. G. *et al.*) 221–229 (Int. Astr. Union, 1995).
21. Latham, D. W., Mathieu, R. D., Milone, A. A. E. & Davis, R. J. in *Binaries as Tracers of Stellar Formation* (eds Duquennoy, A. and Mayor, M.) 132–138 (Cambridge Univ. Press. 1992).
22. Lewis, J. S. & Prinn, R. G. *Planets and their Atmospheres—Origin and Evolution* (Academic, Orlando, 1984).
23. Bonnell, I. A. & Bate, M. R. *Mon. Not. R. astr. Soc.* **271**, 999–1004 (1994).
24. Wolszczan, A. & Frail, D. A. *Nature* **355**, 145–147 (1992).
25. Wolszczan, A. *Science* **264**, 538–542 (1994).

Acknowledgements. We thank G. Burki for analysis of photometric data, W. Benz for stimulating discussions, A. Burrows for communicating preliminary estimates of the radius of Jupiter at different distances from the Sun, and F. Pont for his careful reading of the manuscript. We also thank all our colleagues of Marseille and Haute-Provence Observatories involved in the building and operation of the ELODIE spectrograph, namely G. Adrianzyk, A. Baranne, R. Cautain, G. Knispel, D. Kohler, D. Lacroix, J.-P. Meunier, G. Rimbaud and A. Vin.

1997

(1993).
16. van Kolfschoten, T. Die Vertebraten des Interglazials von Schöningen 12. *Ethnogr.-A*
628 (1993).
17. van Kolfschoten, T. in *Archäologische Ausgrabungen im Braunkohlentagebau Schö Helmstedt* (eds Thieme, H. & Maier, R.) 85–94 (Hahnsche, Hannover, 1995).
18. Schoch, W. H. in *Archäologische Ausgrabungen im Braunkohlentagebau Schön Helmstedt* (eds Thieme, H. & Maier, R.) 73–84 (Hahnsche, Hannover, 1995).
19. Roberts, M. B., Stringer, C. B. & Parfitt, S. A. *Nature* **369**, 311–313 (1994).
20. Roebroeks, W. & van Kolfschoten, T. in *The Earliest Occupation of Europe* (eds Roe Kolfschoten, T.) 297–315 (University of Leiden, 1995).
21. Nitecki, M. H. in *The Evolution of Human Hunting* (eds Nitecki, M. H. & Nitecki, D. New York, 1987).
22. Gamble, C. in *The Pleistocene Old World. Regional Perspectives* (ed. Soffer, O.) 81– York, 1987).

...ents. ...nschweigische Kohlen-Bergwerke AG for t ...nizi... ...pport of the ...ogical excavations in the Schöningen mine; ...tifyin... ...ood; D. Ma... ...he drawings in Figs 1 and 2; P. Pfarr and C. ...ot... ...gs 3–5; a... ...oberts, W. Roebroeks and T. van Kolfschoten fo ...ginal Ger... ...nuscript. The excavation of the spear site was s Deutsche... ...Denkm...

...ce and ... ...or materials should be addressed to the author.

...which is 2.30 m long. The spear is shown to the ... an ...a horse, and the base has been broken ... ...ows a ...pear II. Scale in cm.

...and are at least of stage 9. Correlations of the ...uence to other areas[12,14] suggest that they were ...the fourth-last interglacial, probably at the end ...ging from the mammalian fauna, the Reinsdorf ...unger than the Lower Palaeolithic site of Boxgrove ...K)[19,20].

...scovery, the oldest complete 'spear' known was ...mian deposits at Lehringen (Lower Saxony, Ger- ...ref. 2). Thought to be about 125 kyr old (oxygen- ...5e), this thrusting spear was recovered from ...s of a straight-tusked elephant, and was made ...s). The Schöningen spears are probably three full ...al cycles older than the Lehringen lance (Fig. 2).

...y of spears designed for throwing means that ...evelopment of hunting capacities and subsistence ...dle Pleistocene hominids must be revised, as well- ...ticated hunting weapons were common from an ...he Middle Pleistocene onwards. Accordingly, meat ...ay have provided a larger dietary contribution than ...een acknowledged[21,22]. The Schöningen evidence ...ow little is known about the 'organic' component of ...aterial culture. □

...ccepted 23 December 1996.

...s, P., Keeley, L. H. & Clark, J. D. A reappraisal of the Clacton spearpoint. *Proc.* ...0 (1977).
...Neue Untersuchungen zum eemzeitlichen Elefanten-Jagdplatz Lehringen, Ldkr. ...F. 36, 11–58 (1985).
...& Urban, B. Archäologische Schwerpunktuntersuchungen im Helmstedter ...HB). –Zum Stand der Arbeiten 1983–1986. *Archäol. Korrespondenzbl.* 17, 445–
...& Urban, B. Neue Erkenntnisse zum urgeschichtlichen Siedlungsgeschehen. ..., 26–30 (1992).
...I. & Elsner, H. Biostratigraphische, quartärgeologische und urgeschichtliche ...ebau 'Schöningen', Ldkr. Helmstedt. *Z. Deutsch. Geol. Ges.* 139, 123–154 (1988).
..., Hölzer, A., Mania, D. & Albrecht, B. Eine eem- und frühweichselzeitliche ...Schöningen, Landkreis Helmstedt. *Eiszeit. Gegenw.* 41, 85–99 (1991).
...R., Mania, D. & Albrecht, B. Mittelpleistozän im Tagebau Schöningen, Ldkr. ...*Geol. Ges.* 142, 351–372 (1991).
...R. *Archäologische Ausgrabungen im Braunkohlentagebau Schöningen, Landkreis* ...e, Hannover, 1995).
...ia, D. 'Schöningen 12'—ein mittelpleistozänes Interglazialvorkommen im ...paläolithischen Funden. *Ethnogr.-Archäol. Z.* 34, 610–619 (1993).

Viable offspring derived fr... fetal and adult mammalian... cells

I. Wilmut, A. E. Schnieke*, J. McWhir, A. J. Kin... & K. H. S. Campbell

Roslin Institute (Edinburgh), Roslin, Midlothian EH25 9PS, UK
** PPL Therapeutics, Roslin, Midlothian EH25 9PP, UK*

Fertilization of mammalian eggs is followed by succ... divisions and progressive differentiation, first into ... embryo and subsequently into all of the cell types tha... the adult animal. Transfer of a single nucleus at a speci... development, to an enucleated unfertilized egg, pr... opportunity to investigate whether cellular differen... that stage involved irreversible genetic modification... offspring to develop from a differentiated cell were ... nuclear transfer from an embryo-derived cell line tha... induced to become quiescent[1]. Using the same procedu... report the birth of live lambs from three new cell p... established from adult mammary gland, fetus and em... fact that a lamb was derived from an adult cell con... differentiation of that cell did not involve the irrevers... ification of genetic material required for developmen... The birth of lambs from differentiated fetal and adult... reinforces previous speculation[1,2] that by inducing don... become quiescent it will be possible to obtain norma... ment from a wide variety of differentiated cells.

It has long been known that in amphibians, nuclei t... from adult keratinocytes established in culture suppor... ment to the juvenile, tadpole stage[3]. Although this invol... entiation into complex tissues and organs, no developm... adult stage was reported, leaving open the question of ... differentiated adult nucleus can be fully reprogrammed. ... we reported the birth of live lambs after nuclear tra... cultured embryonic cells that had been induced into quies... suggested that inducing the donor cell to exit the gro...

TWENTY-ONE

Dolly!

Davor Solter

The lamb Dolly was the first mammal to be cloned from an adult cell. Her birth established that when transferred into eggs from which the genetic material has been removed, the nuclei of at least some adult cells can be used to produce sheep or other animals that are genetically identical to the donor. This technology has huge potential but is highly contentious: the scientific and ethical debate continues.

Judging by the astonished reaction of scientists and the public around the world, the announcement on 27 February 1997 of the birth of several cloned lambs caught almost everybody by surprise. In their *Nature* paper[1] of that date, Ian Wilmut and colleagues of the Roslin Institute in Scotland described how they had cloned a total of eight lambs, using the nuclei from several different cell lines. Seven of the lambs survived to adulthood. The cells from which the nuclei—and so the genetic material—came were from a nine-day-old embryo, a twenty-six-day-old fetus, and, most significantly, the udder of a six-year-old ewe. Lamb 6LL3 (known forever after as Dolly) was the only one derived from the nucleus of an adult cell—that is, a cell that was fully "differentiated" and seemingly committed to a particular function.

The method used, nuclear transfer, was fairly standard by that time. The genetic material was removed from egg cells. These were then fused with the cells, growing in cell culture, that provided the nuclei. Wilmut *et al.*[1] considered that their success hinged on the donor cell that provided the nucleus being in a quiescent state (that is, stopped in an inactive phase of the cell-division cycle). After a further period of development in sheep oviducts, those embryos that had reached a stage known as the blastocyst were transferred to the uterus of recipient ewes and allowed to come to term

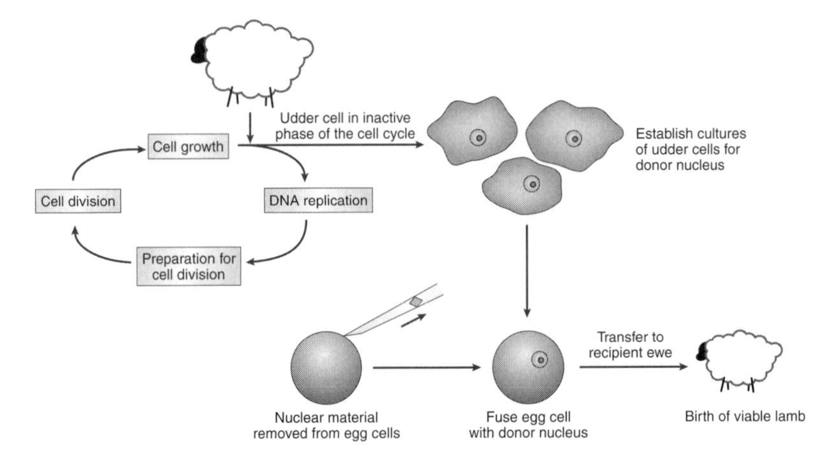

Fig. 21.1. The nuclear transfer technique used to clone Dolly. The nucleus, which contains a cell's genetic material as DNA, was removed from egg cells. In the meantime, a donor cell line was created from cells stopped in an inactive part of the cell cycle. For Dolly, the donor cell came from the udder of a six-year-old ewe. The next step was fusion of the egg cell with a nucleus from the donor cell. Following further development of the embryo, and transfer to the uterus of a recipient ewe, Dolly was born.

(fig. 21.1). The success rate was low (fewer than 1% of manipulated eggs developed to term). And when udder cells were used as nuclear donors, only one live birth ensued from more than 400 manipulated eggs. This did not matter, for with Dolly's birth we had proof of the principle that adult mammals could be cloned. The practical and ethical implications were, and remain, manifold. But first let us look at the scientific background.

The first mammals produced by nuclear transfer were mice, described by Jim McGrath and me[2] in 1983. We established the technique necessary to remove the nucleus from a fertilized mouse egg, and to introduce a new nucleus into that cell (we used an approach involving a virus). These embryos developed to adulthood if the nuclear donor was another fertilized egg, but not if the nuclei were taken from embryos that had divided into two cells or beyond. But although these mice were produced by nuclear transfer, they could not strictly be considered to be clones: the word "clone" presupposes at least the theoretical possibility of more than one genetically identical organism being created.

Using a similar technique to ours, Steen Willadsen[3] produced the first cloned mammal. He removed the genetic material from unfertilized eggs and fused the eggs with cells from sheep embryos in the eight- and sixteen-cell stage of development. Several of these developed to adulthood. At that time it was not clear whether our failure with more advanced nuclear donors and Willadsen's success meant that it is easier to clone sheep than mice, or that unfertilized rather than fertilized egg cells were the better recipients. Over the ensuing ten years, work in many laboratories showed that both of these factors probably applied. Numerous reports have since described the cloning of sheep and cows using nuclei from early embryos. The cloning of mice was also possible using this approach.

The next notable advance came in 1994, with the cloning of calves using nuclei from cells in culture. Michelle Sims and Neal First[4] derived a series of cell lines from the inner cell mass (part of which gives rise to the fetus) of a cow blastocyst. Provided that the cultures were not older than twenty-eight days, live calves were born following nuclear transfer. Crucially, then, at that time the general view was that the cells providing the nuclei should be embryonic, and the younger the better.

So a 1996 report by the Roslin group (Campbell *et al.*),[5] describing the production of sheep "following nuclear transfer from an established cell line," was a real breakthrough. Conceptually, this paper is more important than the report of Dolly's birth. Campbell *et al.* tried to isolate undifferentiated embryonic cells from early sheep embryos (as Sims and First[4] did from the cow), but they did not succeed. They produced cell lines from the embryos, but these cells were obviously differentiated. To their great credit, Campbell and colleagues did not throw these cells away, as they might have done according to accepted wisdom, but proceeded to use them as nuclear donors. The methods were the same as for the cloning of Dolly. Of the five lambs born, three died soon after birth. But two—Megan and Morag—survived and were thus the first clones derived from the genetic material of apparently differentiated cells.

While reading a manuscript of the paper by Campbell *et al.*,[5] I realized that we needed to change our thinking about cloning. It was clear that it might not be necessary to use early embryonic cells as nuclear donors; differentiated and even fully adult cells might work. As I wrote in an article accompanying the paper, "Cloning from adult cells will be considerably harder, but can no longer be considered impossible."[6] Moreover, it was evident that if one could clone from established cells in culture, the addition or elimination of specific genes at that stage would be reflected in the genetic constitution of the cloned adult, with obvious practical potential.

All in all, I was not too surprised when, a year later, I received a call from *Nature* asking me to provide prepublication comments on the paper[1] describing cloning from adult cells. I may not have been surprised, but I was certainly impressed. Wilmut and his colleagues showed remarkable skill, perseverance, and, most of all, a firm belief in their vision of how far cloning could be taken. Being familiar with a subject influences one's outlook, and I (and, it seems, the Roslin group) did not expect the outburst of sound and fury that followed publication of the paper.

Oddly enough, the most obvious application of cloning—the production of genetically modified farm animals—does not depend on cloning from adult cells. Indeed, several calves and sheep have been cloned from cultured fetal cells, called fibroblasts, into which foreign genes have been introduced. We can soon expect herds of animals with all kinds of valuable products in their milk, wool, urine, and who knows what else. Millions of people who are intolerant to lactose, a milk sugar, are probably hoping for lactose-free products from cloned cows that can synthesize and secrete the enzyme lactase (which digests lactose) into their milk. When combined with the genetic manipulation of nuclear donor cells, another potential though contentious use of cloning technology is the production of animal tissues and organs for transplantation into humans. Pigs deficient in an enzyme that results in rejection of pig-to-human transplants have been produced by cloning fibroblasts from which both copies of the gene that encodes the enzyme have been eliminated. Many observers envisage that untold riches will flow from these applications. It is no wonder that there has been a unseemly fight over the significance of small technical differences in the scramble for the ownership of various cloning patents.

While Dolly was the only example of an animal cloned from an adult cell, her provenance, and the very possibility of cloning from adult cells, was doubted in some quarters. But the birth in 1998 of mice cloned from adult cells[7] laid those doubts to rest. Today it is obvious that most, if not all, mammals can be cloned from at least some adult cells, as witnessed by the reports of cloned goats, pigs, cats, and rabbits. Thus it is likely that cloning will succeed in any mammalian species. Despite considerable efforts, at the time of writing no cloning of dogs or primates has been reported, which probably reflects a lack of knowledge about some essential part of their reproduction rather than a biological barrier.

Nonetheless, many basic scientific questions remain unanswered. How does a clonable cell differ from one that cannot be cloned? What is the effect of various technical aspects in determining the success of cloning? Why is the rate of successful cloning so low? What happens during genetic reprogramming (the process in which the donor nucleus falls into step with events in the recipient cell), and how can it be improved? These questions are gradually being addressed, mostly using mice.

It appears that the failure of complete reprogramming can explain both the low success rate and the abnormalities observed in cloned fetuses, newborns, and adults. The incidence of abnormalities is high, and one can justifiably ask: are there any normal cloned mammals?[8] The answer is probably no. Dolly herself was put down on 14 February 2003, suffering from a lung tumor that may or may not have been related to her having been born through cloning.

Although most of the defects seen in cloned animals are probably a consequence of the faulty reprogramming inherent in the nuclear transfer of adult cells into egg cytoplasm, some of them could at least partially be caused by manipulation of the embryo during the cloning process. There are also reports of healthy clones, but there is always the possibility that more extensive analysis would reveal abnormalities in these cases as well. Even if all clones are to some degree abnormal, however, that should not prevent the application of cloning technology in agriculture. Defects that are a consequence of faulty reprogramming, and not inherent in the genetic material, would not be transmitted to offspring.

The cloning of farm animals from adult cells will at least sometimes be used to reproduce "genetically elite" specimens. The difficulties involved are considerable, however, and this application may be limited in scope. Another possibility is the cloning of beloved pets; the cloning of a cat has already been reported and attempts to clone dogs continue unabated. I find it depressing that there are people prepared to spend huge amounts of money on cloning their pet dog in a world in which every day thousands of people are dying of hunger and thousands of unwanted dogs are destroyed. Moreover, the future owners of cloned pets may be disappointed in the results, because cloning does not mean duplicating: the personality of the cloned pet may turn out to be quite different from that of the original.

And how about the cloning of humans? After all, the brouhaha following the birth of Dolly centered on this possibility and everybody, especially representatives of public institutions, chimed in with demands that it should be forbidden. At present there is a moratorium, mostly for safety reasons because of the high failure rate with animals and because some of the clones have abnormalities or die soon after birth. But what if cloning from human adults can be done reliably? It is likely that every society, and ultimately every individual, will have to make a decision about human cloning.

In the meantime, the prospect of using nuclear transfer to produce lines of so-called embryonic stem cells from individual people is likely to be realized in the near future. Embryonic stem cells are undifferentiated, and if produced by cloning they could be used to grow replacement cells or tissues suitable for the person concerned. One can envision the use of cells produced in this way in treating diabetes, heart failure, stroke, degenerative neurological disease, and, in general, any condition for which cell or tissue replacement could be beneficial. As it is possible that the approach will require the creation and subsequent destruction of a human embryo (albeit at a very early stage), this area remains the subject of hot legal, ethical, and theological debate. But the proof of principle using mice has been provided, and one can expect at least some clinical trials in the not too distant future. More technical detail and references about all of the prospects opened up by cloning, and the pitfalls, can be found in two recent publications.[9,10]

Certain scientific papers illuminate previously unforeseen possibilities and stimulate a continuing burst of further research and exploration of applications. The report by Wilmut *et al.*[1] is one of them. We are just starting to walk down the new paths it has opened up.

References

1. Wilmut, I., Schnieke, A. E., McWhir, J., Kind, A. J. & Campbell, K. H. S. Viable offspring derived from fetal and adult mammalian cells. *Nature* **385,** 810–813 (1997).
2. McGrath, J. & Solter, D. Nuclear transplantation in the mouse embryo by microsurgery and cell fusion. *Science* **220,** 1300–1302 (1983).
3. Willadsen, S. M. Nuclear transplantation in sheep embryos. *Nature* **320,** 63–65 (1986).
4. Sims, M. & First, N. L. Production of calves by transfer of nuclei from cultured inner cell mass cells. *Proc. Natl Acad. Sci. USA* **91,** 6143–6147 (1994).
5. Campbell, K. H. S., McWhir, J., Ritchie, W. A. & Wilmut, I. Sheep cloned by nuclear transfer from a cultured cell line. *Nature* **380,** 64–66 (1996).
6. Solter, D. Lambing by nuclear transfer. *Nature* **380,** 24–25 (1996).
7. Wakayama, T., Zuccotti, M., Johnson, K. R., Perry, A. C. F. & Yanagimachi, R. Full term development of mice from enucleated oocytes injected with cumulus cell nuclei. *Nature* **394,** 369–374 (1998).
8. Wilmut, I. Are there any normal cloned mammals? *Nature Med.* **8,** 215–216 (2002).
9. Solter, D. Mammalian cloning: advances and limitations. *Nature Rev. Genet.* **1,** 199–207 (2000).
10. Committee on Science, Engineering, and Public Policy, Board of Life Sciences. *Scientific and Medical Aspects of Human Reproductive Cloning.* (National Academy Press, Washington, 2002). This report can also be read at http://books.nap.edu/books/0309076374/html/index.html.

Further reading

Di Berardino, M. A. *Genomic Potential of Differentiated Cells* (Columbia University Press, New York, 1997).

Kolata, G. B. *Clone: The Road to Dolly, and the Path Ahead* (Morrow, New York, 1998).

McLaren, A. Cloning: Pathways to a pluripotent future. *Science* **288,** 1775–1780 (2000).

Wilmut, I., Campbell, K. & Tudge, C. *The Second Creation: The Age of Biological Control by the Scientists that Cloned Dolly* (Headline, London, 2000).

1997

Viable offspring derived from fetal and adult mammalian cells

I. Wilmut, A. E. Schnieke, J. McWhir, A. J. Kind, and K. H. S. Campbell

Fertilization of mammalian eggs is followed by successive cell divisions and progressive differentiation, first into the early embryo and subsequently into all of the cell types that make up the adult animal. Transfer of a single nucleus at a specific stage of development, to an enucleated unfertilized egg, provided an opportunity to investigate whether cellular differentiation to that stage involved irreversible genetic modification. The first offspring to develop from a differentiated cell were born after nuclear transfer from an embryo-derived cell line that had been induced to become quiescent[1]. Using the same procedure, we now report the birth of live lambs from three new cell populations established from adult mammary gland, fetus and embryo. The fact that a lamb was derived from an adult cell confirms that differentiation of that cell did not involve the irreversible modification of genetic material required for development to term. The birth of lambs from differentiated fetal and adult cells also reinforces previous speculation[1,2] that by inducing donor cells to become quiescent it will be possible to obtain normal development from a wide variety of differentiated cells.

It has long been known that in amphibians, nuclei transferred from adult keratinocytes established in culture support development to the juvenile, tadpole stage[3]. Although this involves differentiation into complex tissues and organs, no development to the adult stage was reported, leaving open the question of whether a differentiated adult nucleus can be fully reprogrammed. Previously we reported the birth of live lambs after nuclear transfer from cultured embryonic cells that had been induced into quiescence. We suggested that inducing the donor cell to exit the growth phase causes changes in chromatin structure that facilitate reprogramming of gene expression and that development would be normal if nuclei are used from a variety of differentiated donor cells in similar regimes. Here we investigate whether normal development to term is possible when donor cells derived from fetal or adult tissue are induced to exit the growth cycle and enter the G0 phase of the cell cycle before nuclear transfer.

Three new populations of cells were derived from (1) a day-9 embryo, (2) a day-26 fetus and (3) mammary gland of a 6-year-old ewe in the last trimester of pregnancy. Morphology of the embryo-derived cells (Fig. 1) is unlike both mouse embryonic stem (ES) cells and the embryo-derived cells used in our previous study. Nuclear transfer was carried out according to one of our established protocols[1] and reconstructed embryos transferred into recipient ewes. Ultrasound scanning detected 21 single fetuses on day 50–60 after oestrus (Table 1). On subsequent scanning at ~14-day intervals, fewer fetuses were observed, suggesting either misdiagnosis or fetal loss. In total, 62% of fetuses were lost, a significantly greater proportion than the estimate of 6% after natural mating[4]. Increased prenatal loss has been reported after embryo manipulation or

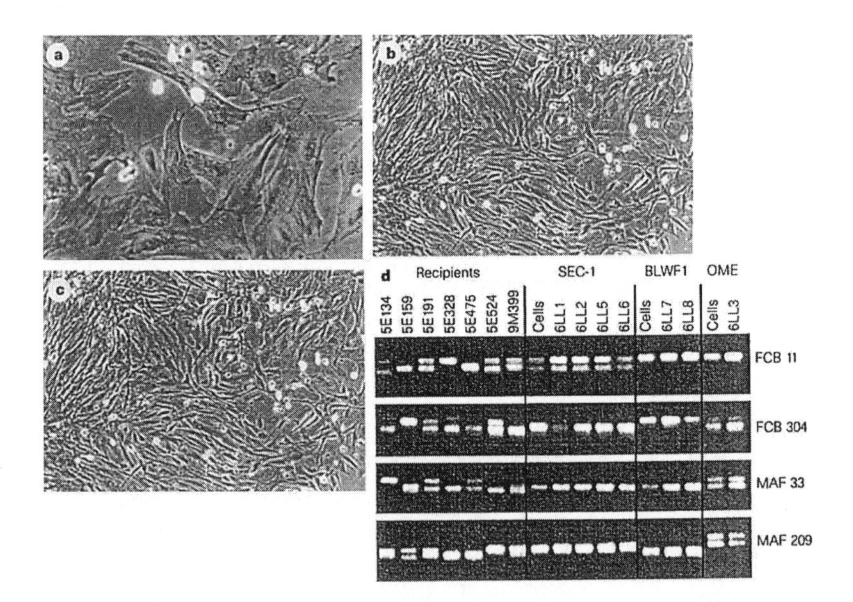

Fig. 1 Phase-contrast photomicrograph of donor-cell populations: **a,** Embryo-derived cells (SEC1); **b,** fetal fibroblasts (BLWF1); **c,** mammary-derived cells (OME). **d,** Microsatellite analysis of recipient ewes, nuclear donor cells and lambs using four polymorphic ovine markers[22]. The ewes are arranged from left to right in the same order as the lambs. Cell populations are embryo-derived (SEC1), fetal-derived (BLW1), and mammary-derived (OME), respectively. Lambs have the same genotype as the donor cells and differ from their recipient mothers.

Table 1 Development of embryos reconstructed with three different cell types

Cell type	No. of fused couplets (%)*	No. recovered from oviduct (%)	No. cultured	No. of morula/ blastocyst (%)	No. of morula or blastocysts transferred†	No. of pregnancies/ no. of recipients (%)	No. of live lambs (%)‡
Mammary epithelium	277 (63.8)[a]	247 (89.2)	—	29 (11.7)[a]	29	1/13 (7.7)	1 (3.4%)
Fetal fibroblast	172 (84.7)[b]	124 (86.7)	—	34 (27.4)[b]	34	4/10 (40.0)	2 (5.9%)
			24	13 (54.2)[b]	6	1/6 (16.6)	1 (16.6%)§
Embryo-derived	385 (82.8)[b]	231 (85.3)	—	90 (39.0)[b]	72	14/27 (51.8)	4 (5.6%)
			92	36 (39.0)[b]	15	1/5 (20.0)	0

* As assessed 1 h after fusion by examination on a dissecting microscope. Superscripts a or b within a column indicate a significant difference between donor cell types in the efficiency of fusion ($P < 0.001$) or the proportion of embryos that developed to morula or blastocyst ($P < 0.001$).

† It was not practicable to transfer all morulae/blastocysts.

‡ As a proportion of morulae or blastocysts transferred. Not all recipients were perfectly synchronized.

§ This lamb died within a few minutes of birth.

culture of unreconstructed embryos[5]. At about day 110 of pregnancy, four fetuses were dead, all from embryo-derived cells, and post-mortem analysis was possible after killing the ewes. Two fetuses had abnormal liver development, but no other abnormalities were detected and there was no evidence of infection.

Eight ewes gave birth to live lambs (Table 1, Fig. 2). All three cell populations were represented. One weak lamb, derived from the fetal fibroblasts, weighed 3.1 kg and died within a few minutes of birth, although post-mortem analysis failed to find any abnormality or infection. At 12.5%, perinatal loss was not dissimilar to that occurring in a large study of commercial sheep, when 8% of lambs died within 24 h of birth[6]. In all cases the lambs displayed the morphological characteristics of the breed used to derive the nucleus donors and not that of the oocyte donor (Table 2). This alone indicates that the lambs could not have been born after inadvertent mating of either the oocyte donor or recipient ewes. In addition, DNA microsatellite analysis of the cell populations and the lambs at four polymorphic loci confirmed that each lamb was derived from the cell population used as nuclear donor (Fig. 1). Duration of gestation is determined by fetal genotype[7], and in all cases gestation was longer than the breed mean (Table 2). By contrast, birth weight is influenced by both maternal and fetal genotype[8]. The birth weight of all lambs was within the range for single lambs born to Blackface ewes on our farm (up to 6.6 kg) and in most cases was within the range for the breed of the nuclear donor. There are no strict control observations for birth weight after embryo transfer between breeds, but the range in weight of

Fig. 2 Lamb number 6LL3 derived from the mammary gland of a Finn Dorset ewe with the Scottish Blackface ewe which was the recipient.

Table 2 Delivery of lambs developing from embryos derived by nuclear transfer from three different donor cell types, showing gestation length and birth weight

Cell type	Breed of lamb	Lamb identity	Duration of pregnancy (days)*	Birth weight (kg)
Mammary epithelium	Finn Dorset	6LL3	148	6.6
Fetal fibroblast	Black Welsh	6LL7	152	5.6
	Black Welsh	6LL8	149	2.8
	Black Welsh	6LL9†	156	3.1
Embryo-derived	Poll Dorset	6LL1	149	6.5
	Poll Dorset	6LL2‡	152	6.2
	Poll Dorset	6LL5	148	4.2
	Poll Dorset	6LL6‡	152	5.3

* Breed averages are 143, 147 and 145 days, respectively for the three genotypes Finn Dorset, Black Welsh Mountain and Poll Dorset.

† This lamb died within a few minutes of birth.

‡ These lambs were delivered by caesarian section. Overall the nature of the assistance provided by the veterinary surgeon was similar to that expected in a commercial flock.

lambs born to their own breed on our farms is 1.2–5.0 kg, 2–4.9 kg and 3–9 kg for the Finn Dorset, Welsh Mountain and Poll Dorset genotypes, respectively. The attainment of sexual maturity in the lambs is being monitored.

Development of embryos produced by nuclear transfer depends upon the maintenance of normal ploidy and creating the conditions for developmental regulation of gene expression. These responses are both influenced by the cell-cycle stage of donor and recipient cells and the interaction between them (reviewed in ref. 9). A comparison of development of mouse and cattle embryos produced by nuclear transfer to oocytes[10,11] or enucleated zygotes[12,13] suggests that a greater proportion develop if the recipient is an oocyte. This may be because factors that bring about reprogramming of gene expression in a transferred nucleus are required for early development and are taken up by the pronuclei during development of the zygote.

If the recipient cytoplasm is prepared by enucleation of an oocyte at metaphase II, it is only possible to avoid chromosomal damage and maintain normal ploidy by transfer of diploid nuclei[14,15], but further experiments are required to define the optimum cell-cycle stage. Our studies with cultured cells suggest that there is an advantage if cells are quiescent (ref. 1, and this work). In earlier studies, donor cells were embryonic blastomeres that had not been induced into quiescence. Comparisons of the phases of the growth cycle showed that development was greater if donor cells were in mitosis[16] or in the G1 (ref. 10) phase of the cycle, rather than in S or G2 phases. Increased development using donor cells in G0, G1 or mitosis may reflect greater access for reprogramming factors present in the oocyte cytoplasm, but a direct comparison of these phases in the same cell population is required for a clearer understanding of the underlying mechanisms.

Together these results indicate that nuclei from a wide range of cell types should prove to be totipotent after enhancing opportunities for reprogramming by using appropriate combinations of these cell-cycle stages. In turn, the dissemination of the genetic improvement obtained within elite selection herds will be enhanced by limited replication of animals with proven performance by nuclear transfer from cells derived from adult animals. In addition, gene targeting in livestock should now be feasible by nuclear transfer from modified cell populations and will offer new opportunities in biotechnology. The techniques described also offer an opportunity to study the possible persistence and impact of epigenetic changes, such as imprinting and telomere shortening, which are known to occur in somatic cells during development and senescence, respectively.

The lamb born after nuclear transfer from a mammary gland cell is, to our knowledge, the first mammal to develop from a cell derived from an adult tissue. The phenotype of the donor cell is unknown. The primary culture contains mainly mammary epithelial (over 90%) as well as other differentiated cell types, including myoepithelial cells and fibroblasts. We cannot exclude the possibility that there is a small proportion of relatively undifferentiated stem cells able to support regeneration of the mammary gland during pregnancy. Birth of the lamb shows that during the development of that mammary cell there was no irreversible modification of genetic information required for development to term. This is consistent with the generally accepted view that mammalian differentiation is almost all achieved by systematic, sequential changes in gene expression brought about by interactions between the nucleus and the changing cytoplasmic environment[17].

Methods

Embryo-derived cells were obtained from embryonic disc of a day-9 embryo from a Poll Dorset ewe cultured as described[1], with the following modifications. Stem-cell medium was supplemented with bovine DIA/LIF. After 8 days, the explanted disc was disaggregated by enzymatic digestion and cells replated onto fresh feeders. After a further 7 days, a single colony of large flattened cells was isolated and grown further in the absence of feeder cells. At passage 8, the modal chromosome number was 54. These cells were used as nuclear donors at passages 7–9. Fetal-derived cells were obtained from an eviscerated Black Welsh Mountain fetus recovered at autopsy on day 26 of pregnancy. The head was removed before tissues were cut into small pieces and the cells dispersed by exposure to trypsin. Culture was in BHK 21 (Glasgow MEM; Gibco Life Sciences) supplemented with L-glutamine (2 mM), sodium pyruvate (1 mM) and 10% fetal calf serum. At 90% confluency, the cells were passaged with a 1:2 division. At passage 4, these fibroblast-like cells (Fig. 1) had modal chromosome number of 54. Fetal cells were used as nuclear donors at passages 4–6. Cells from mammary gland were obtained from a 6-year-old Finn Dorset ewe in the last trimester of pregnancy[18]. At passages 3 and 6, the modal chromosome number was 54 and these cells were used as nuclear donors at passage numbers 3–6.

Nuclear transfer was done according to a previous protocol[1]. Oocytes were recovered from Scottish Blackface ewes between 28 and 33 h after injection of gonadotropin-releasing hormone (GnRH), and enucleated as soon as possible. They were recovered in calcium- and magnesium-free PBS containing 1% FCS and transferred to calcium-free M2 medium[19] containing 10% FCS at 37 °C. Quiescent, diploid donor cells were produced by reducing the concentration of serum in the medium from 10 to 0.5% for 5 days, causing the cells to exit the growth cycle and arrest in G0. Confirmation that cells had left the cycle was obtained by staining with antiPCNA/cyclin antibody (Immuno Concepts), revealed by a second antibody conjugated with rhodamine (Dakopatts).

Fusion of the donor cell to the enucleated oocyte and activation of the oocyte were induced by the same electrical pulses, between 34 and 36 h after GnRH injection to donor ewes. The majority of reconstructed embryos were cultured in ligated oviducts of sheep as before, but some embryos produced by transfer from embryo-derived cells or fetal fibroblasts were cultured in a chemically defined medium[20]. Most embryos that developed to morula or blastocyst after 6 days of culture were transferred to recipients and allowed to develop to term (Table 1). One, two or three embryos were transferred to each ewe depending upon the availability of embryos. The effect of cell type upon fusion and development to morula or blastocyst was analysed using the marginal model of Breslow and Clayton[21]. No comparison was possible of development to term as it was not practicable to transfer all embryos developing to a suitable stage for transfer. When too many embryos were available, those having better morphology were selected.

Ultrasound scan was used for pregnancy diagnosis at around day 60 after oestrus and to monitor fetal development thereafter at 2-week intervals. Pregnant recipient ewes were monitored for nutritional status, body condition and signs of EAE, Q fever, border disease, louping ill and toxoplasmosis. As lambing approached, they were under constant observation and a veterinary surgeon called at the onset of parturition. Microsatellite analysis was carried out on DNA from the lambs and recipient ewes using four polymorphic ovine markers[22].

Roslin Institute (Edinburgh), Roslin, Midlothian EH25 9PS, UK; A.E.S., A.J.K.: *PPL Therapeutics, Roslin, Midlothian EH25 9PP, UK*

Received 25 November 1996; accepted 10 January 1997 [Published 27 February]

References

1. Campbell, K. H. S., McWhir, J., Ritchie, W. A. & Wilmut, I. Sheep cloned by nuclear transfer from a cultured cell line. *Nature* **389**, 64–66 (1996).
2. Solter, D. Lambing by nuclear transfer. *Nature* **380**, 24–25 (1996).
3. Gurdon, J. B., Laskey, R. A. & Reeves, O. R. The developmental capacity of nuclei transplanted from keratinized skin cells of adult frogs. *J. Embryol. Exp. Morph.* **34**, 93–112 (1975).
4. Quinlivan, T. D., Martin, C. A., Taylor, W. B. & Cairney, I. M. Pre- and perinatal mortality in those ewes that conceived to one service. *J. Reprod. Fert.* 11, 379–390 (1966).
5. Walker, S. K., Heard, T. M. & Seamark, R. F. *In vitro* culture of sheep embryos without co-culture: successes and perspectives. *Therio* **37**, 111–126 (1992).
6. Nash, M. L., Hungerford, L. L., Nash, T. G. & Zinn, G. M. Risk factors for perinatal and postnatal mortality in lambs. *Vet. Rec.* **139**, 64–67 (1996).
7. Bradford, G. E., Hart, R., Quirke, J. F. & Land, R. B. Genetic control of the duration of gestation in sheep. *J. Reprod. Fert.* **30**, 459–463 (1972).
8. Walton, A. & Hammond, J. The maternal effects on growth and conformation in Shire horse–Shetland pony crosses. *Proc. R. Soc. B* **125**, 311–335 (1938).
9. Campbell, K. H. S., Loi, P., Otaegui, P. J. & Wilmut, I. Cell cycle co-ordination in embryo cloning by nuclear transfer. *Rev. Reprod.* 1, 40–46 (1996).
10. Cheong, H.-T., Takahashi, Y. & Kanagawa, H. Birth of mice after transplantation of early-cell-cycle-stage embryonic nuclei into enucleated oocytes. *Biol. Reprod.* **48**, 958–963 (1993).
11. Prather, R. S. *et al.* Nuclear transplantation in the bovine embryo. Assessment of donor nuclei and recipient oocyte. *Biol. Reprod.* **37**, 859–866 (1987).
12. McGrath, J. & Solter, D. Inability of mouse blastomere nuclei transferred to enucleated zygotes to support development *in vitro*. *Science* **226**, 1317–1318 (1984).

13. Robl, J. M. *et al.* Nuclear transplantation in bovine embryos. *J. Anim. Sci.* **64**, 642–647 (1987).
14. Campbell, K. H. S., Ritchie, W. A. & Wilmut, I. Nuclear-cytoplasmic interactions during the first cell cycle of nuclear transfer reconstructed bovine embryos: implications for deoxyribonucleic acid replication and development. *Biol. Reprod.* **49**, 933–942 (1993).
15. Barnes, F. L. *et al.* Influence of recipient oocyte cell cycle stage on DNA synthesis, nuclear envelope breakdown, chromosome constitution, and development in nuclear transplant bovine embryos. *Mol. Reprod. Dev.* **36**, 33–41 (1993).
16. Kwon, O. Y. & Kono, T. Production of identical sextuplet mice by transferring metaphase nuclei from 4-cell embryos. *J. Reprod. Fert.* Abst. Ser. **17**, 30 (1996).
17. Gurdon, J. B. The control of gene expression in animal development (Oxford University Press, Oxford, 1974).
18. Finch, L. M. B. *et al.* Primary culture of ovine mammary epithelial cells. *Biochem. Soc. Trans.* **24**, 369S (1996).
19. Whitten, W. K. & Biggers, J. D. Complete development *in vitro* of the preimplantation stages of the mouse in a simple chemically defined medium. *J. Reprod. Fertil.* **17**, 399–401 (1968).
20. Gardner, D. K., Lane, M., Spitzer, A. & Batt, P. A. Enhanced rates of cleavage and development for sheep zygotes cultured to the blastocyst stage in vitro in the absence of serum and somatic cells. Amino acids, vitamins, and culturing embryos in groups stimulate development. *Biol. Reprod.* **50**, 390–400 (1994).
21. Breslow, N. E. & Clayton, D. G. Approximate inference in generalized linear mixed models. *J. Am. Stat. Assoc.* **88**, 9–25 (1993).
22. Buchanan, F. C., Littlejohn, R. P., Galloway, S. M. & Crawford, A. L. Microsatellites and associated repetitive elements in the sheep genome. *Mammal. Gen.* **4**, 258–264 (1993).

Acknowledgements

We thank A. Colman for his involvement throughout this experiment and for guidance during the preparation of this manuscript; C. Wilde for mammary-derived cells; M. Ritchie, J. Bracken, M. Malcolm-Smith, W. A. Ritchie, P. Ferrier and K. Mycock for technical assistance; D. Waddington for statistical analysis; and H. Bowran and his colleagues for care of the animals. This research was supported in part by the Ministry of Agriculture, Fisheries and Food. The experiments were conducted under the Animals (Scientific Procedures) Act 1986 and with the approval of the Roslin Institute Animal Welfare and Experiments Committee.

Contributors

Philip Ball is a Consultant Editor for *Nature.* Among other books he is author of *H_2O: A Biography of Water* (Weidenfeld & Nicolson, 1999) and *Bright Earth: Art and the Invention of Color* (University of Chicago Press, 2002).

C. K. Brain was formerly Director of the Transvaal Museum, Pretoria, South Africa, and a coworker of Raymond Dart. He is author of *The Hunters or the Hunted? An Introduction to African Cave Taphonomy* (University of Chicago Press, 1981).

Sydney Brenner is a Fellow of King's College, Cambridge, and a Distinguished Research Professor at the Salk Institute, La Jolla, California. With Francis Crick and others he identified messenger RNA as the intermediary between DNA and protein, and the code by which the amino acid building blocks of proteins are specified. He was cowinner of the 2002 Nobel Prize in Physiology or Medicine for pioneering use of the nematode worm *Caenorhabditis elegans* as a model organism in biology.

Maurice Goldhaber is a Distinguished Scientist Emeritus at Brookhaven National Laboratory in Upton, New York, and a former director of the laboratory. He started his graduate studies at the University of Cambridge just after James Chadwick had discovered the neutron there, and, with Chadwick, made the first accurate determination of the neutron's mass.

Allan Griffin is a Professor of Physics at the University of Toronto, specializing in the theory of superfluids. He is author of *Excitations in a Bose-Condensed Liquid* (Cambridge University Press, 1993) and editor (with D. W. Snoke and S. Stringari) of *Bose-Einstein Condensation* (Cambridge University Press, 1995).

Jonathan C. Howard is in the Institute for Genetics, University of Cologne. He studies host-pathogen conflict at the cellular level, and is author of *Darwin* (Oxford University Press, reissued 2001).

Peter Little is at the School of Biochemistry and Molecular Genetics, University of New South Wales, Sydney. He is author of *Genetic Destinies* (Oxford University Press, 2002).

Malcolm Longair is Jacksonian Professor of Natural Philosophy and Head of the Cavendish Laboratory at the University of Cambridge. He is author of *High Energy Astrophysics* (Cambridge University Press, 1992, 1994), *Galaxy Formation* (Springer, 1998) and a popular book, *Our Evolving Universe* (Cambridge University Press, 1997).

Dan McKenzie is a Royal Society Research Professor in the Department of Earth Sciences at the University of Cambridge. He is one of the originators of the theory of plate tectonics, which provides a precise mathematical framework for earlier ideas of seafloor spreading and continental drift, and was awarded the 2002 Crafoord Prize by the Royal Swedish Academy of Sciences.

Ginés Morata, a specialist in the developmental genetics of *Drosophila*, is at the Centro de Biología Molecular, Universidad Autónoma de Madrid.

Gregory A. Petsko is Director of the Rosenstiel Basic Medical Sciences Research Center, Brandeis University, Waltham, Massachusetts. His research includes investigations of the structure and function of enzymes.

Marcus E. Raichle is a neurologist and Professor of Radiology at the Washington University School of Medicine, Saint Louis, Missouri.

Joseph Rotblat is an Emeritus Professor of Physics in the University of London, and Emeritus President of the Pugwash Conferences on Science and World Affairs, with which organization he shared the Nobel Peace Prize in 1995. He was one of eleven signatories of the Russell-Einstein Manifesto of 1955, which urged on world governments the imperative to avoid nuclear war.

Fred J. Sigworth is at the Yale University School of Medicine. He was involved in developing the patch-clamp technique, and now studies the structure of ion channels.

Davor Solter is a Director and Member of the Max Planck Institute for Immunobiology, Freiburg, Germany. He was a pioneer of the nuclear transfer technique commonly used in cloning mammals, and of the concept of genomic imprinting.

Richard S. Stolarski is a research scientist at NASA's Goddard Space Flight Center in Greenbelt, Maryland. He has been involved in research on the ozone layer since the early 1970s.

Joseph H. Taylor is a Professor of Physics and former Dean of the Faculty at Princeton University. He shared the 1993 Nobel Prize in Physics for the discovery of a pulsar orbiting another star, which opened up new possibilities for the study of gravitation.

Akira Tonomura is a Fellow of the Hitachi Advanced Research Laboratory, in Saitama, Japan. He is best known for his development of the electron holography microscope, which uses the wave properties of electrons to probe electric and magnetic fields in solids.

Charles H. Townes is a Professor of Physics at the University of California, Berkeley. He shared the 1964 Nobel Prize in Physics for the invention of the maser and laser.

Gordon A. H. Walker is a Professor Emeritus of Astronomy at the University of British Columbia, specializing in the analysis of starlight to look for extrasolar planets, interstellar molecules, and stellar oscillations.

Robin A. Weiss is Professor of Viral Oncology at University College London. He studies retroviruses, including the human immunodeficiency virus, and is coeditor of *RNA Tumor Viruses* (Cold Spring Harbor Laboratory Press, New York, 1985).

Index